Advanced Secure Optical Image Processing for Communications

IOP Series in Imaging Engineering

Series Editor

Leonid P Yaroslavsky
School of Electrical Engineering, Tel Aviv University, Tel Aviv, Israel

Images are the main source of information for humans, and the development of imaging methods and devices has always been at the forefront of technological progress. By the beginning of the 21st century a new branch of modern technology had formed: imaging engineering. The goal of the *IOP Series in Imaging Engineering* is to supply students and practitioners in academia and the imaging industry with the most up-to-date texts and reference books in the field. Encompassing imaging devices and systems, image acquisition, reconstruction and processing and the diverse applications of modern imaging systems, the series consists of compact textbooks and reference books written at a level accessible to graduate students. Authors are encouraged to take advantage of the features made possible by electronic publication to enhance the reader experience through the use of colour, animations, video and incorporating supplementary files in their work.

Advanced Secure Optical Image Processing for Communications

Ayman Al Falou, Editor

IOP Publishing, Bristol, UK

ISBN 978-0-7503-1457-2 (ebook)
ISBN 978-0-7503-1455-8 (print)
ISBN 978-0-7503-1456-5 (mobi)

DOI 10.1088/978-0-7503-1457-2

Version: 20180401

IOP Expanding Physics
ISSN 2053-2563 (online)
ISSN 2054-7315 (print)

British Library Cataloguing-in-Publication Data: A catalogue record for this book is available from the British Library.

Published by IOP Publishing, wholly owned by The Institute of Physics, London

IOP Publishing, Temple Circus, Temple Way, Bristol, BS1 6HG, UK

US Office: IOP Publishing, Inc., 190 North Independence Mall West, Suite 601, Philadelphia, PA 19106, USA

Contents

Preface

Over the last two decades, the optical information processing field has become increasingly popular in optical research communities, thanks to the property of optics allowing us to operate directly on the sensor. Thus, with optical image processing, it is possible to realize some of the pre-processing and/or processing optically before any interactions with the digital world. This is made possible thanks to technological progress, but also and mostly to the numerous research studies carried out in the field as evidenced by a very abundant state of the art in this subject, especially in the field of information security and/or securing people.

In the encryption field (for storage or transmission of information) many research studies have shown the interest and efficiency of optical methods compared to their digital competitors. This was made possible with the optical encryption method called the 'DRP' (Double random encryption method) developed by Javidi and Refregier in 1995. This has been widely commented and studied in the literature as evidenced by:

- The conference dedicated to optical information processing held in Shenzhen city in 2015: *International Symposium on 3D Imaging, Metrology, and Data Security*.
- Alfalou A and Brosseau C 2015 Recent advances in optical image processing *Progress in Optics* vol 60 (Elsevier) pp 119–262.
- Javidi B *et al* 2016 Roadmap on optical security *Journal of Optics* **18** 083001.
- Alfalou A, Situ G, Peng X, He W and Rastogi P 2017 3DIM-DS 2015: Optical image processing in the context of 3D imaging, metrology, and data security *Optics and Lasers in Engineering* **89** 1–202 (Special issue).

Through these publications and many others, it has been proved that optical information processing remains very competitive today. To support this wave of interest in optical information processing, this book presents and discusses the latest optical methods for compression and/or encryption: telecommunications applications. To enlighten the reader, we can divide this book into four parts:

1. The first part '**Tutorial**', which consists of four chapters. These chapters are dedicated to the exploration of the state of the art: encryption/compression/securing of people/commutations.
2. The second part, which consists of four chapters, is dedicated to the development of new optical encryption methods, thus increasing the performance of optical encryption systems.
3. The third part is devoted to the study of the performances of these kinds of methods. In this part several weaknesses have been detected and thus possible improvements have been proposed and validated.
4. This last part is dedicated to real systems and applications using innovative methods to secure information and/or people such as: compressive optical image security, simultaneous compression and encryption, correlation, ….

Chapter 1 A novel gyrator wavelet transform is defined, which is generalized by association with both the wavelet transform and the gyrator transform. The new transform opens new perspectives in optical- and digital-information processing applications. Optical implementation of this transform is suggested. The new transform can be implemented in image encryption, watermarking, fusion and authentication. Numerical simulations demonstrate the feasibility of the applications of the novel transform

Chapter 2 As we know, information security issues combined with free space optics (FSO) technology have been investigated by many scholars in the past two decades, mainly because the optical waveforms possess many complex degrees of freedom such as amplitude, phase, and polarization that can be multiplexed in many ways to make information encryption more secure. However, most of those schemes are designed in accordance with traditional symmetric or asymmetric cryptographic mechanisms so that it is hard for them to meet some common but also special security requirements in the real world, such as checking the integrity or verifying the authenticity of a message. To the best of our knowledge, there are no other solutions except for two recent attempts by us from the perspective of optics, and each of these relies on the construction of a kind of one-way hash function (OWHF) with hybrid opto-electronic means, which we have named optical OWHF.

Chapter 3 In this survey, first a historical background of new challenges in optical communication systems is given. Then, approaches to address recently raised problems by the wireless communication community are introduced, challenges encountered in real-life applications are described, and state of the art advances in optical free space communication (FSO) techniques are explained.

Chapter 4 Holography methods are attractive candidates for security and encryption applications. Digital holography techniques using coherent illumination have been demonstrated for encryption and security. Recently, imaging and digital holography systems using three colors coherent light and incoherent light have proven useful for security applications. Encryption systems that use incoherent light are simpler and cheaper as they do not require lasers. In this line of research, we propose incoherent digital holography systems for securing 3D scenes. Incoherent digital holography systems have been developed for various purposes, such as exploiting the higher spatial bandwidth of incoherent light sources and to achieve super resolution. Incoherent light sources are also used to avoid edge ringing effects and speckle noise during imaging, and sometimes to even avoid the use of laser. In this chapter, we present incoherent digital holography techniques which can also be applied for security applications.

Chapter 5 Image encryption using the method of double random phase encoding (DRPE) has emerged as one of the most prominent methods in optical security applications. The technique required the input image to be bonded with a random phase mask (RPM) and use this as input to a $4f$ (f being the focal length of a converging lens) system. The output of this procedure is a complex-amplitude stationary white noise. The encoded image thus obtained was quite robust to blind convolutions. The $4f$ system practically meant subjecting the input to double Fourier transform, along with RPMs placed in the optical path. The RPMs served as the keys. The decryption involved back tracing the optical path, using either conjugate of the RPMs or conjugate of the encrypted image. It should be mentioned that the interference based encoding technique, like the DRPE method, falls under the category of symmetric cryptosystem. Such cryptosystems have the same encryption and decryption keys. The linearity in these systems made them vulnerable to various kinds of attacks, such as the known plaintext and the chosen plaintext attacks. To remove this anomaly, asymmetric cryptosystems were introduced that had dissimilar encryption and decryption keys.

In this chapter, the overview of optical cryptosystem based on phase-truncated Fourier transform (PTFT) is given and equal modulus decomposition (EMD) with a coherent superposition based scheme is elaborated. The following sections discuss its vulnerability to applicable specific attacks. One of the attack algorithms has been presented to give an idea of formulation of such attacks. This is followed by a discussion on methods devised to overcome the shortcomings of the EMD based cryptosystems.

Chapter 6 Fractional transform is a class of expanded linear canonical transform, such as fractional Fourier transform (FrFT) and gyrator transform (GT). When the parameter in the transforms is changed, the output spectrum has various distribution, which includes amplitude and phase. In mathematics, the FrFT and GT can become Fourier transform, if their transformation angle is equal to $\pi/2$. FrFT can be implemented with a single lens system. A system having six thin cylinder lenses can achieve GT optically. In recent years, the two optical fractional transforms are applied widely for the aim of optical encryption. The parameter is regarded as an additional key in some optical information security methods.

Chapter 7 Privacy protection of biometric data of citizens is a deep concern, because leakage of the biometric database can cause serious damage. We present a method for privacy protection of biometric data by encryption techniques. We have developed a Fourier domain cryptosystem for privacy protection using multimodal biometric encryption. One modality of biometrics (such as fingerprints) is used as the plaintext, which is encrypted by another modality of biometrics (such as the face). A private key is synthesized from the encrypted biometric signature by complex spatial Fourier processing. The encrypted biometric signature can further be encrypted by other biometric modalities, and the corresponding private keys can be synthesized. The resulting biometric signature is privacy protected since the encryption keys are provided by the human, and hence those are private keys. Moreover, the decryption keys are synthesized using those private encryption keys. The encrypted signatures are decrypted using the synthesized private keys and inverse complex spatial Fourier processing. Computer simulations demonstrate the feasibility of the technique proposed.

Chapter 8 In this chapter, we present several joint-transform correlator (JTC) architectures for nonlinear information encryption, decryption and authentication systems. These JTC architectures are developed in different processing domains, such as Fourier, fractional Fourier, Fresnel and gyrator domains. Some of these processing domains can add new security keys in order to improve the security of the image encryption–decryption process. Probability density functions of the random phase masks used as keys in the security system also play a role to increase the security level of the overall system. Additional nonlinear modifications can significantly enhance the quality of the retrieved image after decryption as well as security. We also review recent nonlinear modifications of encryption-authentication systems that permit one to simultaneously encrypt and verify multiple pieces of information for highly secure optical systems

Chapter 9 Past research has demonstrated that single random phase encryption (SRPE) is a simple but highly secure method for encoding an optical image. In applying SRPE, the source image is distorted as it has to be geometrically transformed to prevent the encryption key from being deduced with various forms of attacks. To overcome this shortcoming a novel holographic method, referred to as enhanced single random phase encryption (ESRPE), is proposed in this chapter. The encryption process can be divided into four stages. First the source image to be encrypted is converted into an inline complex hologram. Second, a secondary image that is different from the source image is converted into an off-axis complex hologram. Third, the pair of holograms are summed up, and converted into a phase-only hologram with error diffusion. Finally, a fixed random phase mask is mixed with the phase-hologram, resulting in an encrypted phase-only hologram. In the decryption process the source image, and the secondary image are reconstructed at non-overlapping regions from the encrypted phase-hologram if it is integrated with the correct encryption key. Similar to the SRPE method, the encryption and decryption process of the proposed scheme is simple and highly resistant to various forms of plain-text-attacks. However, the source image is not distorted and can be directly reconstructed with good fidelity.

Chapter 10 In this chapter, current developments about single-pixel optical information encoding and authentication are described. The single-pixel optical information encoding is presented by using an iterative phase extraction scheme, and a typical case is explained. The single-pixel optical information authentication is presented by using signal binarisation method, and a typical case is correspondingly illustrated. Subsequently, some discussions are given to describe the perspective and potential challenges of single-pixel optical information encoding and authentication methods. Finally, a conclusion is made to briefly summarize single-pixel optical information encoding and authentication approaches. It is believed that there will be a long-term benefit for the information security field by optically encoding and authenticating information based on single-pixel imaging structures, and new single-pixel optical information encoding and authentication approaches can be continuously developed and integrated in long-term study.

Chapter 11 Based on compressive sensing, phase-shifting digital holography and optical image security, we study the methods of optical image security based on the single-pixel compressive imaging in fully optical domain to expound the theory and the technique of optical image encryption, optical image hiding, and optical image encryption and hiding in it, by way of theoretical analysis, numerical simulation and experimental validation in this project. These methods will effectively reduce the amount of information recording needed and address the information security issues in the areas of Internet of Things, all-optical network and 2D or 3D scene information security access etc.

Chapter 12 Encryption is the most useful way to solve the privacy problem. However, conventional encryption algorithms are not capable of encrypting a large amount of data for real-time applications. Simultaneous compression and encryption can then be a robust solution that does not result in an increase in computation time while providing some security and keeping the size of the data encoded. In this paper, we propose encryption techniques for compressed video sequences. We will describe them one by one by showing their block diagrams and by explaining the algorithms used. A method of encryption combined with video compression is also presented and will be compared with the simple encryption procedures introduced.

Chapter 13 In the last two decades, face recognition has become an interesting direction of research, and used in different disciplines such as computer vision, image processing, pattern recognition, and machine learning. We have found several methods including local, subspaces and correlation filters of face recognition. The purpose of this project is to establish a comparative study between six approaches, one based on the local feature methods and the second one based on the correlation methods. However, the comparative study between these methods is carried with distance measurements and peak to correlation energy methods. In addition, the evaluation techniques are the yardstick to present the performance and efficiency for each face recognition system. These techniques allow one to interpret, compare, and distinguish the different factors such as images, illumination, location and characteristics of subjects. In this chapter, we review, discuss, and compare some common face recognition algorithms. Our objective of this work is to demonstrate the effectiveness and feasibility of the best methods for face recognition in terms of design, implementation, and application.

Acknowledgments

I would like to thank all the authors who participated with valuable contributions in this project and who showed great interest in the optical information processing. It was a privilege to work with this team. I also thank all the staff of IOP for their help and support throughout this project.

Editor biography

Ayman Al Falou

Ayman Al Falou received his PhD in Telecommunications and signal processing from the French National Telecommunication Graduate Engineering School of Brittany (ENSTB-France) and of the University of Rennes in 1999. He held a Post-Doc position for one year at French National Telecommunications Graduate Engineering School of Brittany (ENSTB-France) (DGA, French Army), consisting in designing and realizing an optical compact and high rate correlator. Since June 2000, he has been a Professor of Telecommunications and signal processing at ISEN-Brest (Institut Supérieur de l'Electronique et du Numérique). At ISEN, he created the Optical Signal and Image Processing Laboratory (VISION). His research interests are signal processing and image processing, telecommunications, optical systems, optical processing, opt-electronics, lasers, and polarization. He is a senior member of OSA, senior member of IEEE, senior member of SPIE and elected member of the Institute of Physics.

List of Contributors

Muhammad Rafiq Abuturab
Maulana Azad College of Engineering and Technology
Patna
India

He Wenqi
Shenzhen University
Shenzhen
China

Mohamed Abaza
Electronics & Communications Engineering Department
Arab Academy for Science, Technology and Maritime Transport
Giza
Egypt

Raed Mesleh
Electrical and Communication Engineering Department
German Jordanian University
Amman
Jordan

Ali Mansour
LabSTICC, UMR 6285 CNRS, ENSTA Bretagne
Brest
France

El-Hadi M Aggoune
Sensor Networks & Cellular System (SNCS)
University of Tabuk
Tabuk
Saudi Arabia

Joseph Rosen
Ben-Gurion University of the Negev
Beer Sheva
Israel

Naveen K Nishchal
Indiand Institute of Technology Patna
Patna
Bihar
India

Zhengjun Liu
Harbin Institute of Technology
Harbin
China

Debesh Choudhury
Infosensys Consulting
Kolkata
India

Elisabet Pérez Cabré
Universitat Politècnica de Catalunya
Terrassa
Spain

Juan Manuel Vilardy
Universidad Popular del Cesar
Valledupar (Cesar)
Colombia

María Sagrario Millán
Universitat Politècnica de Catalunya
Terrassa
Spain

Tsang Wai Ming Peter
City University of Hong Kong
Kowloon
Hong Kong

Wen Chen
The Hong Kong Polytechnic University
Kowloon
Hong Kong

Jun Li
South China Normal University
Guangzhou
China

Miao Lei
South China Normal University
Guangzhou
China

Xiaofang Dai
South China Normal University
Guangzhou
China

Jiaosheng Li
South China Normal University
Guangzhou
China

Ting Zhong
South China Normal University
Guangzhou
China

Hongbing Li
South China Normal University
Guangzhou
China

Bangwen Jia
South China Normal University
Guangzhou
China

Chanxia Yang
South China Normal University
Guangzhou
China

Shangyuan Wang
South China Normal University
Guangzhou
China

Chen Wang
South China Normal University
Guangzhou
China

Nihel Neji
University of Sfax
Sfax
Tunisia

Yassin Kortli
University of Monastir
Monastir
Tunisia

Chapter 1

Optical information security systems based on a gyrator wavelet transform

Muhammad Rafiq Abuturab

1.1 Introduction

With significant advances in communication and information processing technologies, optical information security has attracted widespread attention due to its marked advantages of high parallel processing speed, high flexibility and many degrees of freedom such as phase, amplitude, wavelength, polarization, and time [1]. Réfrégier and Javidi first introduced optical image encryption based on double random phase encoding (DRPE) [2]. A number of DRPE-based optical security systems have been subsequently proposed [3–11]. The cryptanalysis of DRPE-based systems has a security flaw against potential attacks due to the inherent linearity of the cryptosystem [12]. Consequently, an asymmetric cryptosystem based on a phase-truncated Fourier transform has been proposed [13]. However, it has been found to be vulnerable to a specific attack [14]. Recently, numerous asymmetric methods have been suggested to overcome various types of attacks [15–20]. Quantitative techniques for the simultaneous compression and encryption of images have also been developed [21–24].

The discrete wavelet transform (DWT) emerged in the field of image/signal processing as an alternative to the Fourier transform (FT) and its related transforms. The DWT reveals an image's spatial and frequency characteristics, while FT reveals only an image's frequency characteristics. Moreover, DWT is a highly intuitive framework for representation and storage of multi-resolution images. The DWT has been applied efficiently in image encryption and compression in networks, multimedia and communications [25]. Further, a fractional wavelet transform has been introduced, which generalizes the wavelet transform with a generalization based on the fractional Fourier transform [26]. Optical image encryptions based on a fractional wavelet transform [27] and dual fractional Fourier-wavelet domain [28]

doi:10.1088/978-0-7503-1457-2ch1

have been reported. Recently, a new DWT-based image encryption has been proposed where an encrypted original image is transformed into a resulting cover image-like encrypted image instead of a noise-like encrypted image [29]. Further, a gyrator wavelet transform has been introduced and developed to achieve better applications in optical information security [30, 31].

This chapter is organized as follows. Section 1.2 presents the basic theory of the WT, GT, GWT, and logistic map and two-coupled logistic map. Section 1.3 describes applications of GWT in asymmetric multiple-image fusion and authentication, and asymmetric group-multiple-image encryption and watermarking. Section 1.4 concludes the chapter.

1.2 Theory

1.2.1 Wavelet transform

The definition of a 1-D wavelet transform of a signal $f_i(x_i)$ is [32]

$$W(a, b) = \int_{-\infty}^{\infty} f_i(x_i) h_{a,b}^{*}(x_i)\mathrm{d}x_i \tag{1.1}$$

where * represents the complex conjugate and a set of daughter wavelets $h_{a,b}(x_i)$ is derivated from the mother wavelet $h(x_i)$ by dilation/scale a and translation/shift b operations:

$$h_{a,b}(x_i) = \frac{1}{\sqrt{a}} h\left(\frac{x_i - b}{a}\right). \tag{1.2}$$

The definition of a 2-D wavelet transform of a signal $f_i(x_i, y_i)$ is

$$W(a_1, a_2, b_1, b_2) = \int_{-\infty}^{\infty}\int_{-\infty}^{\infty} f_i(x_i, y_i) h_{a_1,a_2,b_1,b_2}^{*}(x_i, y_i)\mathrm{d}x_i\mathrm{d}y_i \tag{1.3}$$

where a set of daughter wavelets $h_{a_1,a_2,b_1,b_2}(x_i, y_i)$ is derived from the mother wavelet $h(x_i, y_i)$ by dilations/scales (a_1 and a_2) and translations/shifts (b_1 and b_2) operations:

$$h_{a_1,a_2,b_1,b_2}(x_i, y_i) = \frac{1}{\sqrt{a_1 a_2}} h\left(\frac{x_i - b_1}{a_1}, \frac{y_i - b_2}{a_2}\right). \tag{1.4}$$

The 2-D WT can be realized in an optical correlator architecture by using a 4 f processor with a multireference-matched filter (MRMF), where each daughter wavelet is encoded with a different reference beam [33].

1.2.2 Gyrator transform

The GT at transformation angle α of a two-dimensional complex field function $f_i(x_i, y_i)$, is expressed as [34]

$$f_o(x_o, y_o) = G^{\alpha}\left[f_i(x_i, y_i)\right](x_o, y_o) = \iint f_i(x_i, y_i) K_{\alpha}(x_i, y_i; x_o, y_o)\mathrm{d}x_i\mathrm{d}y_i. \tag{1.5}$$

The Kernel of GT is defined as

$$K_\alpha(x_i, y_i; x_o, y_o) = \frac{1}{|\sin \alpha|} \exp\left(i2\pi \frac{(x_o y_o + x_i y_i)\cos \alpha - (x_i y_o + x_o y_i)}{\sin \alpha} \right) \quad (1.6)$$

where $G^\alpha[\ ]$ represents a GT operator. (x_i, y_i) and (x_o, y_o) denote the input and output dimensionless coordinates, respectively. The optical setup of GT is constructed by using three generalized lenses L_1, L_2 and L_3 (with $L_3 = L_1$) with a fixed interval z between them. Each generalized lens corresponds to the combination of two convergent thin cylindrical lenses of the same power. The GT at angle α is reached by rotation of these lenses. The GT is applied for space variant filtering, hyperbolic noise reduction, and encryption.

1.2.3 Gyrator wavelet transform

A new 2-D transformation coined the gyrator wavelet transform (GWT) for an input signal $f_i(x_i, y_i)$, defined as [31]

$$W^\alpha(a_1, a_2, b_1, b_2) = \frac{1}{\sqrt{a_1 a_2}} \int_{-\infty}^{\infty}\int_{-\infty}^{\infty}\int_{-\infty}^{\infty}\int_{-\infty}^{\infty} f_i(x_i, y_i) h^*\left(\frac{x_o - b_1}{a_1}, \frac{y_o - b_2}{a_2} \right) \times K_\alpha(x_i, y_i; x_o, y_o) \mathrm{d}x_i \mathrm{d}y_i \mathrm{d}x_o \mathrm{d}y_o. \quad (1.7)$$

For $K_\alpha(x_i, y_i; x_o, y_o) = 1$, the 2-D GWT becomes the conventional 2-D wavelet transform.

The inverse 2-D gyrator wavelet transform for back reconstructing the input signal is

$$f_i(x_i, y_i) = G^{-\alpha}\left[\frac{1}{c} \int_{-\infty}^{\infty}\int_{-\infty}^{\infty}\int_{-\infty}^{\infty}\int_{-\infty}^{\infty} \frac{1}{a_1^3 a_2^3} W^\alpha(a_1, a_2, b_1, b_2) \times h\left(\frac{x_o - b_1}{a_1}, \frac{y_o - b_2}{a_2} \right) \mathrm{d}a_1 \mathrm{d}a_2 \mathrm{d}b_1 \mathrm{d}b_2 \right] \quad (1.8)$$

where c is a constant.

The 2-D GWT can be implemented in a hybrid optical configuration as shown in figure 1.1 The two computer-controlled spatial light modulators SLM_1 and SLM_2 are used to introduce the input and the transformed information, respectively. First, the input information displayed on SLM_1 is illuminated by a collimated beam and then GT at transformation angle α is optically performed by three generalized lenses. Second, the transformed information is imported into SLM_2 and then WT is optically accomplished by Fourier-transforming the information, multiplying by a MRMF, and then performing an inverse Fourier-transform. Fourier transform can be achieved in an optical system having a converging lens L of focal length f. The output signal is registered using a charged couple device (CCD) camera. The reference beam is used to record the complex data. In a similar fashion, the inverse GWT can be optically implemented. This optical system includes the inverse wavelet transform first and then the GT at transformation angle $-\alpha$.

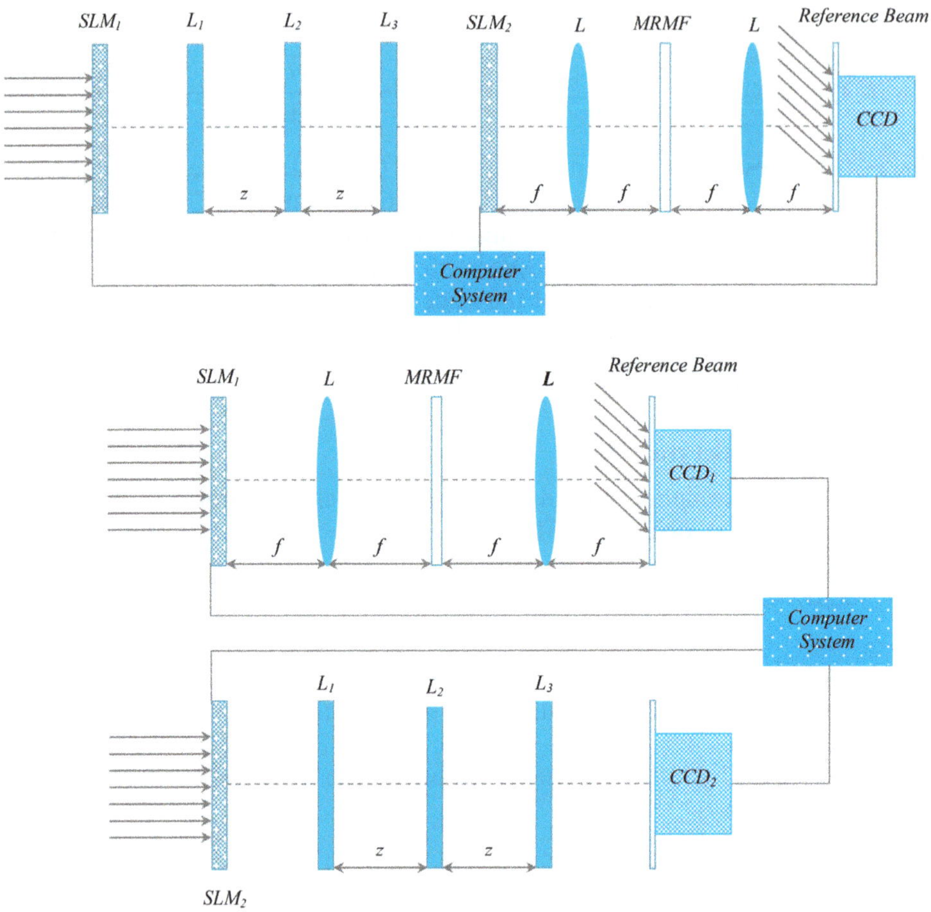

Figure 1.1. (a) Optoelectronic setup for a 2-D gyrator wavelet transform, (b) optoelectronic setup for a decryption system. Reproduced from [31].

1.2.4 Logistic map and two-coupled logistic map

The chaos function is a logistic map. It is a 1-D non-linear chaos function and defined as

$$f(x) = px(1 - x) \tag{1.9}$$

where p is a bifurcation parameter: $0 < p < 4$. The iterative form of equation (1.9) is expressed as

$$x_{n+1} = px_n(1 - x_n) \tag{1.10}$$

where $x_n \in [0, 1]$ is the iterative value and x_0 is the initial value.

However, 1-D chaos function-based image encryption algorithms are broken due to their small key spaces and weakly secure encryption mechanisms. To overcome these problems, a system of two-symmetrically coupled identical logistic maps is proposed [35] and defined as

$$\begin{cases} f(x) = px(1-x) + \varepsilon(y-x), \\ f(y) = py(1-y) + \varepsilon(x-y). \end{cases} \tag{1.11}$$

where ε is the coupling constant: $-2 < \varepsilon < 2$. The iterative form of equation (1.11) is given as

$$x_{n+1} = px_n(1-x_n) + \varepsilon(y_n - x_n), \tag{1.12}$$

$$y_{n+1} = py_n(1-y_n) + \varepsilon(x_n - y_n). \tag{1.13}$$

Suppose that the plaintext image to be encoded consists of $M \times N$ sized pixels. Then the image is permutated by using a sequence of chaotic pairs generated with a system of two-symmetrically coupled identical logistic maps (called the chaotic permutation process) by the following steps: first, the initial values of the initial parameters (x_0, y_0), K, p and ε are set; two random sequences $X = \{x_1, x_2, \ldots,\ldots,x_{M+K}\}$, $x_i \in (0,1)$ and $Y = \{y_1, y_2, \ldots,\ldots,y_{N+K}\}$, $y_i \in (0,1)$ are generated by using equations (1.12) and (1.13). The prior K values are discarded to increase randomness and two sequences $X = \{c(m)|m = 1, 2, \ldots, M\}$ and $Y = \{r(n)|n = 1, 2, \ldots, N\}$ are obtained. Second, the sequences X and Y are sorted in ascending/descending order and two new sequences $X' = \{c[w(m)]|m = 1, 2, \ldots, M\}$ and $Y' = \{r[w(n)]|n = 1, 2, \ldots, N\}$ are found (where w signifies the address code). Finally, the permutated image of $m \times n$ pixels is obtained by using the sequences X' and Y'. Note that the intensity of the $m \times n$ pixel coordinates in the permutated image is equal to the intensity of $w(m) \times w(n)$ pixel coordinates. The parameters x_0, y_0,K,p, and ε can be used as decryption (or private) keys in the proposed system. The inverse chaotic permutation is employed to retrieve the plaintext image [36].

1.3 Applications of a gyrator wavelet transform

In this section, two GWT-based applications for asymmetric optical information security systems have been discussed. First, an asymmetric cryptosystem based on multiple-image fusion and authentication is presented [30]. Second, an asymmetric cryptosystem based on group-multiple-image encryption and watermarking is described [31]. These security systems can be performed by numerical calculations and implemented by hybrid optoelectronic setups.

1.3.1 Asymmetric multiple-image fusion and authentication

In this scheme, the encryption and decryption algorithms have been explained. Then, the corresponding optoelectronic designs have been presented. Finally, numerical simulations have been performed to verify the validity and efficiency of the algorithms.

In the encryption process, the user color image $f_m(x, y)$ is segregated into R, G, and B channels designated as $f_{R_m}(x, y)$, $f_{G_m}(x, y)$ and $f_{B_m}(x, y)$, respectively. The subscript m (= 1,2, ..., N) is the order number of the image. Each user channel is normalized, phase encoded, and then modulated by a random phase function.

The modulated user channels are multiplied together to obtain a single channel user ciphertext.

$$\left.\begin{aligned} U_{R_m}(x, y) &= \exp\left[\pi + \arccos\left[1 - f_{R_m}(x, y)\right] + 2\pi\phi_R(x, y)\right] \\ U_{G_m}(x, y) &= \exp\left[\pi + \arccos\left[1 - f_{G_m}(x, y)\right] + 2\pi\phi_G(x, y)\right] \\ U_{B_m}(x, y) &= \exp\left[\pi + \arccos\left[1 - f_{B_m}(x, y)\right] + 2\pi\phi_B(x, y)\right] \end{aligned}\right\} \tag{1.14}$$

$$U_m(x, y) = U_{R_m}(x, y) \times U_{G_m}(x, y) \times U_{B_m}(x, y). \tag{1.15}$$

Now the secret color image $I_m(x_i, y_i)$ is separated into R, G, and B channels represented as $I_{R_m}(x_i, y_i)$, $I_{G_m}(x_i, y_i)$ and $I_{B_m}(x_i, y_i)$, respectively. Each secret channel is phase encoded and modulated by a random phase function. The corresponding distributions $h_{R_m}(x_i, y_i)$, $h_{G_m}(x_i, y_i)$ and $h_{B_m}(x_i, y_i)$ are independently gyrator transformed at the transformation angles α_{R_m}, α_{G_m}, and α_{B_m}. The transformed secret channels are multiplied together to get a single channel secret ciphertext.

$$H_{R_m}(x, y) = G^{\alpha_{R_m}}\underbrace{\left\{\exp\left[\pi + \arccos\left[1 - I_{R_m}(x_i, y_i)\right] + 2\pi\phi_R(x_i, y_i)\right]\right\}}_{h_{R_m}(x_i, y_i)} \tag{1.16}$$

$$H_{G_m}(x, y) = G^{\alpha_{G_m}}\underbrace{\left\{\exp\left[\pi + \arccos\left[1 - I_{G_m}(x_i, y_i)\right] + 2\pi\phi_G(x_i, y_i)\right]\right\}}_{h_{G_m}(x_i, y_i)} \tag{1.17}$$

$$H_{B_m}(x, y) = G^{\alpha_{B_m}}\underbrace{\left\{\exp\left[\pi + \arccos\left[1 - I_{B_m}(x_i, y_i)\right] + 2\pi\phi_B(x_i, y_i)\right]\right\}}_{h_{B_m}(x_i, y_i)} \tag{1.18}$$

$$H_m(x, y) = H_{R_m}(x, y) \times H_{G_m}(x, y) \times H_{B_m}(x, y). \tag{1.19}$$

The user ciphertext is multiplied with the secret ciphertext to generate a single channel multiplex image $S_m(x, y)$, which is inverse gyrator transformed at the transformation angle β_m and then phase- and amplitude- truncated to produce the encrypted image $E_m(x_i, y_i)$ and the asymmetric key $P_m(x_i, y_i)$, respectively.

$$S_m(x, y) = U_m(x, y) \times H_m(x, y) \tag{1.20}$$

$$E_m(x_i, y_i) = PT\{G^{-\beta_m}[S_m(x, y)]\} \tag{1.21}$$

$$P_m(x_i, y_i) = AT\{G^{-\beta_m}[S_m(x, y)]\} \tag{1.22}$$

where the operators $PT\{\}$ and $AT\{\}$ symbolize phase- and amplitude- truncation, respectively.

The encrypted image $E_m(x_i, y_i)$ is single-level 2-D discrete wavelet transformed. The resultant information is decomposed into four sub-bands W_{LL_m}, W_{HL_m}, W_{LH_m}, and W_{HH_m}, where subscript L refers to low frequency while H stands for high frequency.

$$[W_{LL_m}, W_{HL_m}, W_{LH_m}, W_{HH_m}] = DWT\left[E_m(x_i, y_i)\right]. \tag{1.23}$$

The different decomposed sub-bands of m number of images are independently fused together to construct corresponding fused sub-bands W_{LL}, W_{HL}, W_{LH}, and W_{HH}, which are inverse single-level 2-D discrete wavelet transformed to produce a final encrypted image $E(x_i, y_i)$.

$$W_{LL} = \prod_{m=1}^{N} W_{LL_m},\ W_{HL} = \prod_{m=1}^{N} W_{HL_m},\ W_{LH} = \prod_{m=1}^{N} W_{LH_m},\ W_{HH} = \prod_{m=1}^{N} W_{HH_m}, \tag{1.24}$$

$$E(x_i, y_i) = IDWT(W_{LL}, W_{HL}, W_{LH}, W_{HH}). \tag{1.25}$$

The decryption keys are generated during the encryption process, which are different from the encryption keys. They are the authentication key, secret keys, and sub-band keys in addition to the asymmetric key.

i. $$\text{Authentication key } K_{A_m}(x, y) = \frac{1}{U_m(x, y)} \tag{1.26}$$

ii. $$\text{Secret keys } K_{R_m}(x, y) = \left(\frac{1}{H_{G_m}(x, y) \times H_{B_m}(x, y)}\right) \tag{1.27}$$

$$K_{G_m}(x, y) = \left(\frac{1}{H_{R_m}(x, y) \times H_{B_m}(x, y)}\right) \tag{1.28}$$

$$K_{B_m}(x, y) = \left(\frac{1}{H_{R_m}(x, y) \times H_{G_m}(x, y)}\right) \tag{1.29}$$

iii. Sub-band keys

The individual sub-band keys W_{kLL_n}, W_{kHL_n}, W_{kLH_n}, and W_{kHH_n} are produced by using equation (1.24).

$$W_{kLL_n} = \prod_{n \neq m}^{N} W_{LL_n},\ W_{kHL_n} = \prod_{n \neq m}^{N} W_{HL_n},\ W_{kLH_n} = \prod_{n \neq m}^{N} W_{LH_n},$$
$$W_{kHH_n} = \prod_{n \neq m}^{N} W_{HH_n},\quad (m, n) \in [1, N]. \tag{1.30}$$

The decryption process is simpler than the encryption process. The encrypted image $E(x_i, y_i)$ is single-level 2-D discrete wavelet transformed to obtain sub-bands W_{LL}, W_{HL}, W_{LH}, and W_{HH}. The decomposed sub-bands are divided by corresponding individual sub-band keys W_{kLL_n}, W_{kHL_n}, W_{kLH_n}, and W_{kHH_n}, and then inverse single-level 2-D discrete wavelet transformed.

$$[W_{LL}, W_{HL}, W_{LH}, W_{HH}] = DWT\left[E(x_i, y_i)\right] \tag{1.31}$$

$$W_m(x_i, y_i) = IDWT\left(\frac{W_{LL}}{W_{kLL_n}}, \frac{W_{HL}}{W_{kHL_n}}, \frac{W_{LH}}{W_{kLH_n}}, \frac{W_{HH}}{W_{kHH_n}}\right). \tag{1.32}$$

The resulting image is multiplied with the asymmetric key and gyrator transformed at the transformation angle β_m. The obtained image is multiplied with a secret key of R, G, or B channel and an authentication key of the corresponding user, and then independently inverse gyrator transformed at the transformation angle $-\alpha_{R_m}$, $-\alpha_{G_m}$, or $-\alpha_{B_m}$ to retrieve the corresponding channel.

$$D_m(x, y) = G^{\beta_m}\left[P_m(x_i, y_i) \times W_m(x_i, y_i)\right] \tag{1.33}$$

$$\left.\begin{aligned} I_{R_m}(x_i, y_i) &= |\, G^{-\alpha_{R_m}} [D_m(x, y) \times K_{A_m}(x, y) \times K_{R_m}(x, y)] \,| \\ I_{G_m}(x_i, y_i) &= |\, G^{-\alpha_{G_m}} [D_m(x, y) \times K_{A_m}(x, y) \times K_{G_m}(x, y)] \,| \\ I_{B_m}(x_i, y_i) &= |\, G^{-\alpha_{B_m}} [D_m(x, y) \times K_{A_m}(x, y) \times K_{B_m}(x, y)] \,| \end{aligned}\right\} \tag{1.34}$$

where the operator || stands for the modulus.

The optoelectronic designs for the optical encryption and decryption systems are shown in figures 1.2(a) and (b), respectively. In the encryption process, the digitally

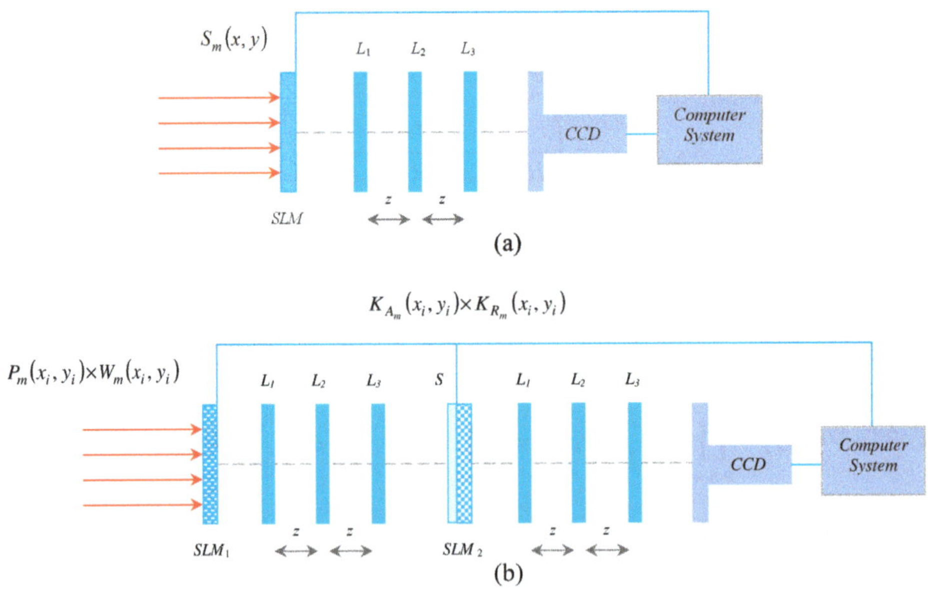

Figure 1.2. Optoelectronic setup for (a) encryption system, and (b) decryption system. Reprinted from [30].

obtained image $S_m(x, y)$ is first displayed on an SLM, illuminated by a uniform plane wave, and then inverse gyrator transformed. The resultant spectrum is recorded by a CCD camera and stored in a computer system. The phase truncations can be accomplished by the CCD camera. The amplitude truncations can be carried out by holographic recording and recovered by phase-shifting interferromentry. The phase truncated image is digitally single-level 2-D discrete wavelet transformed to obtain four sub-bands W_{LL_m}, W_{HL_m}, W_{LH_m}, and W_{HH_m}. The different sub-bands of m number of images are independently fused together to get corresponding fused sub-bands W_{LL}, W_{HL}, W_{LH}, and W_{HH}, which are inverse single-level 2-D discrete wavelet transformed to produce a final encrypted image $E(x_i, y_i)$.

In the decryption process, the digitally obtained image $W_m(x_i, y_i)$ is multiplied with the asymmetric key $P_m(x_i, y_i)$, first displayed on SLM_1, illuminated by a uniform plane wave, and then gyrator transformed at the angle β. The complex wavefront on the screen S modulated by a secret key of each channel and an authentication key of the corresponding user, is displayed on the SLM_2 and then independently inverse gyrator transformed at the angle α. The intensity of each decrypted channel is recorded by CCD in the output plane. Finally, green and blue component images obtained by the same process are combined to a red component image in the computer system to recover the original secret image.

Numerical simulations have been performed on a Matlab 7.11.0 (R2010b) to verify the validity and the efficiency of the proposed encryption technique. The four user color images consist of 512 × 512 × 3 pixels in size: Barbara (user 1), Ali (user 2), Hadi (user 3), and Mahdi (user 4), are shown in figure 1.3. The corresponding four secret color images are 512 × 512 × 3 pixels in size: Olive fruits, Honey, Date tree, and Camel, are depicted in figure 1.4. The authentication keys and asymmetric keys for user 1, user 2, user 3, and user 4 are illustrated in figures 1.5 and 1.6, respectively. For brevity, the secret individual keys of user 1 for R channel, G channel, B channel and fused sub-band key are demonstrated in figures 1.7(i)(a)–(d). The secret sub-band keys: LL sub-band key, HL sub-band key, LH sub-band key, and HH sub-band key, are displayed in figures 1.7(ii)(a)–(d). The ciphertext of user 1, ciphertext of secret image 1, and multiplex image of ciphertexts of user 1 and the secret image 1 are, respectively, presented in figures 1.8(i)(a)–(c). The final encrypted image for all users with the correct parameters is shown in figure 1.8(i)(d). It can be seen that the encrypted image appears to be a noisy pattern, which has no visual relation to the original secret color image. In addition, the gray ciphertexts can confuse unauthorized users. The decrypted secret images with all the correct keys for user 1, user 2, user 3, and user 4 are, respectively, depicted in figures 1.8(ii)(a)–(d).

The decrypted images without the asymmetric key, authentication key, and secret key for user 1, user 2, user 3, and user 4 are, respectively, illustrated in figures 1.9–1.11. The reconstructed images without the LL sub-band key, HL sub-band key, LH sub-band key, and HH sub-band key, for user 1, user 2, user 3, and user 4 are, respectively, demonstrated in figures 1.12(i)–(iv). Figures 1.9–1.12 are unrecognizable signifying that the original color images cannot be retrieved if any one decryption key is wrong.

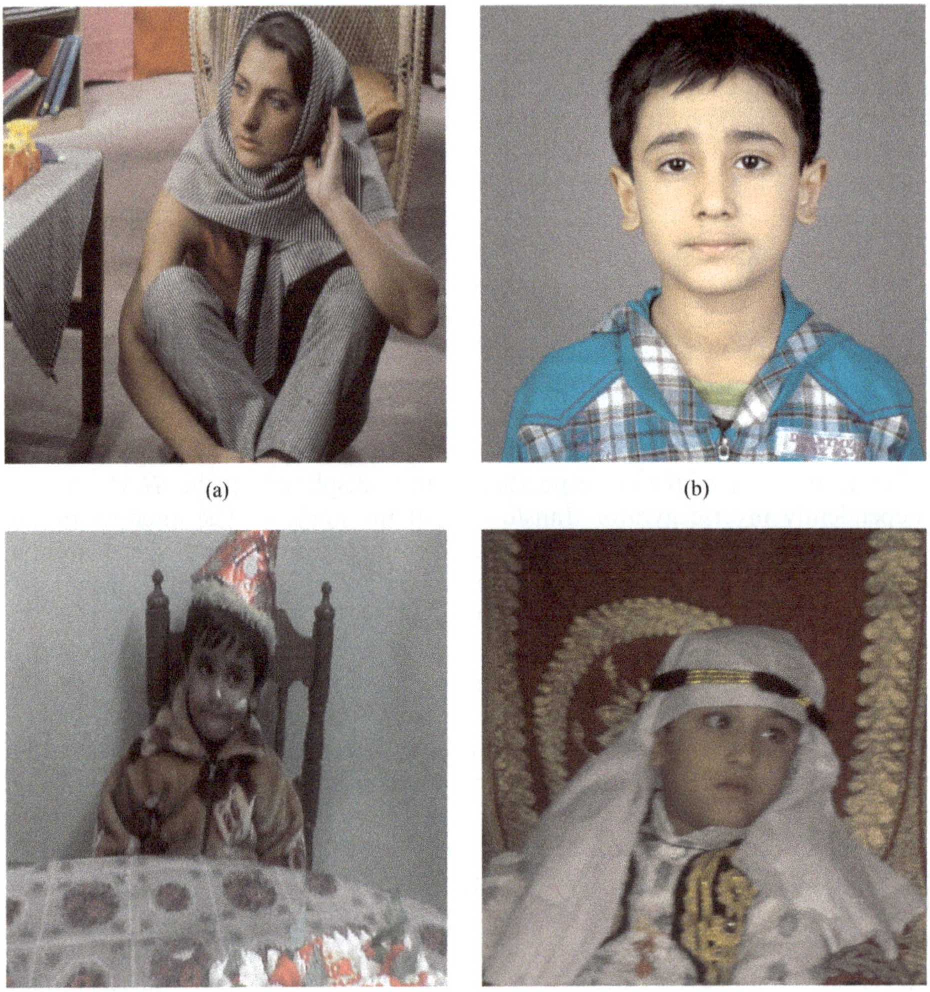

(a) (b) (c) (d)

Figure 1.3. Users color images: (a) User 1, (b) User 2, (c) User 3, and (d) User 4. Reprinted from [30]. Copyright 2015, with permission from Elsevier B.V. All rights reserved.

For simplicity, the CC curves for the R, G, and B channels of the original secret image of user 1 as a function of the first transformation angle ($\alpha_1 = 0.50°$) and the second transformation angle ($\beta_1 = 0.50°$) are plotted as shown in figure 1.13(a) and (b). The CC approaches unity when the images are decrypted with correct transform orders, whereas it increases sharply in the case of a slight departure from the correct orders. From these CC curves, it can be concluded that the cryptosystem is very sensitive to a slight deviation in orders of GT.

In order to investigate the encryption system resisting statistical analysis, the auto-correlation coefficients of the original secret image of user 1, as depicted in figure 1.4(a) and that of the encrypted image as displayed in figure 1.8(i)(d), have been analyzed, and shown in figures 1.14(a) and (b), respectively. Figure 1.14(a)

(a) (b) (c) (d)

Figure 1.4. Secret color images: (a) Olive fruits, (b) Honey, (c) Date tree, and (d) Camel. Reprinted from [30].

shows that the correlation between the original image pixels is very strong, whereas figure 1.14(b) shows that correlation of the encrypted image is very weak. Thus, the proposed encryption system has strong decorrelation power, and resists statistic attacks. The multiplexing capacity of the proposed algorithm has been investigated. The CCs of the decrypted color images against the number of encrypted images ($N = 50$) are illustrated in figure 1.15. It can be observed that the CC values remain constant with an increase in the number of secret images. Thus, it can be concluded that the cryptosystem has high storage capacity and has no cross-talk noise effect.

The PTFT-based asymmetric cryptosystem can endure chosen-plaintext and known-plaintext attacks owing to employment of a nonlinear operation of phase truncation. However, a specific attack, which is based on a two-step iterative

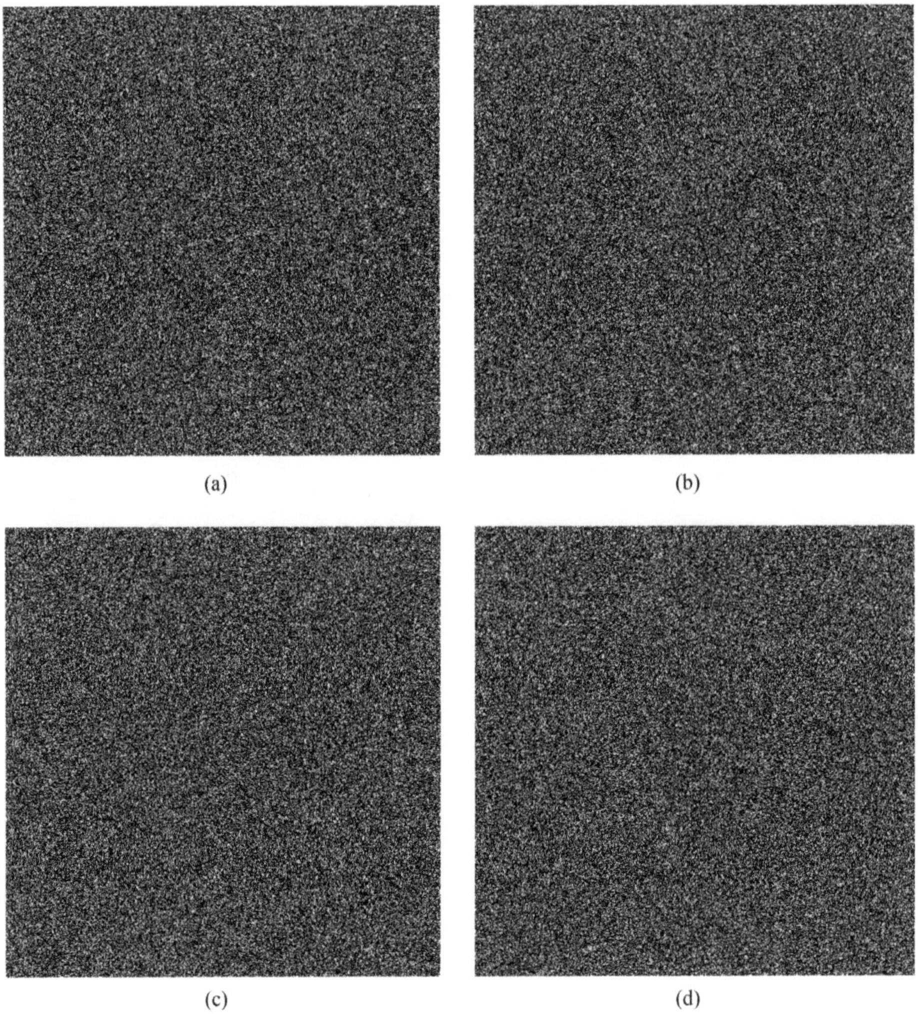

Figure 1.5. Authentication keys for: (a) User 1, (b) User 2, (c) User 3, and (d) User 4. Reprinted from [30]. Copyright 2015, with permission from Elsevier B.V. All rights reserved.

amplitude retrieval approach and works by using public keys and ciphertexts, would allow an attacker to reveal the encrypted information and the decryption keys that were generated in the encryption procedure [14]. The simplest solution is to keep the encryption keys as private keys or apply different phase keys for different plaintexts during the encryption to avoid a known public key attack as well as a specific attack. Since the single channel image $S_m(x, y)$ is treated as input information, attackers do not have enough constraints to establish retrieval even if they have obtained the exact GT orders. Additionally, different ciphertext has different individual decryption (/private) keys as the authentication key, asymmetric key, secret keys, and subband keys in the proposed cryptosystem. This means that the known-plaintext attack, chosen-plaintext attack and specific attack will not work.

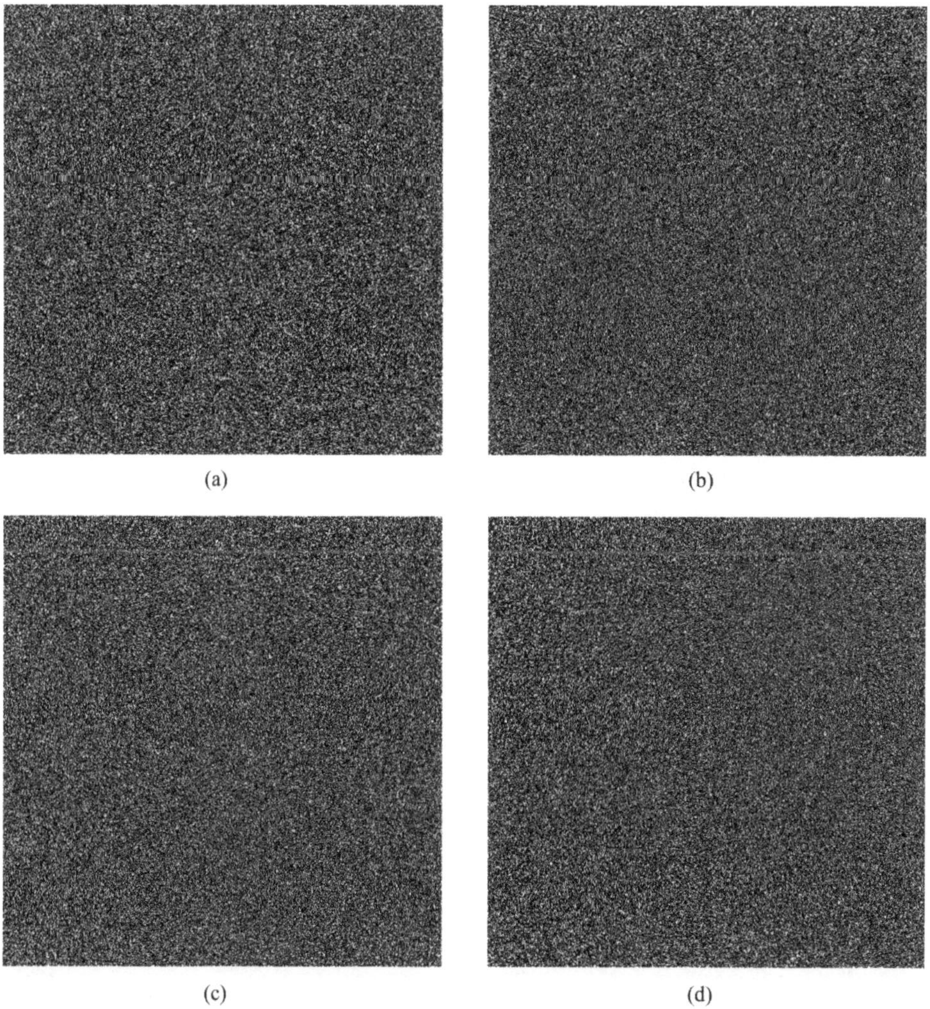

Figure 1.6. Asymmetric keys for: (a) User 1, (b) User 2, (c) User 3, and (d) User 4. Reprinted from [30]. Copyright 2015, with permission from Elsevier B.V. All rights reserved.

1.3.2 Asymmetric group-multiple-image encryption and watermarking

In this system, the encryption and decryption algorithms have been described. Then, the corresponding optoelectronic architectures have been proposed. Finally, numerical simulation results confirm the viability and security of the proposed scheme.

The group-multiple-image encoding and watermarking processes involve the following steps:

Step 1: A color image $f_m^1(x_i, y_i)$ of the first group is divided into R, G, and B channels, represented as $f_R^1(x_i, y_i)$, $f_G^1(x_i, y_i)$ and $f_B^1(x_i, y_i)$, respectively. The subscript m is the serial number of image. Each channel is first independently permutated by using a chaotic permutation process and then gyrator transformed.

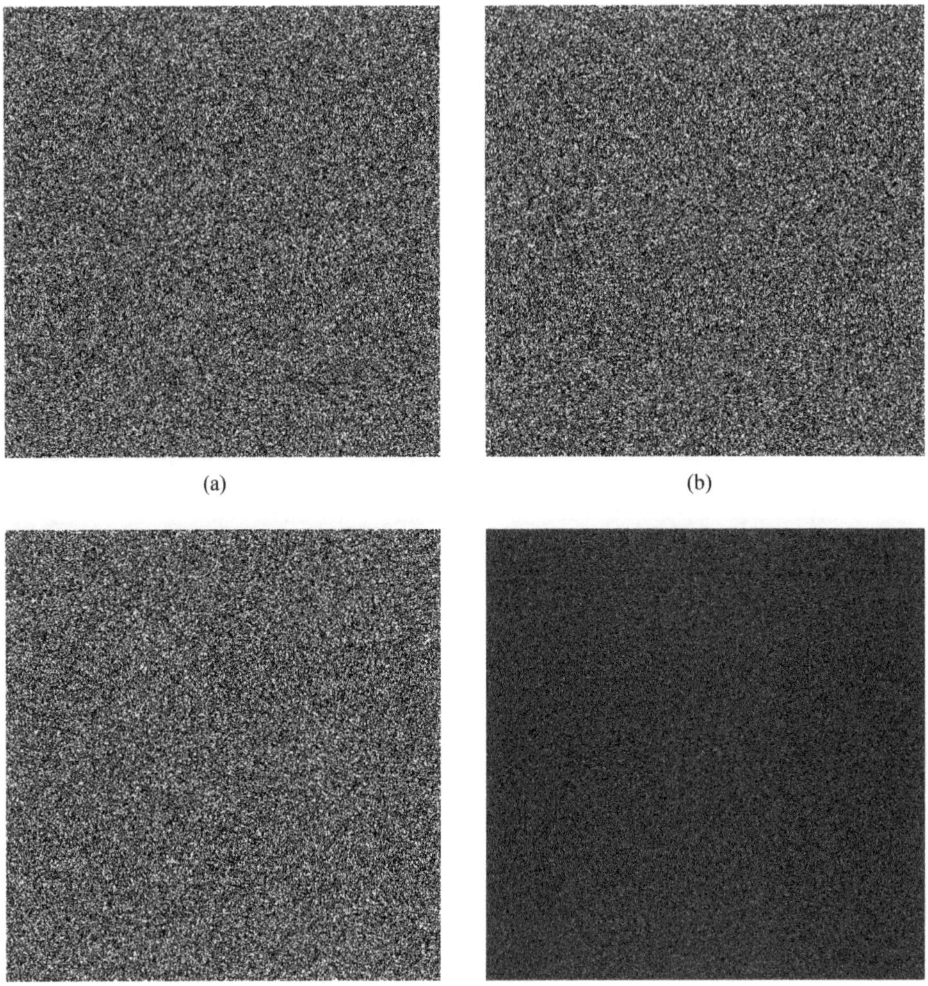

Figure 1.7. (i). Secret keys of user 1 for: (a) R channel, (b) G channel, (c) B channel, and (d) fused sub-band key. Reprinted from [30]. Copyright 2015, with permission from Elsevier B.V. All rights reserved.

$$\left.\begin{aligned} g_{R_m}^1(x, y) &= G^{\alpha_m^1}\left\{C\left(x_m^1, y_m^1, k_m^1, p_m^1, \varepsilon_m^1\right)\left[f_{R_m}^1(x_i, y_i)\right]\right\} \\ g_{G_m}^1(x, y) &= G^{\alpha_m^1}\left\{C\left(x_m^1, y_m^1, k_m^1, p_m^1, \varepsilon_m^1\right)\left[f_{G_m}^1(x_i, y_i)\right]\right\} \\ g_{B_m}^1(x, y) &= G^{\alpha_m^1}\left\{C\left(x_m^1, y_m^1, k_m^1, p_m^1, \varepsilon_m^1\right)\left[f_{B_m}^1(x_i, y_i)\right]\right\} \end{aligned}\right\} \tag{1.35}$$

where $C(\,)$ represents an operator of the chaotic permutation process.

Step 2: Each transformed channel is individually multiplied together and then modulated by a random phase mask to get its multiplex channel.

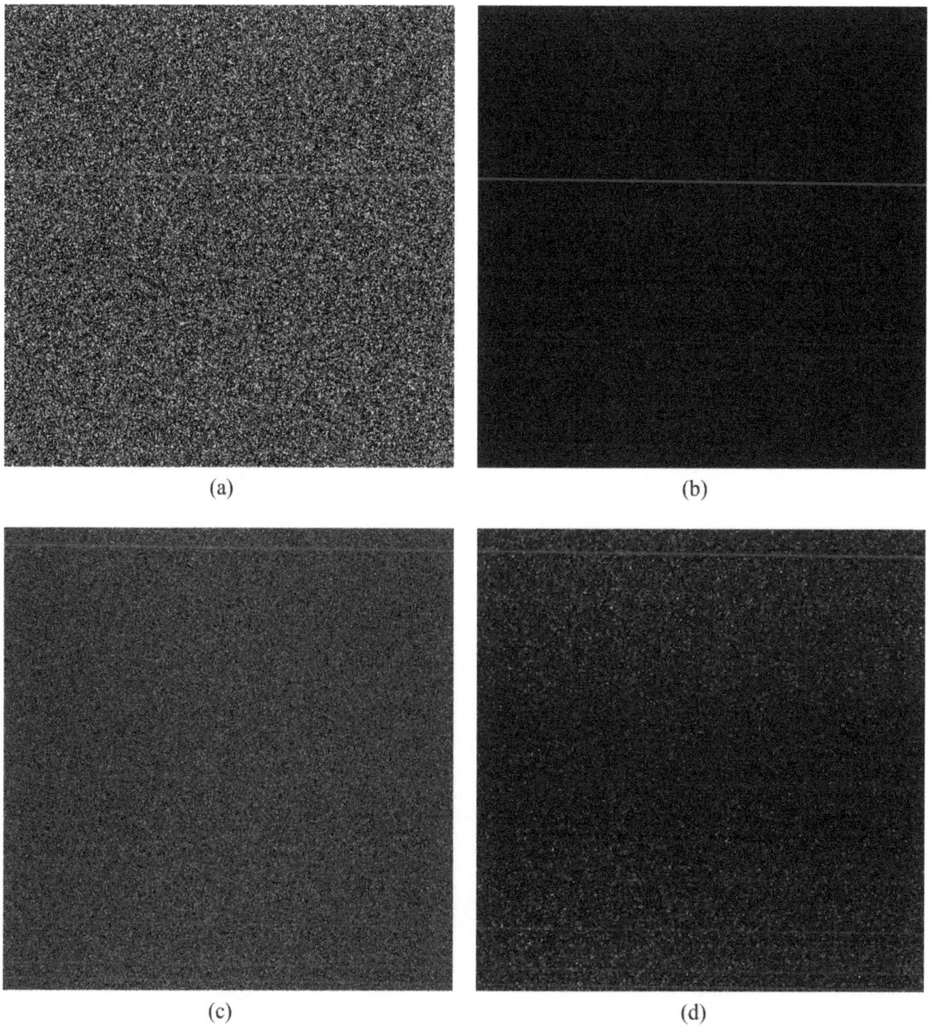

Figure 1.7. (ii). Individual sub-band keys for user 1: (a) *LL* sub-band key, (b) *HL* sub-band key (c) *LH* sub-band key, and (d) *HH* sub-band key. Reprinted from [30]. Copyright 2015, with permission from Elsevier B.V. All rights reserved.

$$\left.\begin{aligned} g_R^1(x, y) &= \prod_{n=1}^{m} g_{R_n}^1(x, y)\exp\left\{i\left[2\pi\phi_R(x, y)\right]\right\} \\ g_G^1(x, y) &= \prod_{n=1}^{m} g_{G_n}^1(x, y)\exp\left\{i\left[2\pi\phi_G(x, y)\right]\right\} \\ g_B^1(x, y) &= \prod_{n=1}^{m} g_{B_n}^1(x, y)\exp\left\{i\left[2\pi\phi_B(x, y)\right]\right\} \end{aligned}\right\} \quad (1.36)$$

Figure 1.8. (i). Encrypted images with all the correct parameters: (a) Ciphertext of User 1, (b) ciphertext of secret image 1, (c) multiplex image of ciphertext of User 1 and secret image 1, and (d) final encrypted image (for all users). Reprinted from [30]. Copyright 2015, with permission from Elsevier B.V. All rights reserved.

Step 3: The first group of encoded multiplex images is formed by the concatenation of $g_R^1(x, y), g_G^1(x, y)$, and $g_B^1(x, y)$ multiplex channels. Correspondingly, steps 1 and 2 are followed to construct the second and third groups of encoded multiplex images.

$$\left.\begin{aligned} g^1(x, y) &= \left[g_R^1(x, y), g_G^1(x, y), g_B^1(x, y)\right] \\ g^2(x, y) &= \left[g_R^2(x, y), g_G^2(x, y), g_B^2(x, y)\right] \\ g^3(x, y) &= \left[g_R^3(x, y), g_G^3(x, y), g_B^3(x, y)\right] \end{aligned}\right\} \tag{1.37}$$

(a) (b) (c) (d)

Figure 1.8. (ii). Decrypted secret images with all the correct parameters: (a) User 1, (b) User 2, (c) User 3, and (d) User 4. Reprinted from [30].

Step 4: The phase and amplitude functions of $g^1(x, y)$, $g^2(x, y)$, and $g^3(x, y)$ are expressed as:

$$P^1(x, y) = \arg[g^1(x, y)], \quad A^1(x, y) = |g^1(x, y)| \tag{1.38}$$

$$P^2(x, y) = \arg[g^2(x, y)], \quad A^2(x, y) = |g^2(x, y)| \tag{1.39}$$

$$P^3(x, y) = \arg[g^3(x, y)], \quad A^3(x, y) = |g^3(x, y)| \tag{1.40}$$

Step 5: The host image is a single-level 2-D discrete wavelet transformed to decompose into four sub-bands W_{LL}, W_{HL}, W_{LH}, and W_{HH}, where

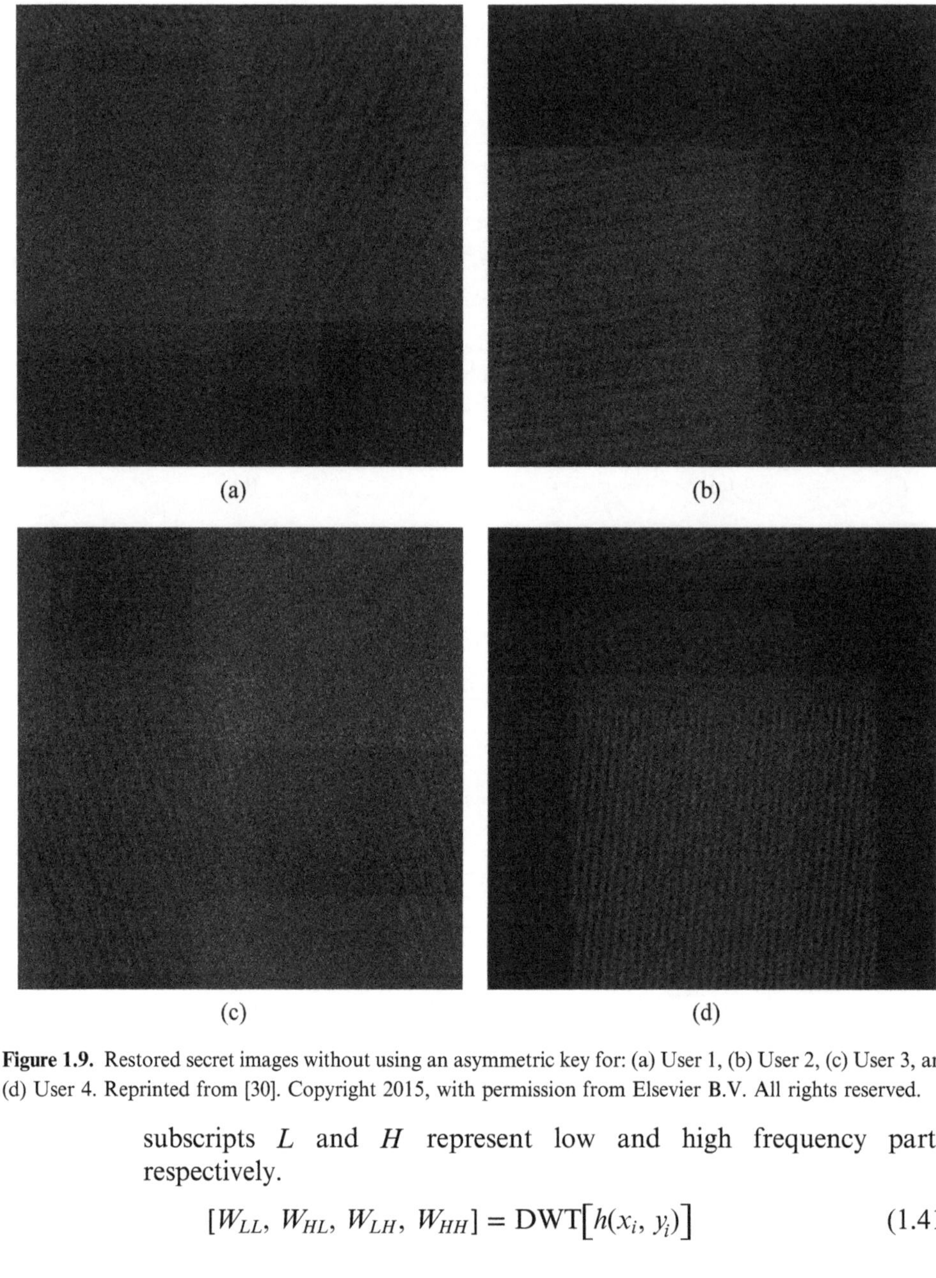

Figure 1.9. Restored secret images without using an asymmetric key for: (a) User 1, (b) User 2, (c) User 3, and (d) User 4. Reprinted from [30]. Copyright 2015, with permission from Elsevier B.V. All rights reserved.

subscripts L and H represent low and high frequency parts, respectively.

$$[W_{LL}, W_{HL}, W_{LH}, W_{HH}] = \mathrm{DWT}\left[h(x_i, y_i)\right] \tag{1.41}$$

Step 6: Since most of the energy is concentrated on the W_{LL} sub-band of the host image, W_{HL}, W_{LH}, and W_{HH} sub-bands of the host image are replaced with $P^1(x, y)$, $P^2(x, y)$, and $P^3(x, y)$, respectively. Then, an inverse single-level 2-D discrete wavelet is transformed to attain a watermarked image $E(x_i, y_i)$ as the final encrypted image, which is visually the same as the host image.

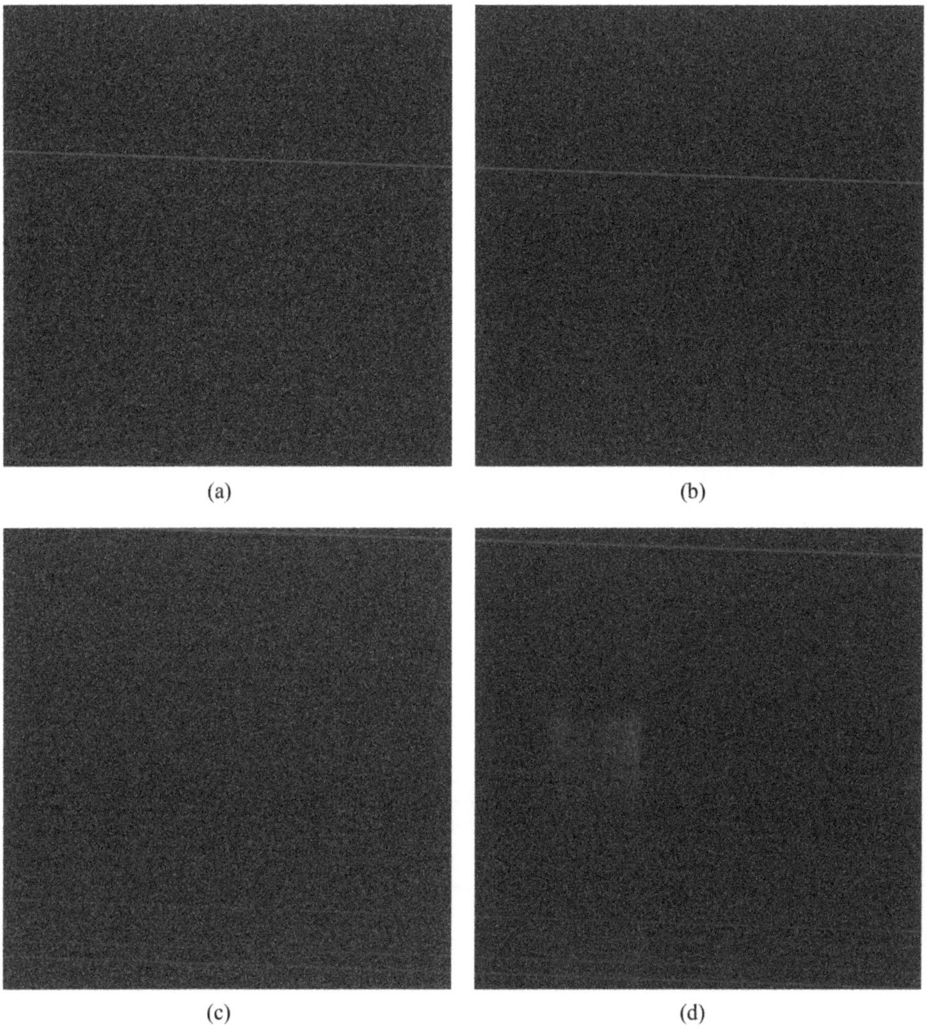

Figure 1.10. Recovered secret images without using an authentication key for: User 1, (b) User 2, (c) User 3, and (d) User 4. Reprinted from [30]. Copyright 2015, with permission from Elsevier B.V. All rights reserved.

$$E(x_i, y_i) = \mathrm{IDWT}(W_{LL}, P^1(x, y), P^2(x, y), P^3(x, y)). \tag{1.42}$$

Individual decryption keys of Group 1 multiple images are produced as

$$I^1_{R_n}(x, y) = \left\{ \prod_{\substack{k=1 \\ k \neq n}}^{m} g^1_{R_k}(x, y) \exp\left\{ \mathrm{i}\left[2\pi\phi_R(x, y)\right]\right\} \right\}^{-1} \tag{1.43}$$

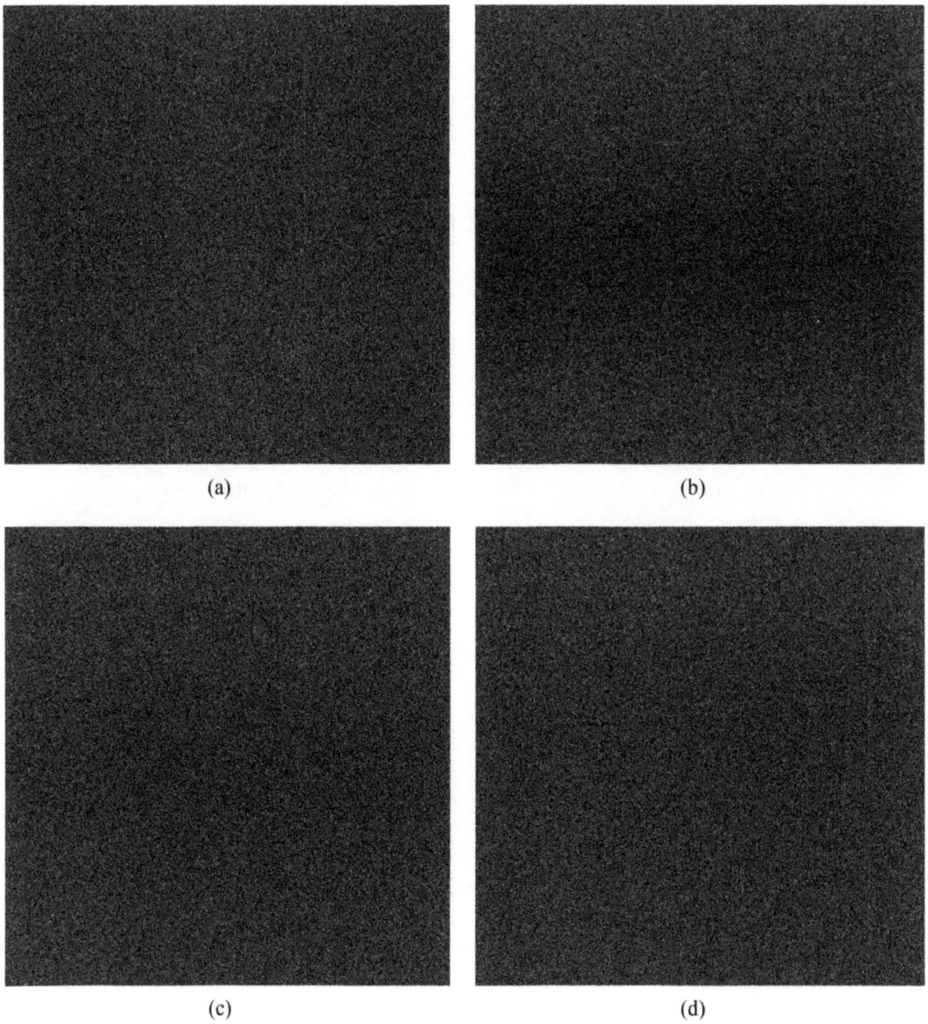

Figure 1.11. Reconstructed secret images without secret keys for: (a) User 1, (b) User 2, (c) User 3, and (d) User 4. Reprinted from [30].

$$I^1_{G_n}(x, y) = \left\{ \prod_{\substack{k=1 \\ k\neq n}}^{m} g^1_{G_k}(x, y) \exp\left\{ \mathrm{i}\left[2\pi\phi_G(x, y)\right]\right\} \right\}^{-1} \tag{1.44}$$

$$I^1_{B_n}(x, y) = \left\{ \prod_{\substack{k=1 \\ k\neq n}}^{m} g^1_{B_k}(x, y) \exp\left\{ \mathrm{i}\left[2\pi\phi_B(x, y)\right]\right\} \right\}^{-1}. \tag{1.45}$$

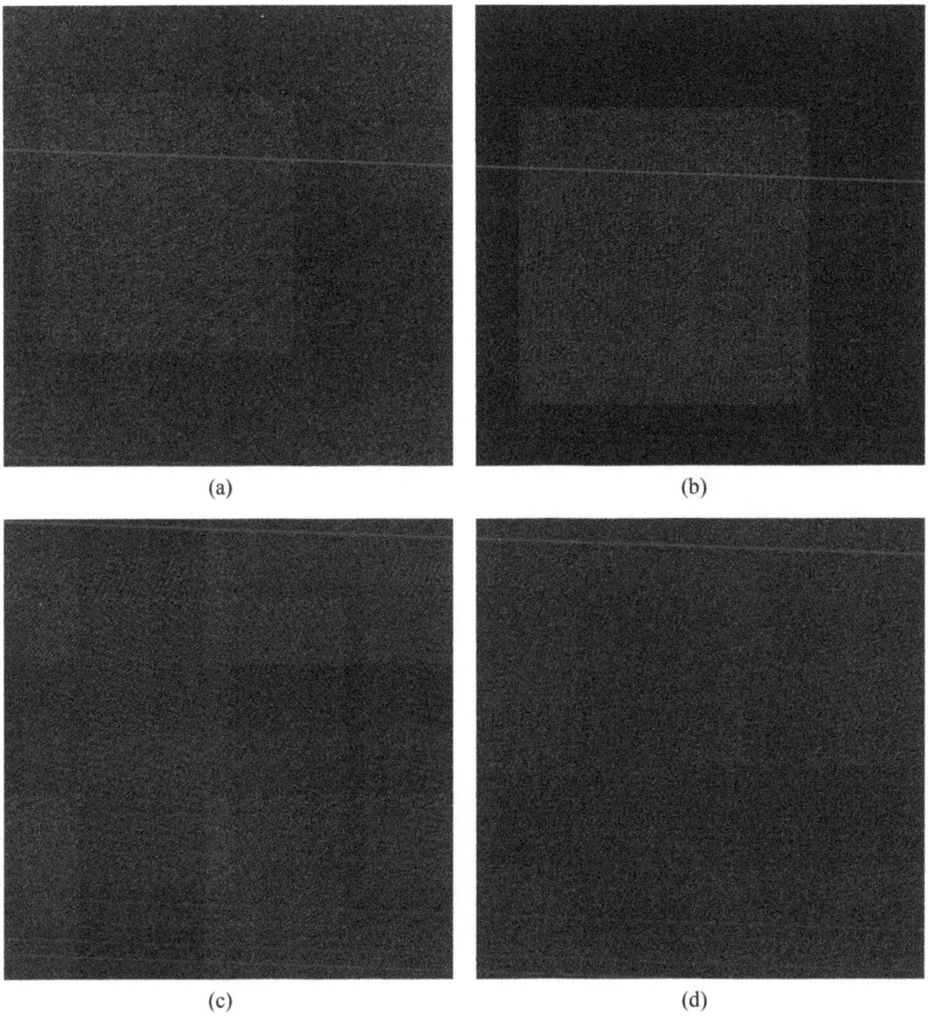

Figure 1.12. (i). Retrieved secret images without the LL sub-band key for: (a) User 1, (b) User 2, (c) User 3, and (d) User 4. Reprinted from [30].

The watermark extraction and group-multiple-image decoding processes imply the following steps:

Step 1: The encrypted image $E(x_i, y_i)$ is a single-level 2-D discrete wavelet transformed to obtain W_{LL}, W_{HL}^1, W_{LH}^2, and W_{HH}^3 sub-bands.

$$[W_{LL}, W_{HL}^1, W_{LH}^2, W_{HH}^3] = \mathrm{DWT}\left[E(x_i, y_i)\right] \tag{1.46}$$

Step 2: W_{HL}^1, W_{LH}^2, and W_{HH}^3 sub-bands are phase encoded, then respectively, multiplied by the amplitudes of the encoded multiplex image as group decryption keys $A^1(x, y)$,$A^2(x, y)$, and $A^3(x, y)$ to get corresponding $D^1(x, y)$, $D^2(x, y)$, and $D^3(x, y)$.

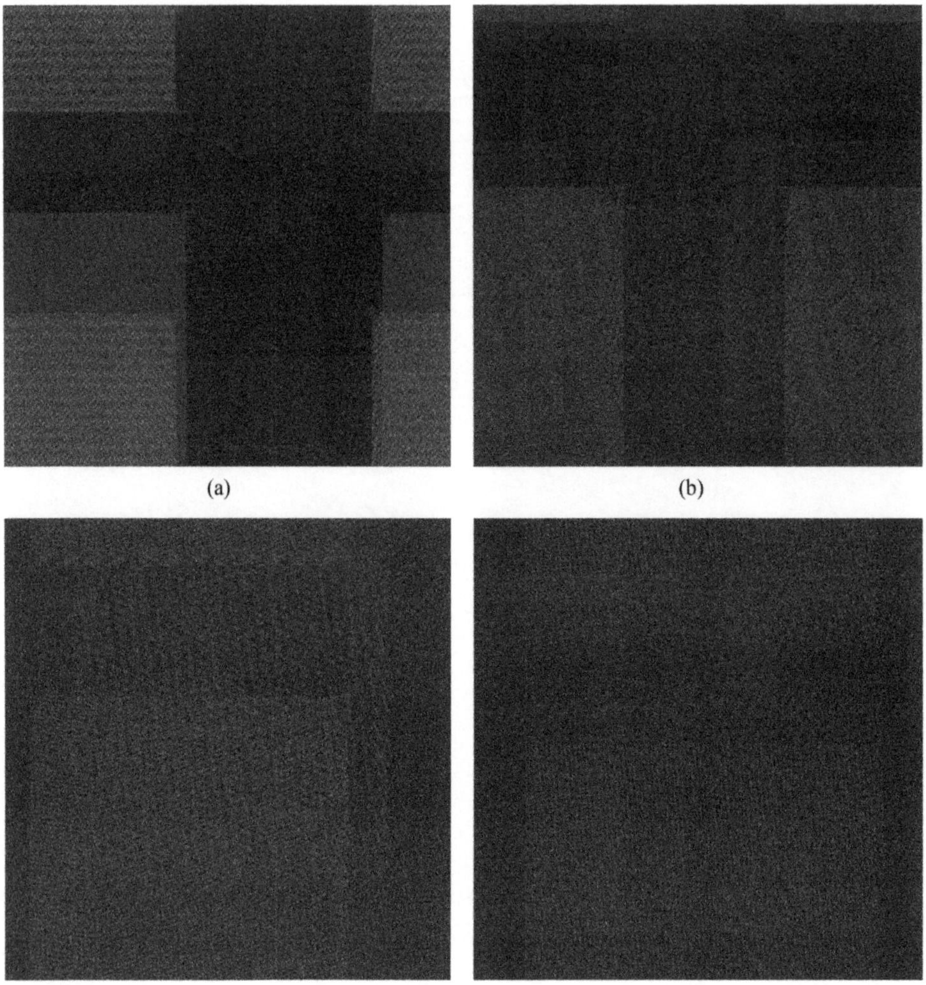

(a) (b) (c) (d)

Figure 1.12. (ii). Restored secret images without the *HL* sub-band key for: (a) User 1, (b) User 2, (c) User 3, and (d) User 4. Reprinted from [30].

$$\left.\begin{aligned} D^1(x, y) &= \exp[i(W_{HL}^1)] \times A^1(x, y) \\ D^2(x, y) &= \exp[i(W_{LH}^2)] \times A^2(x, y) \\ D^3(x, y) &= \exp[i(W_{HH}^3)] \times A^3(x, y) \end{aligned}\right\} \tag{1.47}$$

Step 3: $D^1(x, y)$, $D^2(x, y)$, and $D^3(x, y)$ are independently decomposed into R,G, and B images.

$$\left.\begin{aligned} D^1(x, y) &= [D_R^1(x, y), D_G^1(x, y), D_B^1(x, y)] \\ D^2(x, y) &= [D_R^2(x, y), D_G^2(x, y), D_B^2(x, y)] \\ D^3(x, y) &= [D_R^3(x, y), D_G^3(x, y), D_B^3(x, y)] \end{aligned}\right\} \tag{1.48}$$

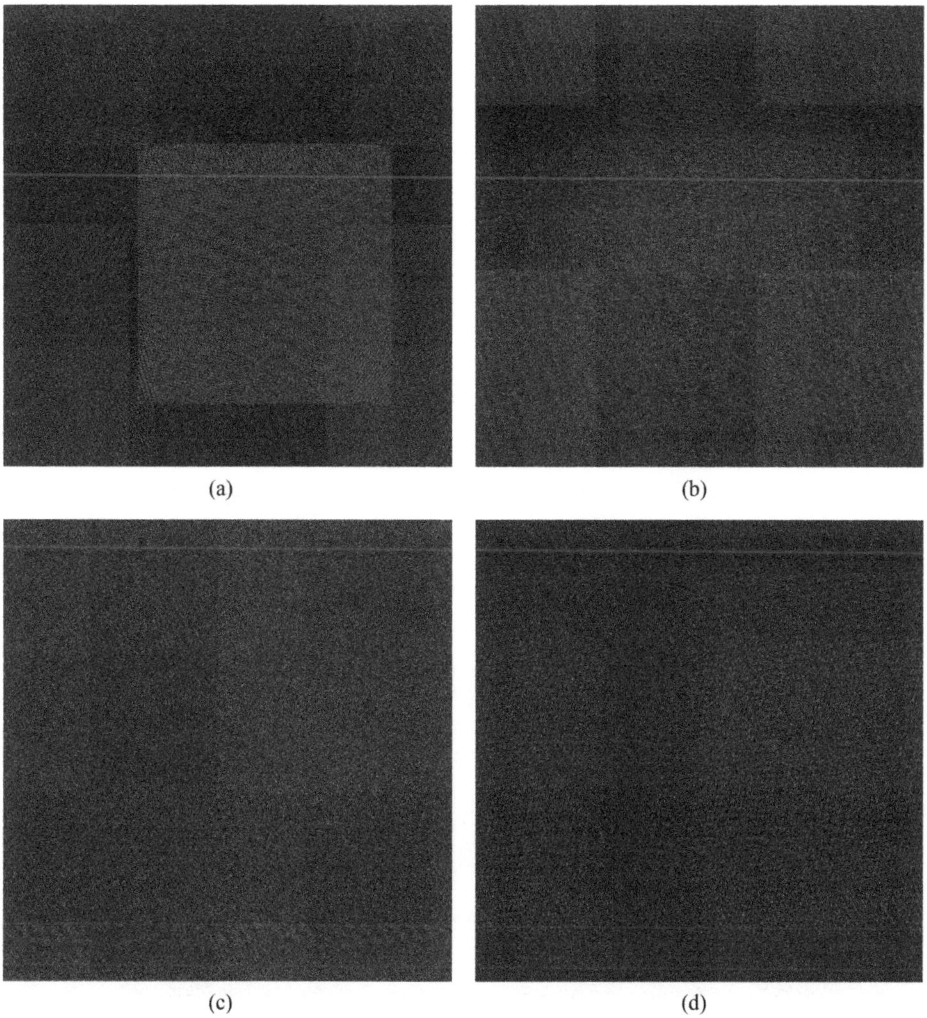

Figure 1.12. (iii). Recovered secret images without the *LH* sub-band key for: (a) User 1, (b) User 2, (c) User 3, and (d) User 4. Reprinted from [30].

Step 4: To decrypt the first group of multiple images, $D_R^1(x, y)$, $D_G^1(x, y)$, and $D_B^1(x, y)$ are first multiplied with corresponding individual decryption keys $I_{R_n}^1(x, y)$, $I_{G_n}^1(x, y)$, and $I_{B_n}^1(x, y)$, and then inverse gyrator transformed. Finally, inverse chaotic permutations processes are separately performed on the resulting image to retrieve R,G, and B channels of the first group of multiple images.

$$\left.\begin{aligned} f_{R_m}^1(x, y) &= C^{-1}\left(x_m^1, y_m^1, k_m^1, p_m^1, \varepsilon_m^1\right)\left\{G^{-\alpha_m^1}\left[D_{R_m}^1(x_i, y_i)I_{R_n}^1(x, y)\right]\right\} \\ f_{G_m}^1(x, y) &= C^{-1}\left(x_m^1, y_m^1, k_m^1, p_m^1, \varepsilon_m^1\right)\left\{G^{-\alpha_m^1}\left[D_{G_m}^1(x_i, y_i)I_{G_n}^1(x, y)\right]\right\} \\ f_{B_m}^1(x, y) &= C^{-1}\left(x_m^1, y_m^1, k_m^1, p_m^1, \varepsilon_m^1\right)\left\{G^{-\alpha_m^1}\left[D_{B_m}^1(x_i, y_i)I_{B_n}^1(x, y)\right]\right\} \end{aligned}\right\}. \tag{1.49}$$

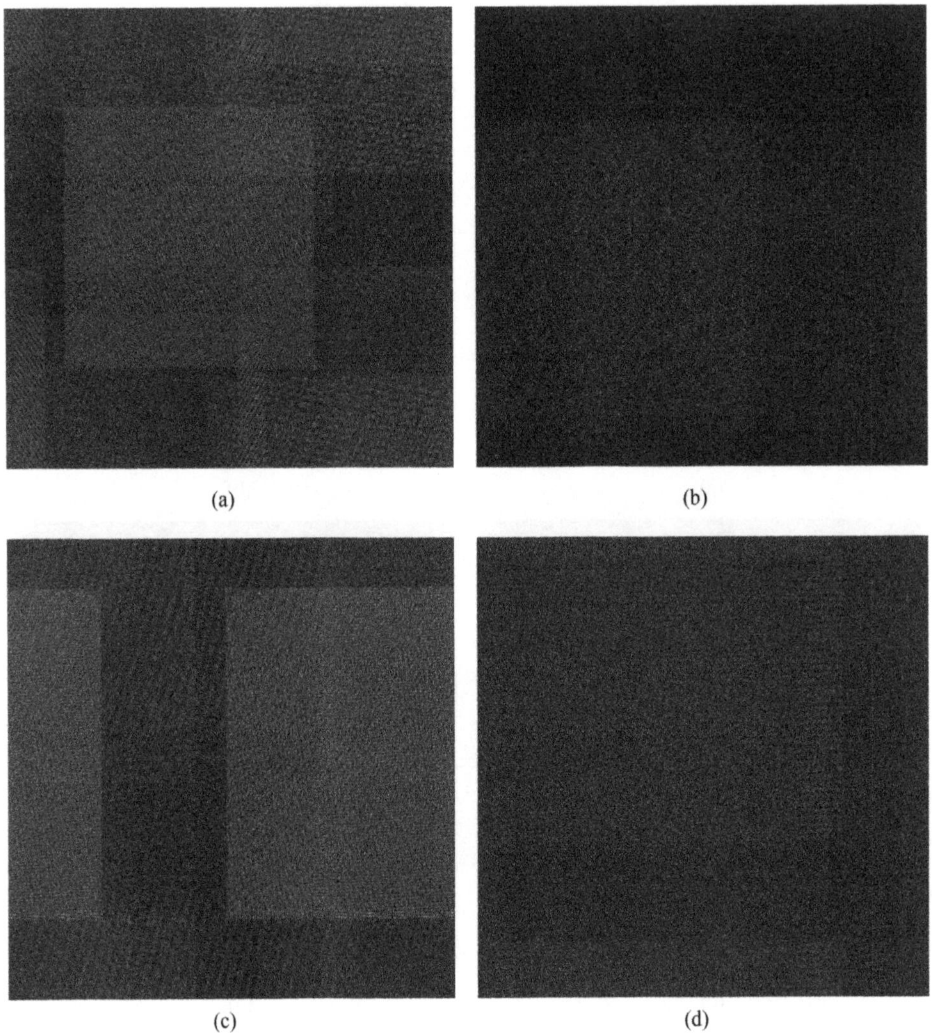

Figure 1.12. (iv). Reconstructed secret images without the *HH* sub-band key for: (a) User 1, (b) User 2, (c) User 3, and (d) User 4. Reprinted from [30].

Similarly, steps 1–4 are repeated to recover *R*, *G*, and *B* channels of the second and third groups of multiple images.

It is worth noting that the encryption process can be realized digitally on a computer system and the decryption process can be implemented using optoelectronic devices for a 2-D GWT decryption system as shown in figure 1.1(b).

For simplicity, decryption of only the first group is illustrated. The digitally obtained watermarked image $E(x_i, y_i)$ is displayed on the first spatial light modulator (SLM_1) and illuminated by a uniform plane wave, and then WT is optically accomplished by Fourier-transforming the information, multiplying by a MRMF, and then performing an inverse Fourier-transform. Each daughter wavelet

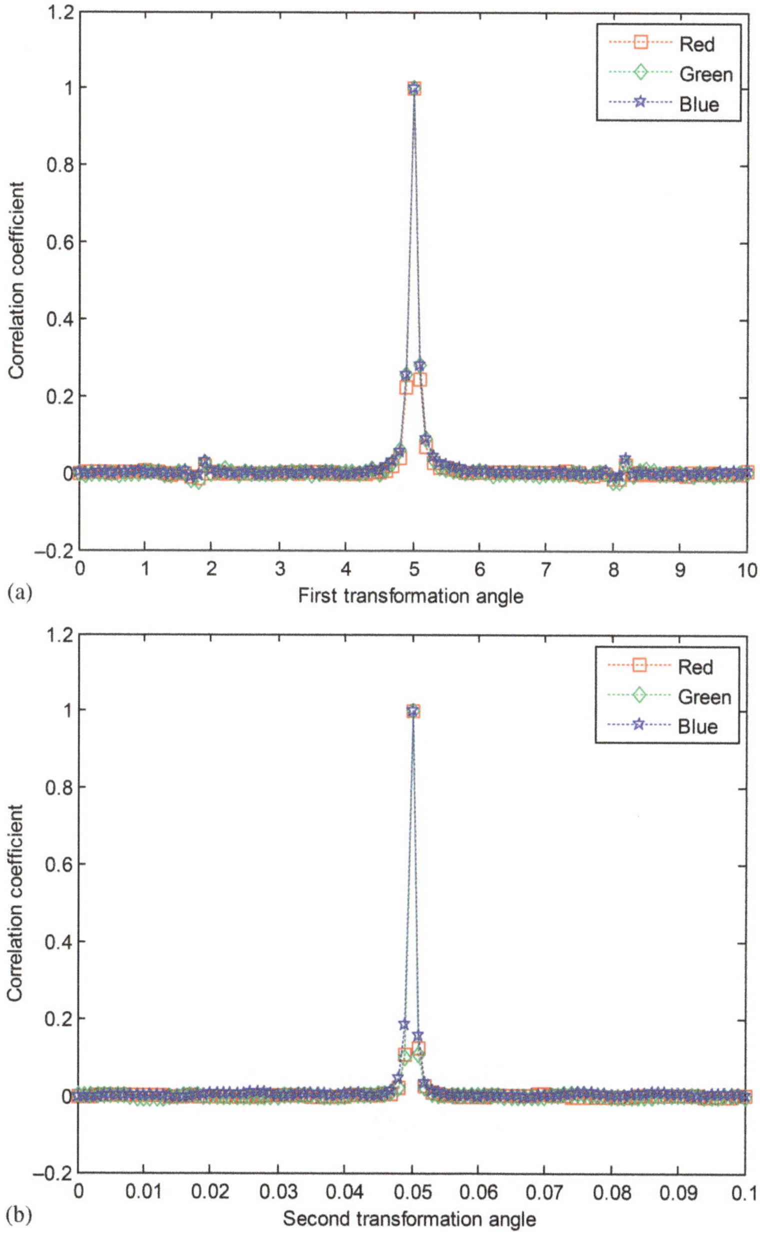

Figure 1.13. (a) Correlation coefficient against deviation of the first transformation angle for the secret image of user 1. (b) Correlation coefficient against deviation of the second transformation angle for the secret image of user 1. Reprinted from [30]. Copyright 2015, with permission from Elsevier B.V. All rights reserved.

is obtained with a different reference beam. The retrieved W_{HL}^1 is phase encoded, multiplied by $A^1(x, y)$, decomposed into $D_R^1(x, y)$, $D_G^1(x, y)$, and $D_B^1(x, y)$ and then independently multiplied with corresponding individual decryption keys $I_{R_n}^1(x, y)$, $I_{G_n}^1(x, y)$, and $I_{B_n}^1(x, y)$ in a computer system. Now $D_R^1(x, y) \times I_{R_n}^1(x, y)$

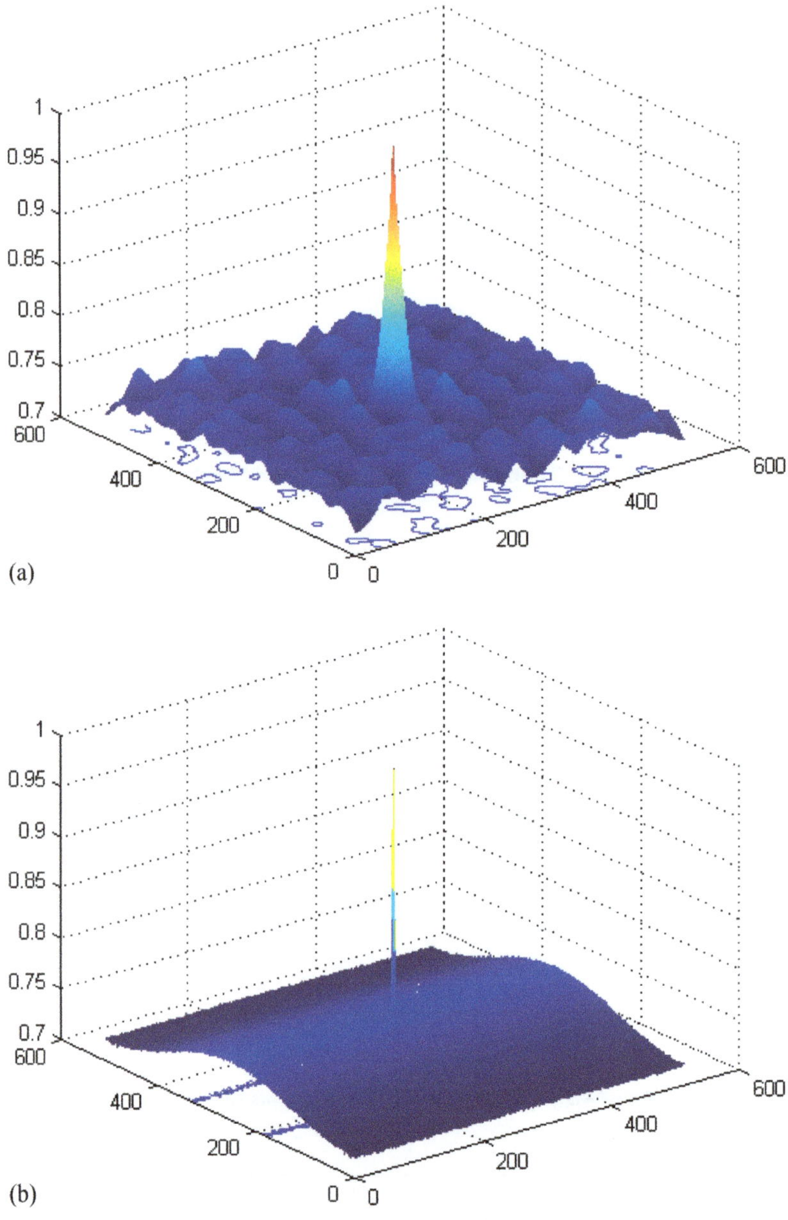

Figure 1.14. The auto-correlation coefficients of (a) the original secret image as shown in figure 1.3(a), and (b) the final encrypted image as shown in figure 1.8(i)(d). Reprinted from [30]. Copyright 2015, with permission from Elsevier B.V. All rights reserved.

is imported into SLM_2 and then GT is optically performed by using three generalized lenses L_1, L_2 and L_3 at transformation angle $-\alpha$. The output image is registered using a CCD_2 camera. Finally, the inverse chaotic permutation process is performed digitally on the output image to decrypt the R channel of the first group of multiple images. In the same process, the G and B channels are recovered.

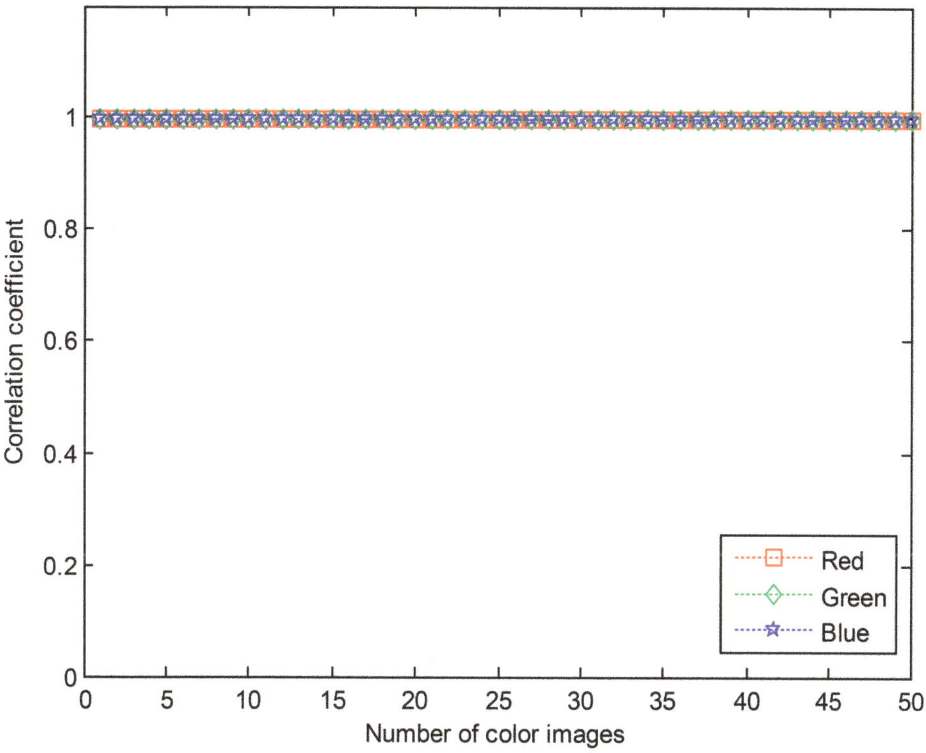

Figure 1.15. Correlation coefficient against the number of color images. Reprinted from [30]. Copyright 2015, with permission from Elsevier B.V. All rights reserved.

Numerical simulations on a Matlab 7.11.0 (R2010b) platform have been conducted to prove the feasibility of the proposed idea. Each secret color image is 256 × 256 × 3 pixels in size. For simplicity, each group consists of two secret color images. The groups I, II, and III include images of (Barbara and Father), (Hadi and Date), and (Mahdi and Honey) as shown in figures [1.16(a) and 1.16(b)], [1.16(c) and 1.16(d)], and [1.16(e) and 1.16(f)], respectively. The encrypted groups I, II, and III are depicted in figures 1.17(a)–(c), respectively. As can be seen, the resulting images are noisy-like appearances and thus the original image can not be perceived. The decryption keys of groups I, II, and III are displayed in figures 1.17(d)–(f), respectively. The decrypted groups I, II, and III with the correct parameters illustrated in figures [1.18(a) and 1.18(b)], [1.18(c) and 1.18(d)], and [1.18(e) and 1.18(f)], respectively. It can be observed that the original images are reconstructed only with the correct decrypting keys. The host image of 512 × 512 × 3 sized pixels is shown in figure 1.19(a). Its single-level 2-D discrete wavelet decomposition is *LL*, *HL*, *LH*, and *HH* sub-bands as demonstrated in figure 1.19(b). So, each sub-band will be 256 × 256 × 3 pixels in size. The resulting image obtained by replacing encrypted I, II, and III groups of multiple-images with corresponding *HL*, *LH*, and *HH* sub-bands of the host image, is illustrated in figure 1.19(c). The encrypted

Figure 1.16. Secret color images, each 256 × 256 × 3 pixels in size, to be encrypted: Group I [(a) Barbara, (b) Father], Group II [(c) Hadi, (d) Date], and Group III [(e) Mahdi, (f) Honey]. (a)–(c) reproduced from [31]; (d)–(f) reprinted from [30], copyright 2015, with permission from Elsevier B.V, all rights reserved.

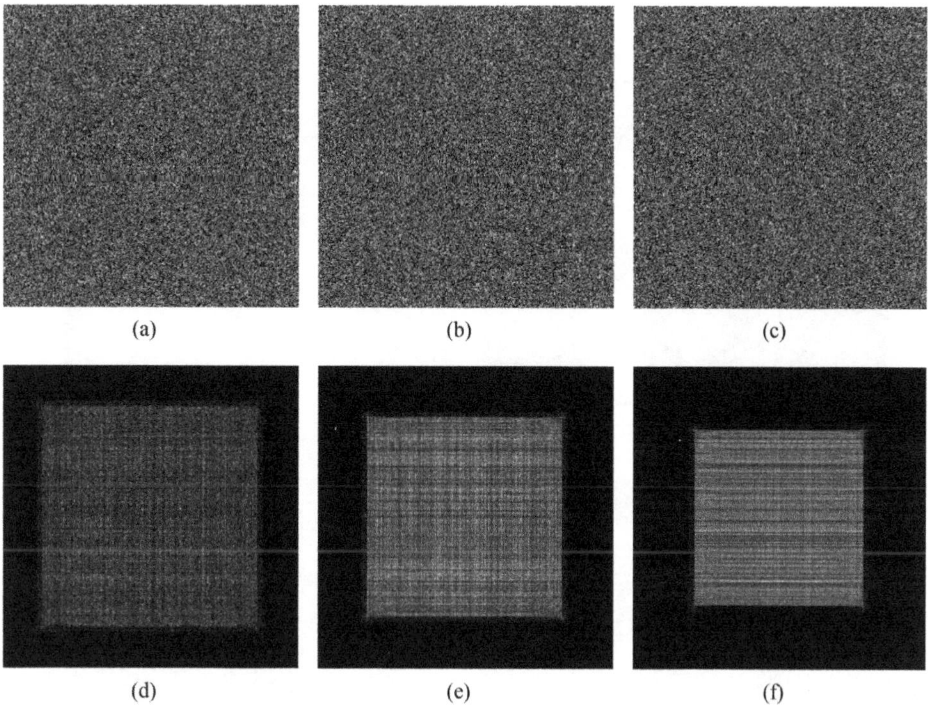

Figure 1.17. (a) Encrypted group I (b) encrypted group II, (c) encrypted group III, (d) decryption key of group I, (e) decryption key of group II, and (f) decryption key of group III. Reproduced from [31].

watermark image is shown in figure 1.19(d). It can be seen that the encrypted groups of multiple-images are fully disguised as the encrypted image visually resembles the host image.

The robustness of the decryption keys (/or asymmetric keys) of the security system for two images of group I has been analyzed during image decryption. Figures 1.20(a) and 1.21(a) show the decrypted images without decryption keys. The corresponding MSE values are $(2.793 \times 10^8, 1.8214 \times 10^8, 1.7763 \times 10^8)$ and $(3.3339 \times 10^8, 1.8237 \times 10^8, 1.7680 \times 10^8)$. The MSE values are quite high and no constructive information about the primary images could be observed from it. Thus, the decryption keys introduced in the proposed system are essential to the decryption process in order to recover the original images of the group.

The sensitivity of the decrypted results with respect to the error of the parameters of the proposed system for two images of group I has also been studied. Figures 1.20(b)–(f) display the restored images for image 1 of group 1 with, x_1 changed by 1×10^{-15}, y_1 changed by 1×10^{-16}, p_1 changed by 1×10^{-15}, ε_1 changed by 1×10^{-16}, and α_1 changed by 0.04. The corresponding MSE values are $(5.6383 \times 10^3, 4.3507 \times 10^3, 5.1641 \times 10^3)$, $(5.5510 \times 10^3, 4.3006 \times 10^3, 5.1105 \times 10^3)$, $(5.5232 \times 10^3, 4.4323 \times 10^3, 5.2417 \times 10^3)$, $(6.3149 \times 10^3, 4.4584 \times 10^3, 5.2779 \times 10^3)$, and $(6.4571 \times 10^3, 4.4992 \times 10^3, 4.5880 \times 10^3)$. These results confirm that the sensitivities of the initial parameters, bifurcation parameter and coupling constant are of the order of 10^{-15} to 10^{-16}, whereas

Figure 1.18. Decrypted secret color images with all the correct keys: Group I [(a) Barbara, (b) Father], Group II [(c) Hadi, (d) Date], and Group III [(e) Mahdi, (f) Honey]. Reproduced from [31].

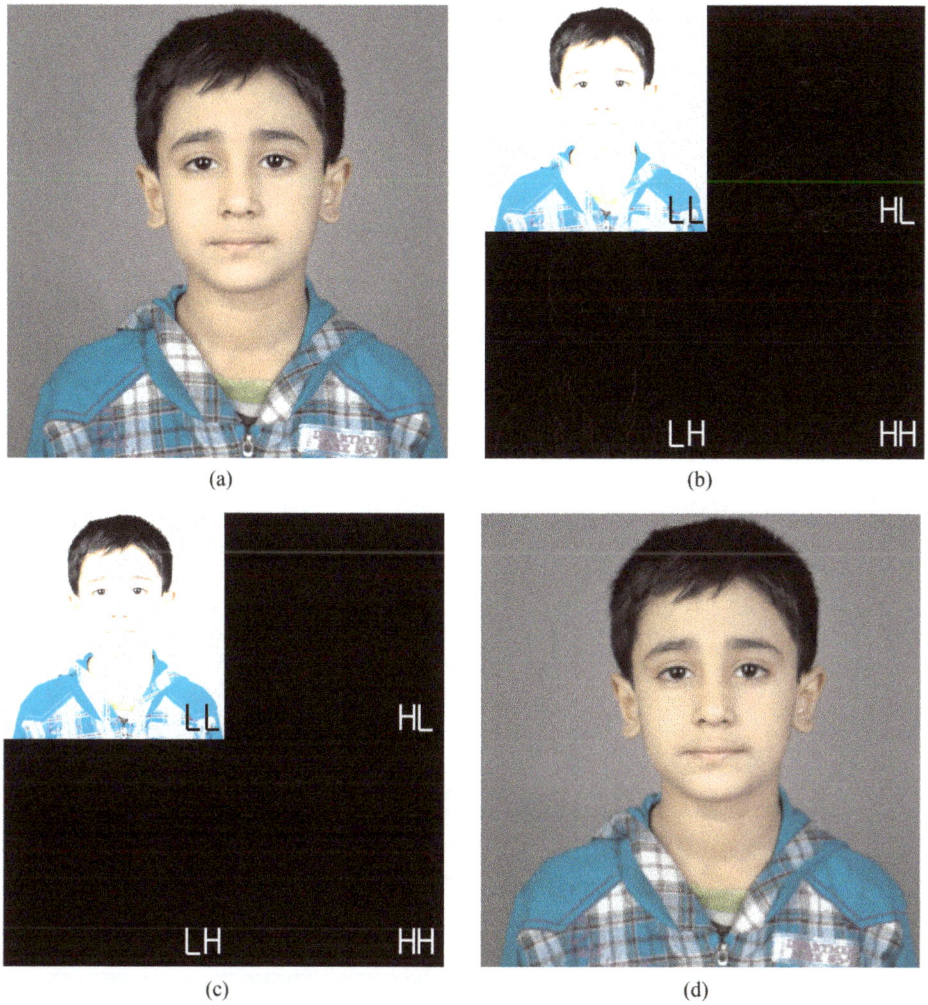

Figure 1.19. (a) Host image (size 512 × 512 × 3 pixels), (b) single-level 2-D discrete wavelet decomposition of the host image, (c) resultant image obtained by fusing encrypted I, II, and III groups of multiple-images with corresponding HL, LH, and HH sub-bands of the host image, (d) encrypted watermark image. Reproduced from [31].

the sensitivity of the transformation angle is 10^{-2}. Therefore the multilayered security levels of image 1 of group 1 are very sensitive with very small deviations from the encryption values.

Figures 1.21(b)–(f) demonstrate the reconstructed images for image 2 of group 1 with, x_2 changed by 1×10^{-16}, y_2 changed by 1×10^{-16}, p_2 changed by 1×10^{-16}, ε_2 changed by 1×10^{-16}, and α_2 changed by 0.04. The corresponding MSE values are (9.3058×10^3, 9.9444×10^3, 6.5783×10^3), (9.3058×10^3, 9.9444×10^3, 6.5783×10^3), (9.2357×10^3, 9.8952×10^3, 6.5458×10^3), (9.3124×10^3, 9.9379×10^3, 6.5894×10^3), and (7.7653×10^3, 7.4104×10^3, 4.2864×10^3). These results also verify that the sensitivities of the initial parameters, bifurcation parameter and

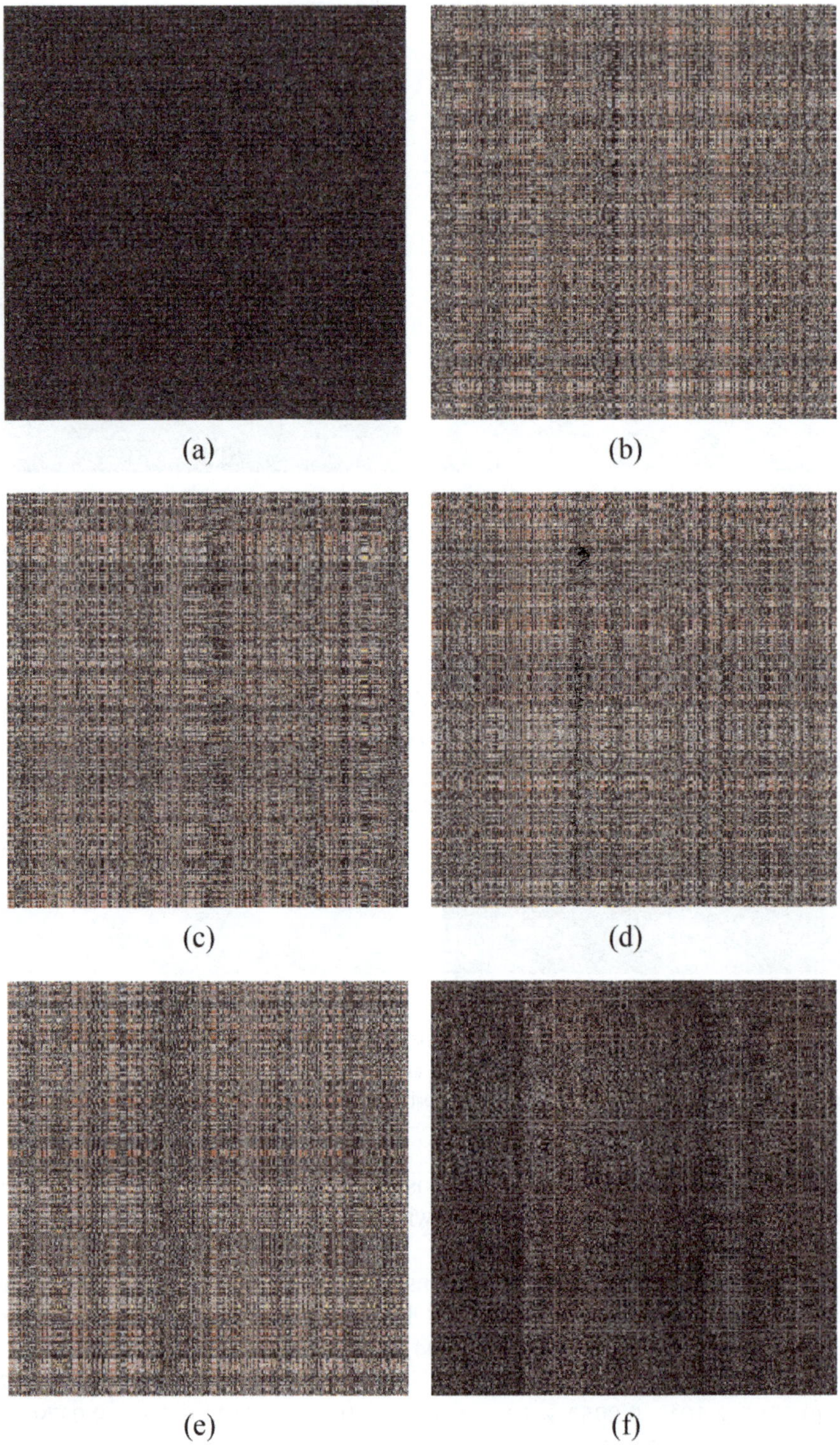

Figure 1.20. Decrypted results for image 1 of group 1 with: (a) no decryption key, (b) x_1 changed by 1×10^{-15}, (c) y_1 changed by 1×10^{-16} (d) p_1 changed by 1×10^{-15}, (e) ε_1 changed by 1×10^{-16},(f) α_1 changed by 0.04. Reproduced from [31].

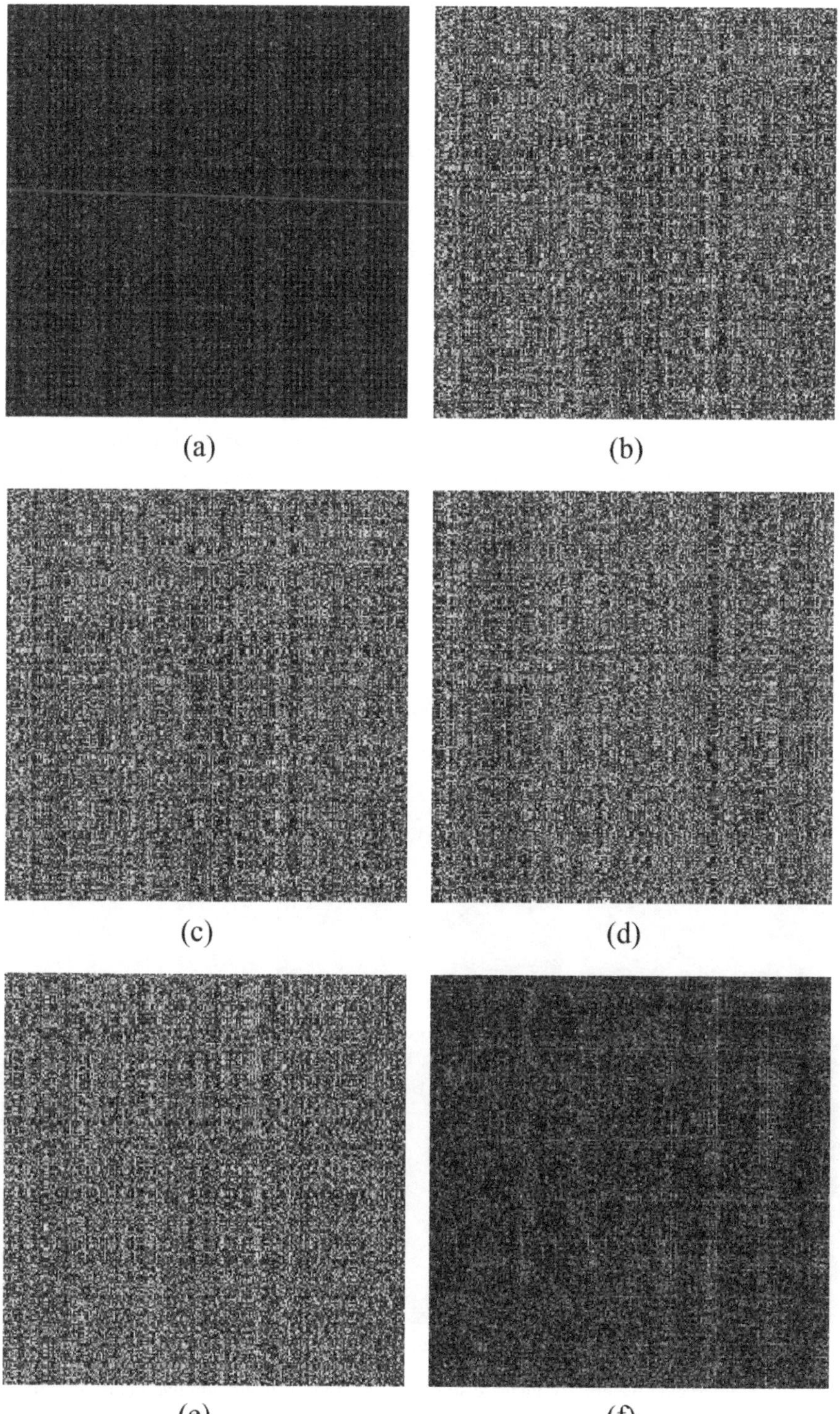

Figure 1.21. Decrypted results for image 2 of group 1 with: (a) no decryption key, (b) x_2 changed by 1×10^{-16}, (c) y_2 changed by 1×10^{-16} (d) p_2 changed by 1×10^{-16}, (e) ε_2 changed by 1×10^{-16}, (f) α_2 changed by 0.04. Reproduced from [31].

coupling constant are of the order of 10^{-16}, while the sensitivity of the transformation angle is 10^{-2}. Thus, the multilayered security levels of image 2 of group 1 are also very sensitive with very small variations from the encryption values.

The robustness against noise perturbation of the security scheme has been researched. Figure 1.22(a) is the encrypted watermark image distorted by zero-mean white additive Gaussian noise with a standard deviation of 0.01. Figures 1.22(b) and (c) are the decrypted images recovered from the noised watermark image with all the correct decryption keys. The tolerance against the loss of encrypted data has been studied. Figure 1.22(d) is the encrypted watermark image with 50% occlusion. Figures 1.22(e) and (f) are the decrypted images retrieved from the occluded watermark image with all the correct decryption keys. From these results, one can still discern the main information of the original images of the group and thus the system can resist against a certain level of noise perturbation or occlusion.

The dependence of MSE for R, G, and B channels of images 1 and 2 of group 1 on the variation of transformation angles α_1 and α_2 is displayed in figures 1.23(a) and (b), respectively. The MSE values reach zero only when α_1 and α_2 are at the correct encryption values. The MSE values increase sharply for a very narrow deviation from the correct values.

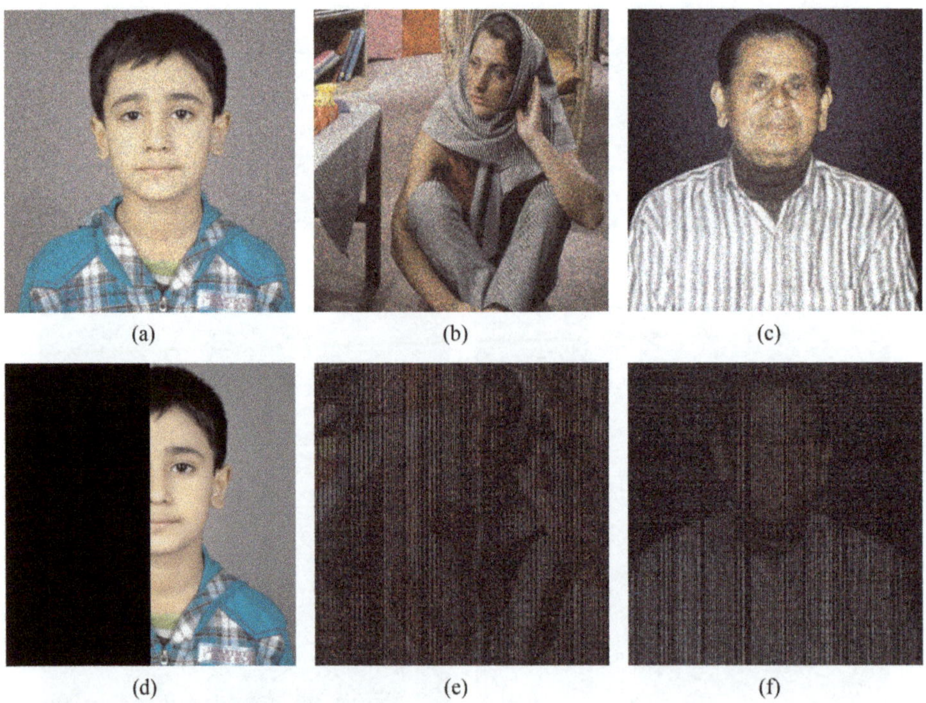

Figure 1.22. (a) Encrypted watermark image distorted by zero-mean white additive Gaussian noise with a standard deviation of 0.01 and corresponding, (b) decrypted image 1 (Barbara) of group 1, (c) decrypted image 2 (Father) of group 1; (d) encrypted watermark image with 50% occlusion and corresponding, (e) retrieved image 1 (Barbara) of group 1, and (f) retrieved image 2 (Father) of group 1. Reproduced from [31].

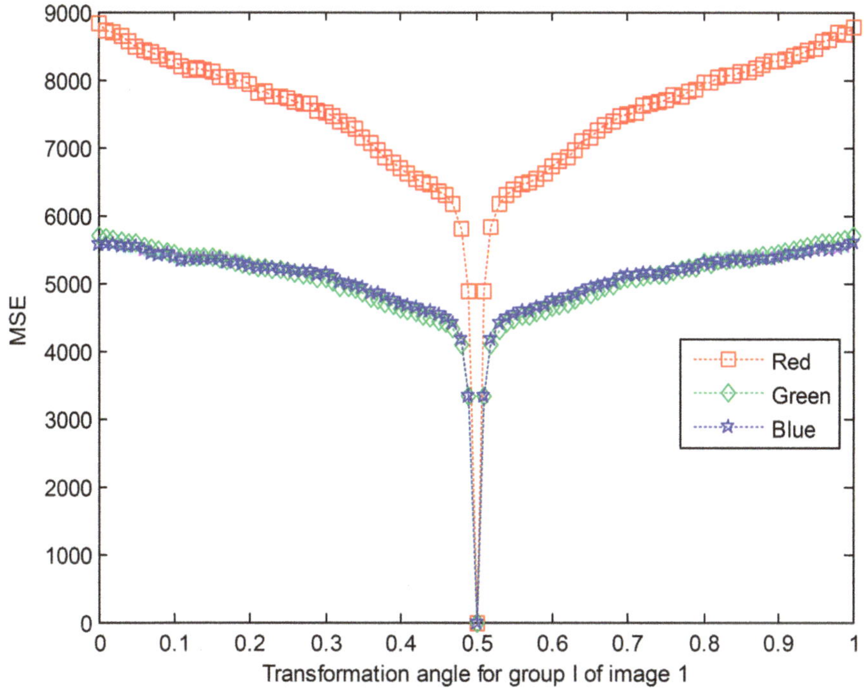

(a)

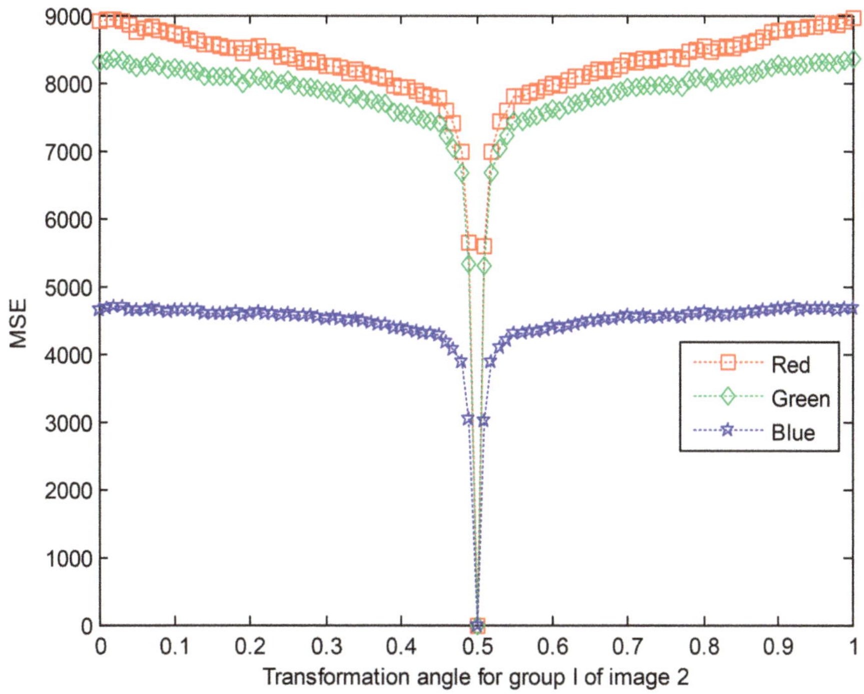

(b)

Figure 1.23. MSE against deviation of the transformation angle for group I: image 1, and (b) image 2. Reproduced from [31].

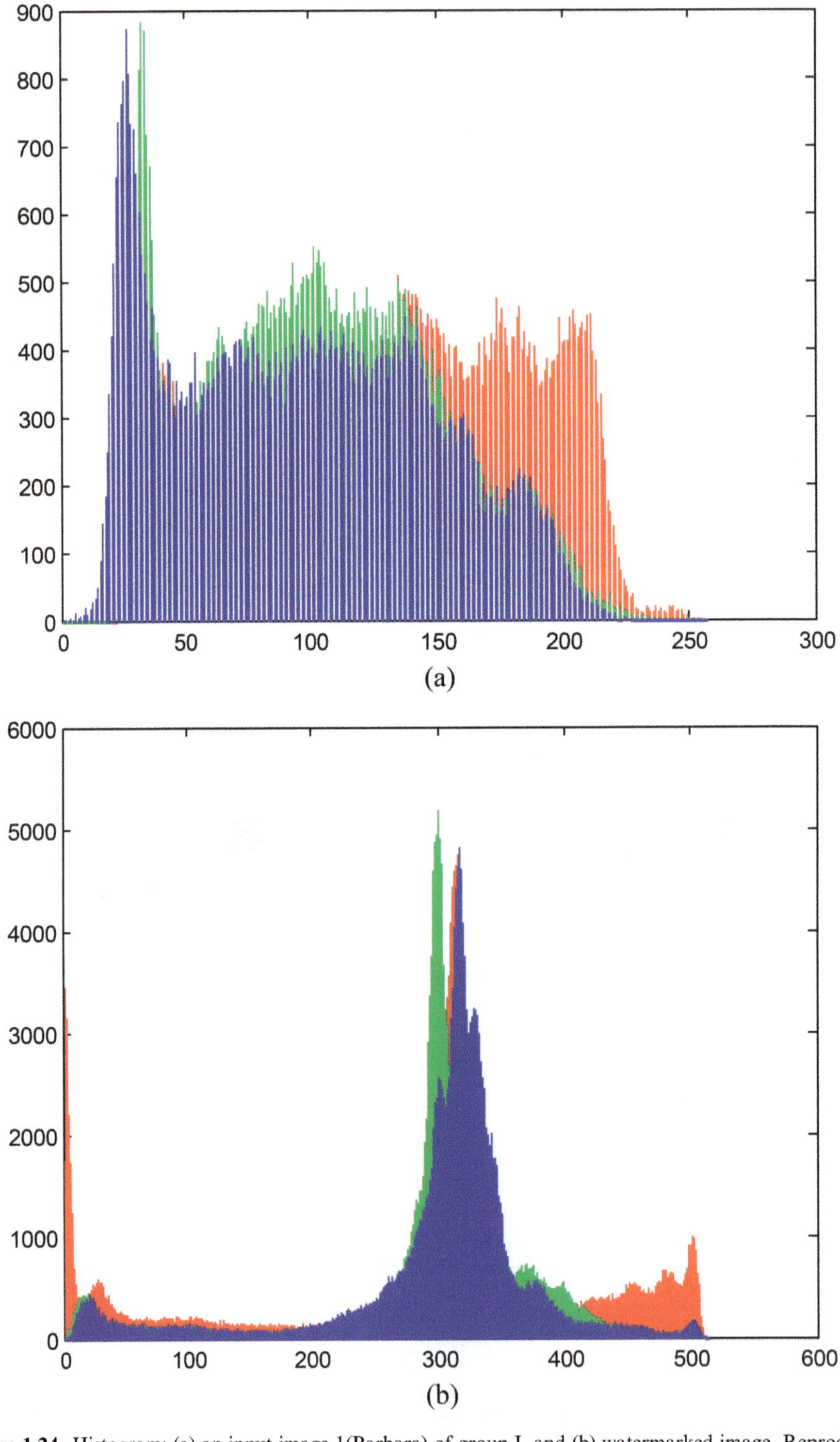

Figure 1.24. Histogram: (a) an input image 1(Barbara) of group I, and (b) watermarked image. Reproduced from [31].

The security of the cryptosystem is discussed. The generations of decryption keys during the encryption stage are non-linear; hence the system will resist the known plaintext and chosen-plaintext attacks. The special attack cracks PTFT-based asymmetric cryptosystem by using two public phase keys and ciphertext as three constraints based on the iterative phase retrieval algorithm [14]. The cryptosystem exploits a sequence of chaotic pairs generated with a system of two-symmetrically coupled identical logistic maps to permutate input information, so unauthorized users do not have sufficient constraints to set up the retrieval, however they identify the correct transformation angle of GT orders.

Statistical analysis of image 1 group I has been exemplified. The histogram of the original image 1 is shown in figure 1.24(a). The histogram of the watermarked image is common to all the three groups, as depicted in figure 1.24(b). It can be observed that there is a marked difference between the histogram of the original image 1 and the watermarked image. The histogram of the original image has different pixel value distributions from that of the watermarked image. Consequently, it is difficult for attackers to find significant information from the statistic characteristic.

1.4 Conclusion

In this chapter, a novel GWT and its applications for asymmetric multiple-image and group-multiple-image security systems have been reviewed. In an asymmetric multiple-image security system, each authorized user has one authentication key, one asymmetric key, three secret keys, and four sub-band keys as individual decryption keys. Moreover, the orders of gyrator transform provide extra degrees of freedom. In an asymmetric group-multiple-image security system, the three groups of multiple images are protected not only in a physical way with an encryption algorithm, but also by way of visually disguising hidden encrypted multiple images in the host image. Each group contains group decryption keys and each image of the group consists of individual decryption keys besides parameters of coupled logistic maps and gyrator transform. So, this is a great advantage for optical encryption techniques using multiple decryption keys, which not only expand the key space but also provide asymmetric characteristics to control system security and thus prevent potential attacks. The numerical simulation results authenticate the viability of the applications of the GWT.

Acknowledgements

The author is indebted to Muhammad Waizul Haque and Mehr-un-nisa for their inspiring support. Furthermore, I am grateful to Elsevier B. V. and the Optical Scociety of America for the permission to reprint the paper referred to in [30] and [31], respectively.

References

[1] Alfalou A and Brosseau C 2009 Optical image compression and encryption methods *Adv. Opt. Photon.* **1** 589–636

[2] Refregier P and Javidi B 1995 Optical image encryption based on input plane encoding and Fourier plane random encoding *Opt. Lett.* **20** 767–9

[3] Unnikrishnan G, Joseph J and Singh K 2000 Optical encryption by double-random phase encoding in the fractional Fourier domain *Opt. Lett.* **25** 887–9
[4] Situ G and Zhang J 2004 Double random-phase encoding in the Fresnel domain *Opt. Lett.* **29** 1584–6
[5] Liu Z and Liu S 2007 Random fractional Fourier transform *Opt. Lett.* **32** 2088–90
[6] Clemente P, Durn V, Tajahuerce E and Lancis J 2010 Optical encryption based on computational ghost imaging *Opt. Lett.* **35** 2391–3
[7] Chen W, Chen X and Sheppard C J R 2010 Optical image encryption based on diffractive imaging *Opt. Lett.* **35** 3817–19
[8] Abuturab M R 2012 Color image security system using double random-structured phase encoding in gyrator transform domain *Appl. Opt.* **51** 3006–16
[9] Shi Y, Li T, Wang Y, Gao Q, Zhang S and Li H 2013 Optical image encryption via ptychography *Opt. Lett.* **38** 1425–27
[10] Barrera J F, Mira A and Torroba R 2013 Optical encryption and QR codes: Secure and noise-free information retrieval *Opt. Express* **21** 5373–78
[11] Liu J, Xu X, Wu Q, Sheridan J T and Situ G 2015 Information encryption in phase space *Opt. Lett.* **40** 859–62
[12] Frauel Y, Castro A, Naughton T J and Javidi B 2007 Resistance of the double random phase encryption against various attacks *Opt. Express* **15** 10253–65
[13] Qin W and Peng X 2010 Asymmetric cryptosystem based on phase-truncated Fourier transforms *Opt. Lett.* **35** 118–20
[14] Wang X and Zhao D 2012 A special attack on the asymmetric cryptosystem based on phase-truncated Fourier transforms *Opt. Commun.* **285** 1078–81
[15] Abuturab M R 2013 Information authentication system using interference of two beams in gyrator transform domain *Appl. Opt.* **52** 5133–42
[16] Liu W, Liu Z and Liu S 2013 Asymmetric cryptosystem using random binary phase modulation based on mixture retrieval type of Yang–Gu algorithm *Opt. Lett.* **38** 1651–3
[17] Liu W, Liu Z, Wu J and Liu S 2013 Asymmetric cryptosystem by using modular arithmetic operation based on double random phase encoding *Opt. Commun.* **301** 56–60
[18] Sui L, Duan K, Liang J and Hei X 2014 Asymmetric double-image encryption based on cascaded discrete fractional random transform and logistic maps *Opt. Express* **22** 20605–21
[19] Abuturab M R 2015 Generalized Arnold map-based optical multiple color-image encoding in gyrator transform domain *Opt. Commun.* **343** 157–71
[20] Abuturab M R 2017 Securing multiple information using chaotic spiral phase encoding with simultaneous interference and superposition methods *Opt. Lasers Eng.* **98** 1–16
[21] Alfalou A and Brosseau C 2010 Exploiting root-mean-square time-frequency structure for multiple-image optical compression and encryption *Opt. Lett.* **35** 1914–6
[22] Alfalou A, Brosseau C, Abdallah N and Jridi M 2011 Simultaneous fusion, compression, and encryption of multiple images *Opt. Express* **19** 24023–29
[23] Alfalou A, Brosseau C, Abdallah N and Jridi M 2013 Assessing the performance of a method of simultaneous compression and encryption of multiple images and its resistance against various attacks *Opt. Express* **21** 8025–43
[24] Aldossari M, Alfalou A and Brosseau C 2014 Simultaneous compression and encryption of closely resembling images: application to video sequences and polarimetric images *Opt. Express* **22** 22349–68

[25] Dang P P and Chau P M 2000 Image encryption for secure internet multimedia applications *IEEE Trans. Consum. Electron.* **46** 395–403
[26] Mendlovic D, Zalevsky Z, Mas D, Garcia J and Ferreira C 1997 Fractional wavelet transform *Appl. Opt.* **36** 4801–6
[27] Chen L and Zhao D 2005 Optical image encryption based on fractional wavelet transform *Opt. Commun.* **254** 361–7
[28] Chen L and Zhao D 2009 Color image encoding in dual fractional Fourier-wavelet domain with random phases *Opt. Commun.* **282** 3433–8
[29] Bao L, Zhou Y and Chen C L P 2013 Image encryption in the wavelet domain *Proc. SPIE* **8755** 875502
[30] Abuturab M R 2015 Gyrator wavelet transform based non-linear multiple single channel information fusion and authentication *Opt. Commun.* **355** 462–78
[31] Abuturab M R 2015 Group-multiple-image encoding and watermarking using coupled logistic maps and gyrator wavelet transform *J. Opt. Soc. Am.* A **32** 1811–20
[32] Daubechies I 1990 The wavelet transform time–frequency localization and signal analysis *IEEE Trans. Inf. Theory* **36** 961–1005
[33] Mendlovic D and Konforti N 1993 Optical realization of the wavelet transform for two-dimensional objects *Appl. Opt.* **32** 6542–6
[34] Rodrigo J A, Alieva T and Calvo M L 2007 Application of gyrator transform for image processing *Opt. Commun.* **278** 279–84
[35] Zhusubaliyev Z T and Mosekilde E 2009 Multilayered tori in a system of two coupled logistic maps *Phys. Lett.* A **373** 946–51
[36] Sui L, Duan K and Liang J 2015 Double-image encryption based on discrete multiple-parameter fractional angular transform and two-coupled logistic maps *Opt. Commun.* **343** 140–9

IOP Publishing

Advanced Secure Optical Image Processing for Communications

Ayman Al Falou

Chapter 2

Optical one-way hash function

Wenqi He and Xiang Peng

Part A. Introduction to the one-way hash function

Unlike symmetric or asymmetric cryptosystems, the one-way hash function (OWHF), also called the cryptographic hash function, is a special cryptographic technique with the chief purpose to guarantee the integrity of a message (text, audio, or video), but not protect its secrecy [13]. Moreover, if a secret key is involved in the construction process of the OWHF, one can then not only check a message's integrity, but also verify its authenticity. This type of OWHF (with a secret key) is also called a message authentication code (MAC) or keyed OWHF [13]. The rest of this section will give a brief introduction to the OWHF (section A.1) and MAC (section A.2), respectively.

A.1 One-way hash function—acquire the 'fingerprint' of a message

To begin, we would like to show you a practical case about Alice, and how she might use the OWHF to facilitate her work.

Alice is a software engineer. One night, she finally finished all of her software programming work. Normally, all she needs to do next is package the source codes and write it on to a CD. However, at that moment, because she was really exhausted and sleepy, she decided to go home and leave the rest of the work until tomorrow. The next day, just before she was ready to package her codes, a doubt flashed in her mind: had anyone (say Mallory) tampered with her source codes? This is indeed a valid question since digital data are easy to tamper with nowadays, without any trail due to various attackers and specific software. At this moment, Alice is eager to know whether the source codes, which she has spent a lot of energy on, are 'real', or say, exactly the same as the ones she completed before she went home last night; we also call this requirement 'integrity'. That is to say, Alice wants to check the integrity of her source codes.

doi:10.1088/978-0-7503-1457-2ch2

A quick thought about the solution to this problem: before Alice went home, she could have transferred her computer to a safe place, or copied the source codes on to a flash drive and kept it secretly. Unfortunately, this would be inconvenient or even impossible in most practical application cases, especially when the data is large and thus the copy manipulation would be very time-consuming and inefficient (another user case will be provided to further illustrate this issue at the end of section A.1).

Thanks to the OWHF, Alice could easily get the 'fingerprint' (hash value) of her original source codes, then she would just need to preserve it and keep this unique fingerprint secret. On the following day, by comparing it with the new fingerprint of the suspicious source codes, she can easily tell whether the suspicious source codes have integrity or not. A flowchart is presented below to clarify the working principle of OWHF in the aforementioned case (figure 2.1).

One formal definition of OWHF goes like this, an OWHF is any function that can be used to map an arbitrary-length message, M, to a fixed-length output, h, i.e. $h = \mathrm{H}(M)$. The output returned by an OWHF is called the hash value, digests, or digital fingerprint. An ideal OWHF would have several main properties:

(1) it is easy to generate the hash value for any given message;
(2) it is hard to generate a message from its hash value;
(3) it is deterministic so the same message always leads to the same hash value;
(4) it is hard to find two different messages with the same hash value;
(5) a tiny change to a message should change the hash value so extensively that the new hash value appears uncorrelated with the old hash value.

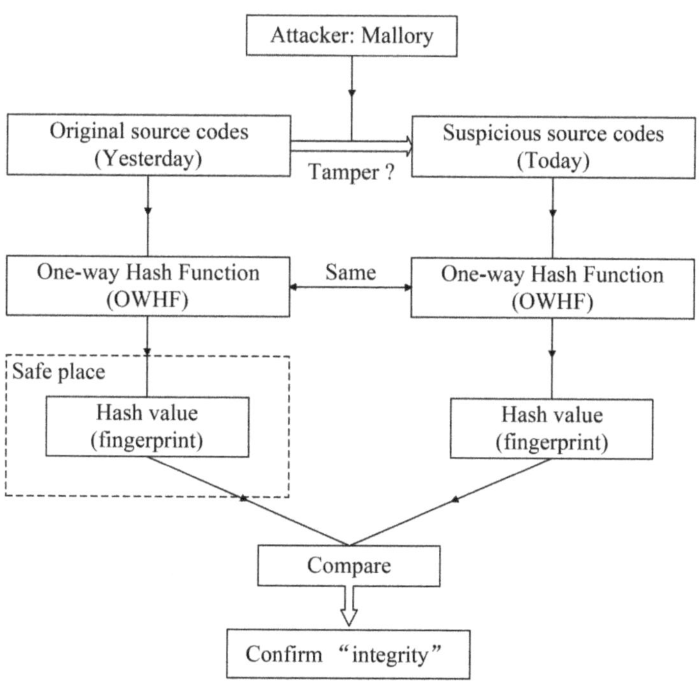

Figure 2.1. Flowchart of the working principle of the OWHF in the aforementioned case.

In the above, term (4) is also referred to as 'strong collision resistance', and term (5) is also known as the 'avalanche effect'; these two properties actually connect with each other to some extent.

Obviously, it is not easy to design a function that allows an input with arbitrary length, let alone make it one-way. In the real world, OWHFs are built on the idea of a compression function (CF), which serves as the key component. This CF could convert an input with a size of m bits into an output with a size of n bits ($m>n$). A series of CFs working together in a cascaded way could then convert an input (size of p bits) into an output (size of n bits) where $p \gg n$. A typical structure of OWHF can be depicted as follows (f represents the CF) (figure 2.2):

Since the invention of the first traditional OWHF, MD4 by Rivest, in 1990 [14], a series of OWHFs such as MD5, SHA-0, SHA-1, PIPEMD-160, SHA-256, SHA-512, etc, have been successively proposed to achieve the purpose of pursuing higher security strength [13–15]. Among those OWHFs, the MD5 and SHA-1 are the most commonly used at present, even though recently their strong-collision-resistance properties have proven to be vulnerable [16–18].

To complete this section, we share one more user case regarding the OWHF: it continues with the case that we have mentioned at the beginning, with the software engineer Alice who developed software one day. She then put the software on her own website to be downloaded by users for free. Meanwhile, in order to release the heavy-load traffic of the server, she decided to distribute her software on several mirror sites as well.

At this moment, a question was raised in her mind: how do I know whether the software copies in the mirror sites have been tampered with or even contain an implanted virus by attackers? Furthermore, how do I know whether the software will be tampered with or even implanted with a virus by attackers during the downloading process? Again, the OWHF is regarded as a perfect way to solve this kind of issue. One standard solution is, Alice computes the hash value h for the software before she sends it to those mirror sites, and then makes h public on her own website. Then the users would be aware of any changes by comparing the computed new hash value h' for the downloaded software with the original hash

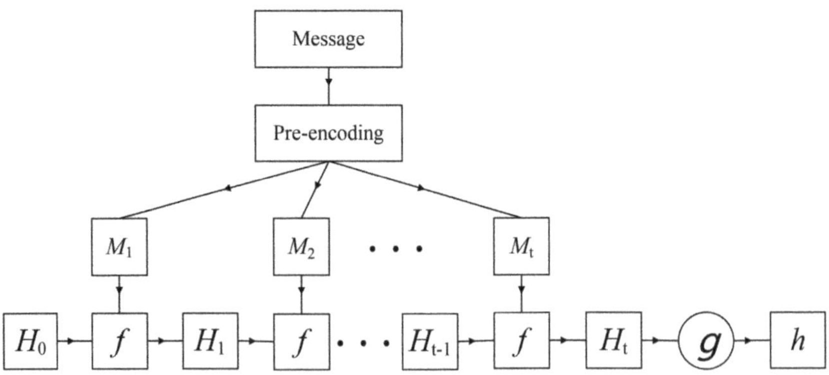

Figure 2.2. A typical structure of an OWHF.

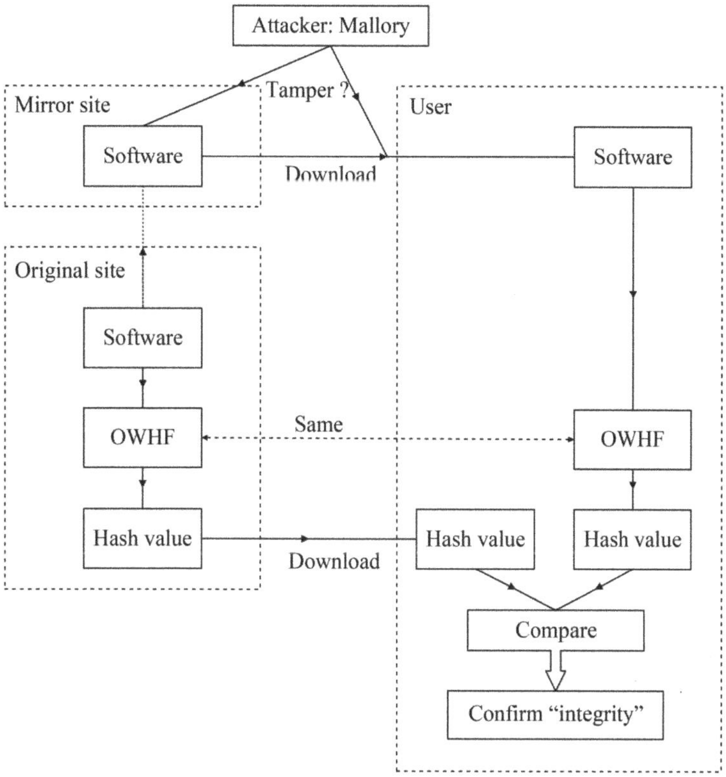

Figure 2.3. Flowchart of the integrity checking process for software.

value *h* published on Alice's website. A flowchart is provided for ease of understanding (figure 2.3).

One last concern for the above user case: if the original software *M*, together with its hash value *h*, were published on the same mirror site, could the user distinguish whether the downloaded software was modified by an attacker or not? The answer is absolutely no, because an attacker could tamper with the software and the hash value at the same time, in other words, the attacker could generate a similar software M' (probably with a virus) and meanwhile compute the corresponding hash value h', and then put both of them on the mirror site, which would mislead users. This is also called counterfeit or forged. How can this issue be solved? Please read the next section A.2.

A.2 Message authentication code—a message coming from the right sender

Again, to start, we would like to present another practical case about Alice and Bob; they represent two commercial banks this time. One day, 'bank Alice' sent a request to 'bank Bob' using the internet, it read: 'Transfer 1 million from account A-111 to account B-111'. Just before bank Bob was ready to transfer the money, he thought: what if someone (say, Mallory) intercepted the original request and re-sent a tampered one? For example, Mallory could capture and tamper with the original request so that it reads: 'Transfer 1 million from account A-111 to account M-111', Mallory would then receive 1 million if bank Bob was not aware of this counterfeit message.

Inspired by the technique of OWHF mentioned in section A.1, someone may have an idea like this: bank Alice computes the hash value for the original request and puts it on a safe website, then sends the request to bank Bob. When Bob receives the request (it might have been tampered with), he also first computes its hash value, then compares the computed hash value with the one published on the safe website, and he would immediately know whether the received request was the original one. Yes, this sounds good but not feasible, at least it is inefficient in reality, especially when bank Alice and bank Bob communicate a lot. Someone may also think: what if bank Alice sends the original request together with its hash value to bank Bob? This way is even worse since Mallory could forge both the original request and its hash value as long as he captures the original ones.

Obviously, the aforementioned conundrum requires the ability to not only integrity check but also use identity authentication. That is to say, bank Bob needs to know whether the request (or message) has integrity and whether it comes from bank Alice (right sender). To handle this issue, we need the message authentication code (MAC), which is also regarded as a type of keyed OWHF (K-OWHF). The difference between MAC (or K-OWHF) and OWHF is shown in figure 2.4.

The detailed process of using a MAC to solve the problem encountered by bank Alice and bank Bob could be depicted as follows (also see figure 2.5):

(1) sender Alice and receiver Bob share a secret key in advance;
(2) sender Alice computes the MAC value for the request to transfer money (using the shared secret key);
(3) sender Alice sends the request together with its MAC value to receiver Bob;
(4) receiver Bob computes the MAC value for the received request (using the shared secret key);

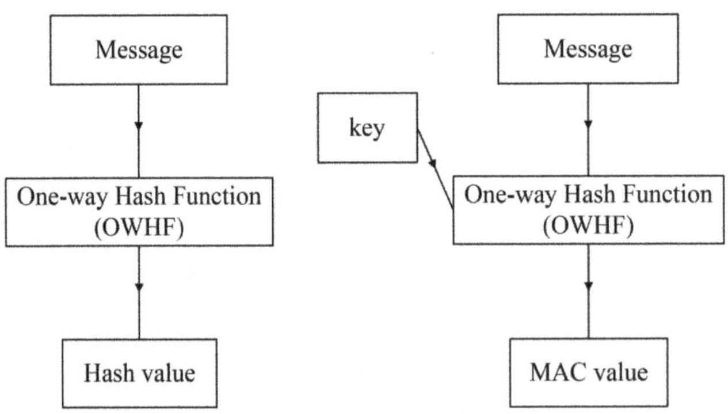

Figure 2.4. A brief comparison of the OWHF and MAC (or K-OWHF).

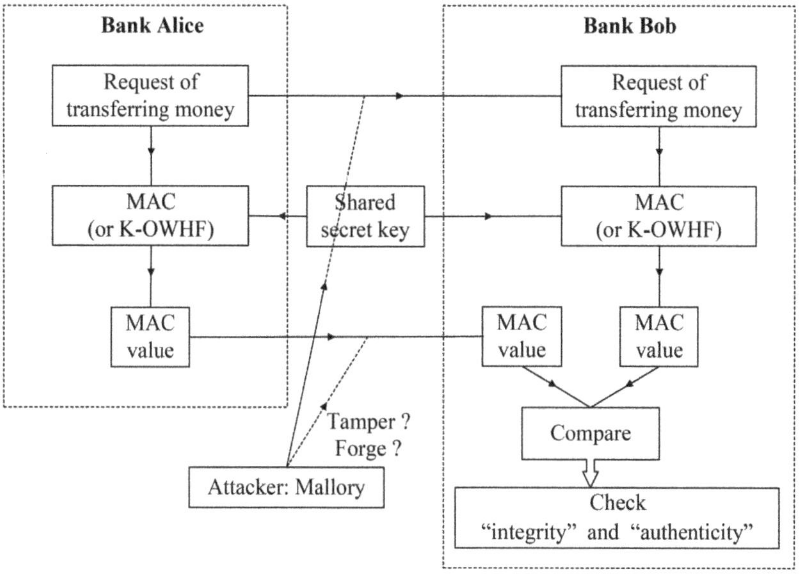

Figure 2.5. Flowchart of using the MAC for integrity and authenticity checking.

(5) receiver Bob compares the MAC value computed by himself with the received MAC value;
(6) if the two MAC values are exactly the same, it means that the received request is complete and indeed from Alice; if not, it implies the received request is not authentic, it has either been tampered with or forged.

A MAC (or K-OWHF) can convert an arbitrary-length message, M, and a secret key, K, into a fixed-length MAC value, h, i.e. $h = H(M, K)$. The MAC should at least fulfill the following four necessary conditions [13]:
(1) given M and K, it is easy to compute h;
(2) given h and M, it is hard to compute K such that $H(M, K) = h$;
(3) given M and K, it is hard to find another message, M', and another secret key, K', such that $H(M, K) = H(M', K')$;
(4) a slight change in M or K will result in a prominent change in h, which is the so-called avalanche effect.

Up to now, is anyone wondering why does Alice not just directly encrypt the message and send it to Bob by employing a traditional symmetric encryption scheme? Can the MAC be replaced by a symmetric encryption scheme instead? This is a good question, because there are indeed some types of MAC which are designed based on a symmetric encryption algorithm. But these types of MAC have some limitations in practical applications. They can only authenticate messages with a certain structure; for example, sentences with a clear meaning. However, they are useless for messages without a special meaning, for example, random numbers/characters themselves.

Part B. Construction of the optical one-way hash function

In part A, we learned that the key component of the OWHF and MAC is a compression function (CF), which therefore becomes our priority task when we try to design an optical OWHF. Keeping this in mind, we have made two attempts based on optical information processing techniques [11, 12]. The first one proposed a concept of 'phase-truncated Fourier transform' (PTFT); it serves as the CF, while the interference phenomenon itself serves as the CF in the second attempt. It should be noted that both of the schemes inevitably involved many digital manipulations, so we call them hybrid opto-electronic systems or prototypes.

B.1 Optical OWHF based on phase-truncated Fourier transform

In this section, we provide a method to optically construct an OWHF [11]. This method largely relies on the concept of phase-truncated Fourier transform (PTFT), in which the one-way and nonlinear compression mechanisms originate from the operation of phase-truncation by a charge coupled device (CCD). By using the PTFT in a cascaded way, we can then get an optical OWHF that satisfies all the basic requirements of an OWHF. One possible optical/digital hybrid setup to implement so-called cascaded PTFTs (C-PTFTs) is shown in figure 2.6.

Prior to explaining how this hybrid system works, we need a pre-processing step for the message to be handled. A message with arbitrary length should be divided into a number of data blocks of 512-bit length, and then each of them is further encoded into 8×8 sub-images with 8-bit (256 gray levels) referred to as the 'information plane'. Suppose that we have a message with $512 \times N$ bits that could be divided into N numbers of sub-image matrices (M_1, M_2, …, M_N).

Next, we will describe how a 128-bit hash value is generated for the aforementioned N numbers of sub-image matrices in detail. As shown in figure 2.6, the spatial light modulator, SLM_1 is an amplitude-only (AO) modulation device, while SLM_2 is a phase-only (PO) modulation device. Electronically, we can first write M_1 and M_2

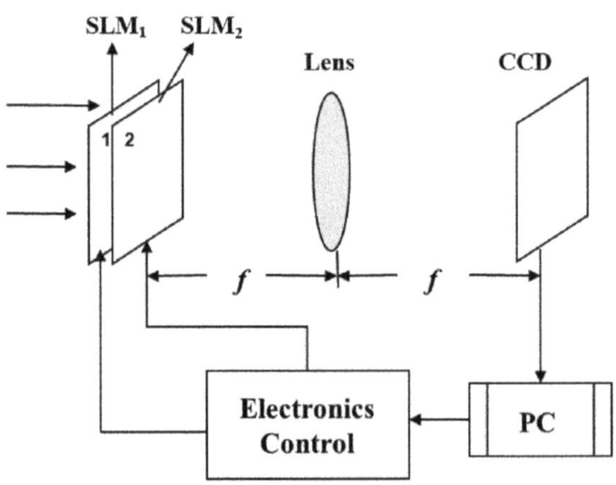

Figure 2.6. Schematic diagram of an optical OWHF generator.

onto SLM_1 and SLM_2, respectively, so that M_1 is AO modulated, whereas M_2 is PO modulated. Combining SLM_1 and SLM_2 leads to the initial complex amplitude t_1, which can be written as

$$t_1 = M_1 \cdot \exp(j2\pi M_2/255). \tag{2.1}$$

This synthesized complex transmittance is then illuminated with a plane wave. A CCD camera with 12-bit quantized accuracy on the back focal-plane of the Fourier-transformation (FT) lens detects only the power spectrum of t_1 whereas the phase information is automatically truncated. We denote this process as PTFT $[M_i, M_j]$, which is also regarded as a CF because the phase part of the signal has been removed. Numerically, we are able to quantize the power spectrum to 12-bit, and the lower 8-bit-plane is chosen and stored as an intermediate result, H_1. This process can be mathematically expressed as:

$$H_1 = Q_{12\to 8}\{|\mathrm{FT}(t_1)|^2\}, \tag{2.2}$$

where FT(*) denotes the Fourier transform, |*| stands for taking the modulus operation, $|*|^2$ can lead to the power spectrum of FT(*), and $Q_{12\to 8}\{*\}$ represents an operation of extracting the lower 8-bit-plane from the quantized power spectrum with 12-bit grey levels (see figure 2.7).

The combination of equations (2.1) and (2.2) could be regarded as a fundamental processing unit. Substituting (2.1) for (2.2) we have

$$\begin{aligned} H_1 &= \mathrm{PTFT}[M_1, M_2] \\ &= Q_{12\to 8}\{|\mathrm{FT}[M_1 \cdot \exp(j2\pi M_2/255)]|^2\}. \end{aligned} \tag{2.3}$$

Next, we update SLM_1 with AO modulation of the H_1, while updating SLM_2 with PO modulation of the M_3, making up a new complex amplitude, t_2, at the input plane of figure 2.6, i.e.

$$t_2 = H_1 \cdot \exp(j2\pi M_3/255). \tag{2.4}$$

By repeating the procedure described by (2.1)–(2.2) (or (2.3)) as imposed on t_1, but now on t_2, we obtain the second intermediate result H_2, which can be written as:

$$\begin{aligned} H_2 &= \mathrm{PTFT}[H_1, M_3] \\ &= Q_{12\to 8}\{|\mathrm{FT}[H_1 \cdot \exp(j2\pi M_3/255)]|^2\}. \end{aligned} \tag{2.5}$$

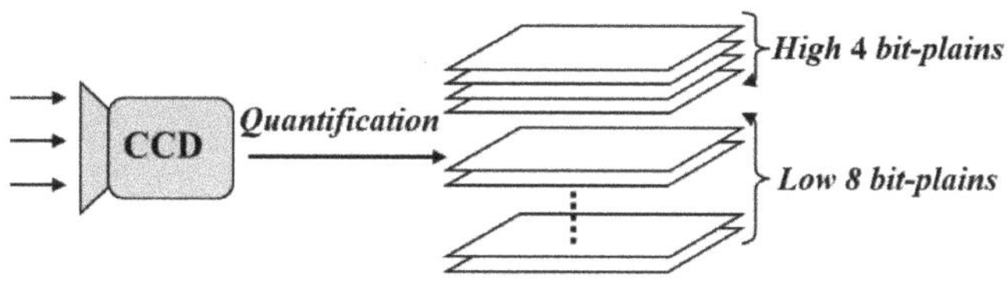

Figure 2.7. Illustration of the function of operator $Q_{12\to 8}\{*\}$.

The above process continues until all the sub-image blocks (M_1, M_2, ..., M_N) are traversed. It is obvious that the process described above is achieved by the C-PTFTs, whose structure diagram is shown in figure 2.8.

So, we rewrite the notation PTFT as a more indicative operator C-PTFTs [M_1, M_2, ..., M_{N-1}, M_N], with which we can formulate the process mentioned above as more general expressions:

$$\begin{aligned} H_{k-1} &= \mathrm{PTFT}[H_{k-2}, M_k] \quad (1 < k \le N) \\ &= Q_{12\to 8}\left\{|\mathrm{FT}[H_{k-2} \cdot \exp(j2\pi M_k/255)]|^2\right\}. \end{aligned} \tag{2.6}$$

$$\begin{aligned} M' &= H_{N-1} \\ &= \text{C-PTFTs}[M_0, M_1, M_2, \ldots, M_N] \\ &= \mathrm{PTFT}[\ldots \mathrm{PTFT}[\mathrm{PTFT}[M_0, M_1], M_2], \ldots, M_{N-1}], M_N]. \end{aligned} \tag{2.7}$$

Equations (2.6) and (2.7) give a general formulation of an algorithm to create an OWHF in an iterative fashion, with which we can make all the N sub-image blocks correlative with the final output M' by using the CF (or PTFT) N times repeatedly.

To reinforce the capability of diffusion and confusion of our OWHF, we divide M' into 4 sub-matrices, (M_a, M_b, M_c, M_d), each 4×4 pixels in size. Then we expand each sub-matrix to another 4 sub-matrices (M_A, M_B, M_C, M_D), each with a larger size of 8×8 pixels, by duplicating every element of M' four times. This operation is illustrated in figure 2.9. Thereafter, making an operation of C-PTFTs on the expanded matrices to generate another 8×8 image matrix, M'':

$$M'' = C\text{-PTFTs}[M_A, M_B, M_C, M_D]. \tag{2.8}$$

Numerically, M'' is a matrix with 8×8 elements from which we pick up eight elements (64 bits), for example, along the diagonal direction of this matrix as part of the final hash value, we call it $hash_1$. It should be noted that we need to take a zero-removing process to each numerical matrix before it is modulated to the amplitude in SLM_1.

The reason is explained as follows: if the value of any element in any numerical matrix modulated in the SLM as the amplitude of a complex wavefront is zero, any changes to the value of the corresponding element in the numerical matrix modulated as the phase of the complex wavefront will not change the final output

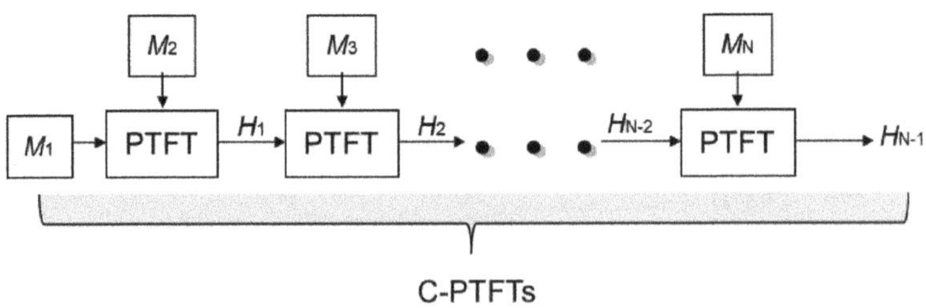

Figure 2.8. Structure diagram of the C-PTFTs.

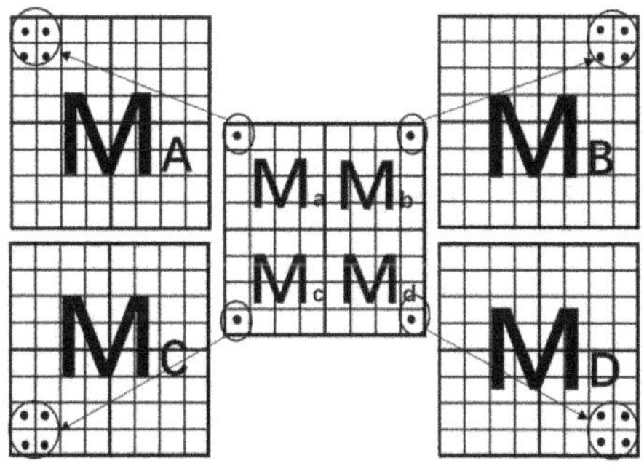

Figure 2.9. Schematic diagram of an expanding numerical matrix.

hash value; in this case, a collision occurs. The implementation of the zero-removing process can be described as:

$$\begin{cases} M(i,j)' = M(i,j) + k, & M(i,j) <= (255 - k) \\ M(i,j)' = M(i,j) + k - 255, & M(i,j) > (255 - k) \end{cases} \tag{2.9}$$

where $M(i, j)$ and $M(i, j)'$ represent a numerical matrix before and after the process, respectively, k stands for an integral number between 0~255, while 3 is selected in the following simulation.

The last 'big' and important step: it is just like a duplication of all the previous steps except for a pre-processing of the N sub-image blocks $M_1, M_2, \ldots, M_N$, that is, each element of the sub-image blocks is circularly shifted leftward with 4-bit and as a result we obtain more N sub-image data blocks, which are called 'auxiliary information planes'.

For example, the values 150 (10010110) in the original sub-image blocks will be changed to 105 (01101001) in the auxiliary ones. In such a way, we obtain another 64-bit output $hash_2$. Combining $hash_2$ with $hash_1$ end-to-end, we finally obtain a 128-bit hash value $hash$ from a message with $512 \times N$ bits. A complete block diagram of our optical OWHF algorithm is shown in figure 2.10.

To evaluate the avalanche performance of our optical OWHF, we define an avalanche effect coefficient (AEC) with the following expression

$$AEC = diff(hash, hash')/128, \tag{2.10}$$

where $hash$ is the original hash value, formed from the original message; $hash'$ represents the hash value formed from the message with a slight change to the original one; and $diff[.]$ denotes an operation accounting for how many different bits there are between the two hash values, $hash$ and $hash'$. Obviously, the larger the value of AEC, the better the avalanche effect of the OWHF.

A set of elementary simulation tests were taken according to the following steps:

(1) arbitrarily generate three messages with ten, a hundred, and a thousand kilobits, respectively;

(2) compute hash values, h_1, h_2, and h_3 for three generated messages, respectively;

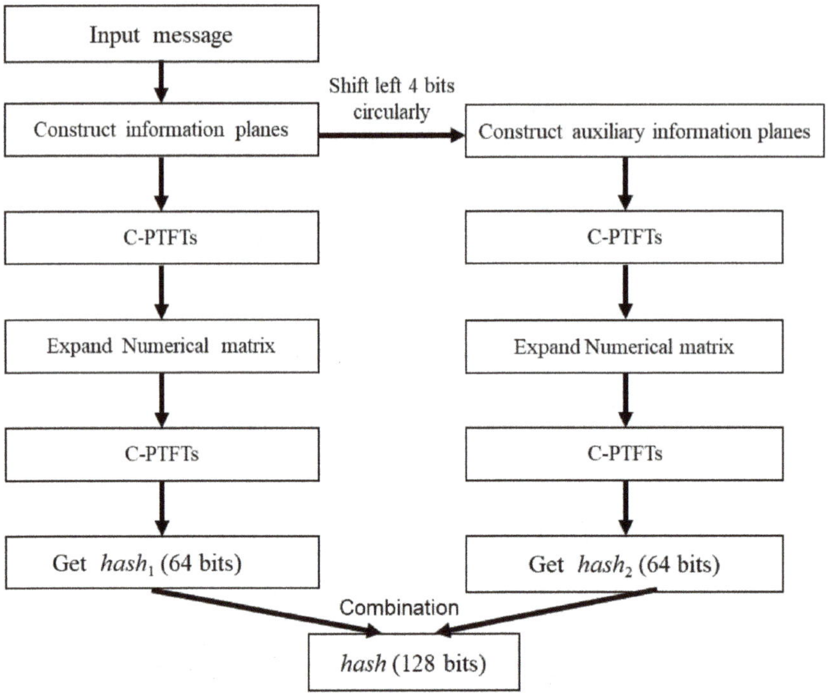

Figure 2.10. Flowchart showing the construction of the optical OWHF.

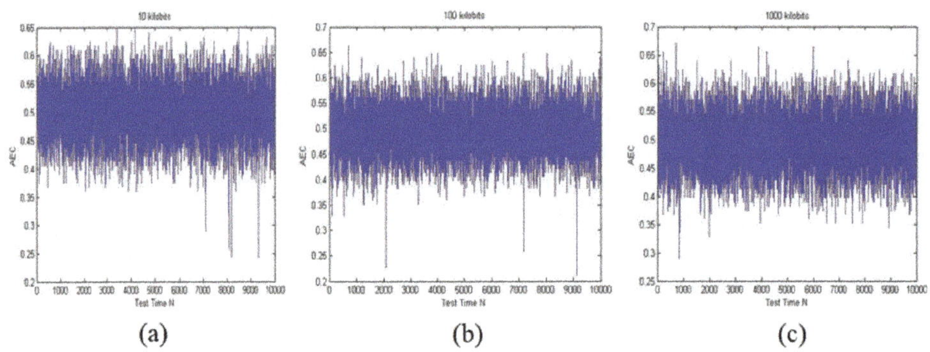

Figure 2.11. Distribution of the values of *AEC* in the tests for messages with (a) 10 kilobits, (b) 100 kilobits, (c) 1000 kilobits, respectively.

(3) randomly choose and change only one bit in three messages and calculate the corresponding hash values h_1', h_2' and h_3';
(4) calculate the *AEC* value according to (2.10);
(5) repeat the steps (2.3)–(2.4) 10 000 times.

The test results are shown as figures 2.11(a)–(c) and tabulated in table 2.1. From the results, we can see that the average values of *AEC* are very close to that of MD5 and SHA-1, which are the most famous traditional OWHFs in reality [13].

Table 2.1. The results of testing the avalanche effect of the optical OWHF.

	AEC value			
Scale of data(kilobits)	Min	Max	Average	Run-time for single optical OWHF
10	0.24	0.67	0.49	0.0207 s
100	0.22	0.66	0.49	0.0716 s
1000	0.29	0.67	0.49	1.7803 s

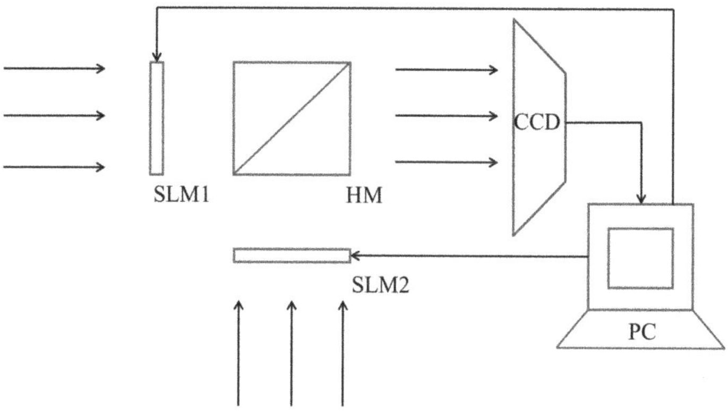

Figure 2.12. Optoelectronic architecture of realizing the CF based on two-beam interference. Reproduced from [12].

Meanwhile, the average value of the standard deviation of *AEC* values for 3 kinds of data scale is 0.0441, while that of the MD5 and SHA-1 are 0.0442 and 0.0488 [19], respectively. Therefore, we can conclude that the constructed optical OWHF has a similar performance to conventional OWHFs. Here, the standard deviation of the *AEC* values is defined as:

$$std = \sqrt{\frac{\sum_{i=1}^{N}(AEC_i - \overline{AEC})^2}{N - 1}}, \tag{2.11}$$

where N stands for the number of attack tests (here $N = 10\,000$), and $\overline{AEC}$ represents the average value of *AEC* values in the tests.

B.2 Optical OWHF based on two-beam interference

In this section, we provide another alternative method to optically construct an OWHF [12]. The CF of this optical OWHF is designed based on the concept of two-beam interference, as shown in figure 2.12, where CCD, HM and PC represent the charge-coupled device with 14-bit quantized accuracy (it is commercially available), half mirror and personal computer, respectively. The SLM1 and SLM2 represent two amplitude-only SLMs, which are employed to load the images electronically. The function of the CF based on two-beam interference is to convert TWO images of 8 × 8 pixels and 8-bit quantization levels (256 gray levels) into ONE output image of the same size. It means that the two inputs of the CF (with a total amount of 1024 bits) can be compressed into one output with only 512 bits.

Now, we explain the working principle of the CF based on two-beam interference in detail. In advance, two images of 8 × 8 pixels and 8-bit quantization levels are written and electronically addressed by the SLM1 and SLM2, respectively. Then, two coherent light beams modulated separately by two amplitude-only SLMs will combine on the HM and then interfere with each other at the output plane, where an intensity detector is placed to record the interferogram. It should be noted that the employed intensity detector is a high-accuracy CCD with 14-bit quantization levels.

Furthermore, we carry out post-processing to the acquired interferogram. By extracting the lower 8-bit-plane of the middle portion (8 × 8 pixels) of the interferogram, we will get the final output of the CF. Up to now, we can see that the two input images—a total amount of 1024 bits—have been converted into one image with 512 bits.

Here, we would like to emphasize that the construction of an OWHF could also be regarded as a one-way encryption process for a message. In this method, the message to be handled should first be pre-encoded before it is processed by using the aforementioned CF based on two-beams interference. Specifically, the message is divided into a number of data blocks with 512-bit for each, they are then encoded into a series of sub-images with 8 × 8 pixels and 8-bit quantization levels. Note that if the length of the message is not the integer multiple of the default data block (512 bits) in length, an extra compensating operation will be required, and the method of padding bits with '1' is the most frequently used way, just as in a traditional OWHF. Therefore, it is reasonable to suppose that any message could be pre-encoded into N sub-images with 8 × 8 pixels and 8-bit quantization levels; here, N is a positive integral. Meanwhile, a precoding process is illustrated in figure 2.13.

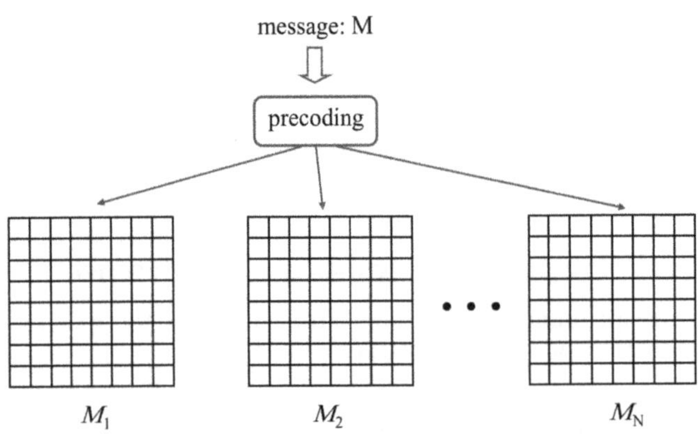

Figure 2.13. Precoding process for a message. Reproduced from [12].

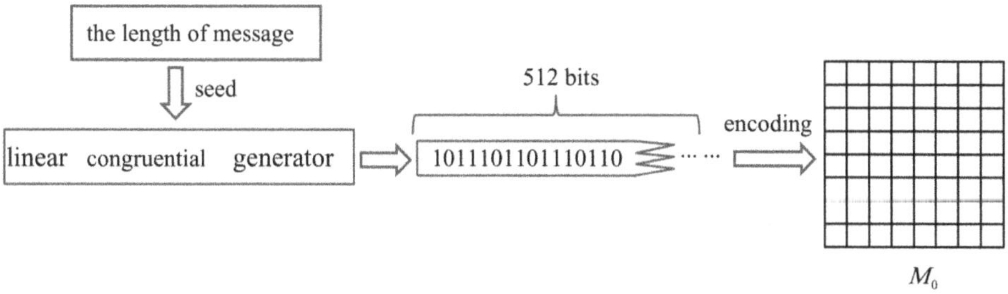

Figure 2.14. A flowchart for creating a pseudo random image.

Next, we need to create an initial pseudo random image for the following cascade compression procedure. Here, a linear congruential generator is adopted to generate 512 bits pseudo random numbers while the length of the message to be handled is taken as the 'seed'. A pseudo random image (M_0) of 8 × 8 pixels and an 8-bit quantization level can thus be digitally created, and a flowchart can be seen in figure 2.14. Meanwhile, it is worth pointing out that the introduction of M_0 could effectively avoid potential collision. This is because all the sub-images will be the same if the differences of two or more messages to be encrypted only exist before the padding bits operation.

Up to now, we have an initial pseudo random image (M_0) as well as all the sub-images (M_1, …, M_N) of the message to be processed. A detailed description of the construction process of our optical OWHF is presented as follows:

(1) the images M_0 and M_1 are separately written into two amplitude-only SLMs, SLM1 and SLM2;

(2) two coherent parallel light beams modulated by SLM1 and SLM2 combine on HM and then interfere with each other at the output plane, leading to an interferogram that is captured by the CCD camera (14-bit quantized accuracy). The interference field can be expressed as:

$$o(m, n) = M_0 * h(x, y, l) + M_1 * h(x, y, l), \tag{2.12}$$

where

$$h(x, y, l) = \frac{\exp(i2\pi l/\lambda)}{il\lambda}\exp\left[\frac{i\pi}{l\lambda}(x^2 + y^2)\right], \tag{2.13}$$

O (m, n), h (x, y, l) represent the distribution of the complex interference field and the point spread function (PSF) of the Fresnel diffraction, respectively. l and * are the distance between any a SLM with the CCD camera and the convolution operation.

(3) By simple manipulation of the intensity distribution of the interference field, obtained in the last step, we can obtain an intermediate value H_1. This process can be formulated as (2.14):

$$H_1 = Q_{14\to 8}\left\{Q_{14}(|o(m, n)|^2)\right\}, \tag{2.14}$$

where the operator $|*|^2$ represents the operation of taking the modulus, $Q_{14}(*)$ means quantifying the intensity distribution with 14-bit accuracy, $Q_{14\to 8}\{*\}$ implies extracting the lower 8-bit-plane from an image with 14-bit accuracy. For convenience, we would like to further simplify (2.14) as:

$$\begin{aligned} H_1 &= f[M_0, M_1] \\ &= Q_{14\to 8}\{Q_{14}(|o(m, n)|^2)\}. \end{aligned} \tag{2.15}$$

(4) Substitute the image M_0 and M_1 written on SLM1 and SLM2 with the intermediate values H_1 and the subsequent sub-image M_2, respectively. By repeating steps (2.2)–(2.3), we can get another intermediate value:

$$H_2 = f[H_1, M_2]. \tag{2.16}$$

(5) Repeat the steps (2.1)–(2.3) N times. All the sub-images of the message will be involved and the final output (H_N) of the cascaded compression process can be obtained.

For the convenience of understanding, two more images are shown: figures 2.15 and 2.16:

So far, we have compressed a message with $N \times 512$ bits into an image with only 512 bits. To further turn the 512-bit output into a final 128-bit hash value, an output transformation is then required as in the construction of traditional OWHF.

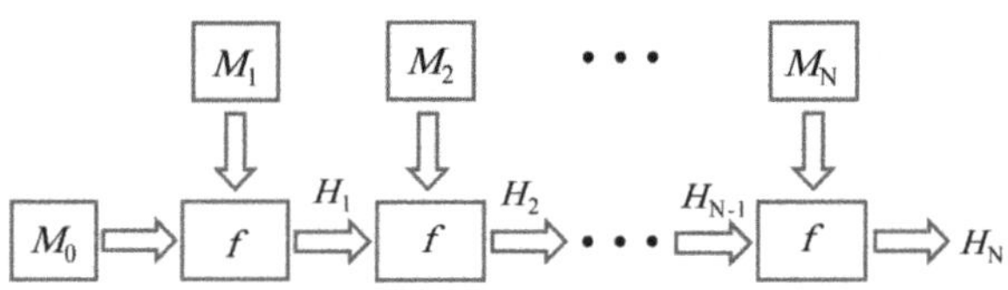

Figure 2.15. Illustration of the cascaded compression process. Reproduced from [12].

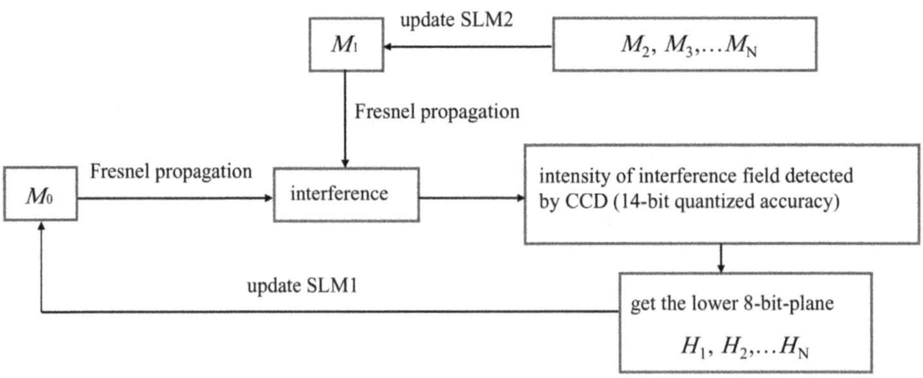

Figure 2.16. Flowchart of the cascade compression procession. Reproduced from [12].

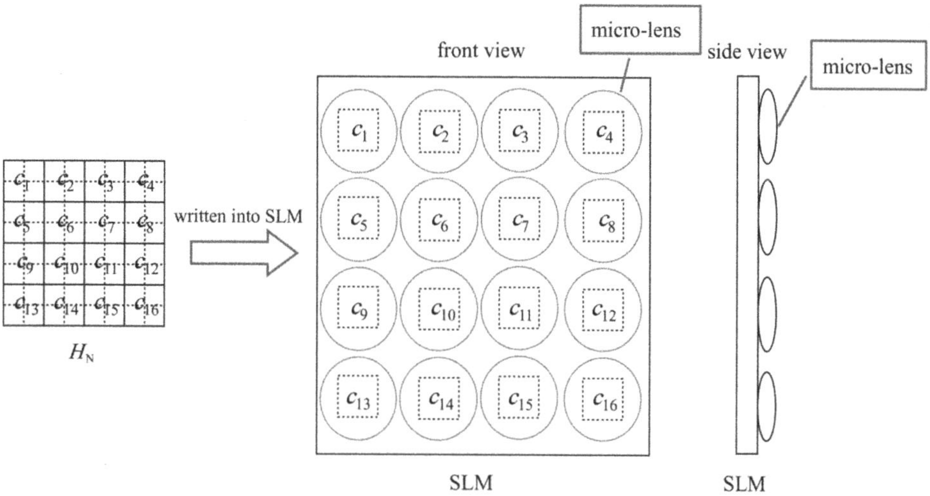

Figure 2.17. The relative positions of the micro-lens array and sub-images written on an SLM. Reproduced from [12].

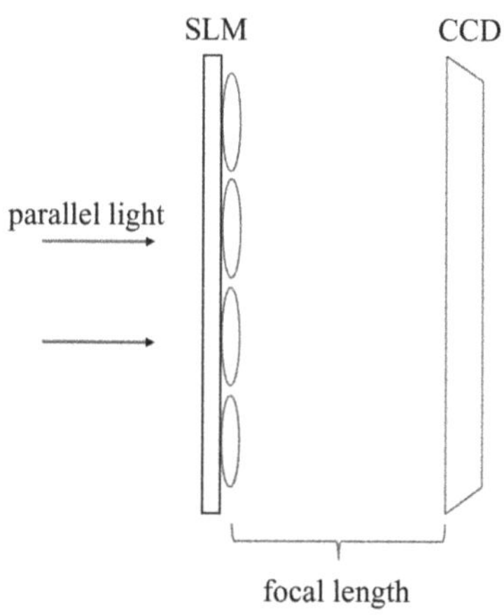

Figure 2.18. Optical realization of the output transformation. Reproduced from [12].

The details of the output transformation are depicted as follows: first, the obtained output H_N is further divided into 16 sub-images (c_1, c_2, …, c_{16}, as shown in figure 2.17) with the size 2×2 pixels for each. These sub-images are then written into an amplitude-only SLM, which sticks closely with a micro-lens array, noting that there exists a fixed distance between all the sub-images in the SLM, and it largely depends on the size of the micro lens.

By illuminating the SLM shown in figure 2.18, the sub-spectrums of c_1 to c_{16} can be observed and integrated at the back focal plane of the micro-lens array, where a CCD camera is placed [20], as expressed by (2.17). In the CCD plane, we can observe 16 bright light spots in 16 sub-spectrums, separately. They are quantified into 14 bits, and the lower 8-bit data of each one is selected to directly combine together into a 128-bit (16 × 8) final output. It is regarded as the hash value of the message.

$$C_n(u, v) = \frac{\exp\left[i\frac{k}{2f}(u^2 + v^2)\right]}{i\lambda f} \int_{-\infty}^{+\infty}\int_{-\infty}^{+\infty} c_n(x', y')\exp\left[-i\frac{2\pi}{\lambda f}(x'u + y'v)\right]dx'dy'. \quad (2.17)$$

Next, we introduce several typical parameters [21, 22] to evaluate the avalanche effect of our optical OWHF. They are defined as equations (2.8)–(2.12):

$$B_i = diff(hash, hash'). \quad (2.18)$$

$$\overline{B} = \frac{1}{N}\sum_{i=1}^{N} B_i. \quad (2.19)$$

$$P = (\overline{B}/128) \times 100\%. \quad (2.20)$$

$$\Delta B = \left[\frac{1}{N-1}\sum_{i=1}^{N}(B_i - \overline{B})^2\right]^{1/2}. \quad (2.21)$$

$$\Delta P = \left[\frac{1}{N-1}\sum_{i=1}^{N}(B_i/128 - P)^2\right]^{1/2} \times 100\%. \quad (2.22)$$

In (2.18), *hash* and *hash'* mean the hash values of the original message to be handled and a slightly modified one, operator *diff*(*) calculates the number of different bits between the two hash values (*hash* and *hash'*) and the subscript i represents the ith test. In (2.19) and (2.20), N denotes the test times, $\overline{B}$ and P are the changed bit number and probability of each bit on average, respectively. The standard variance of numbers of the changed bit number and probability of each bit are indicated by ΔB and ΔP in (2.21) and (2.22), respectively. Obviously, the larger the three parameters (B_i, B and P), the more distinct the avalanche effect. Meanwhile, the smaller the other two parameters (ΔB and ΔP), the better the stability of the avalanche effect.

The test process is performed as follows:

(1) the hash value of an arbitrary message is generated by using the constructed optical OWHF;
(2) one arbitrarily selected bit of the message is modified;
(3) the different bit number between the two hash values can be counted according to (2.18).

Repeat the aforementioned procedures (2.2)–(2.3) 10 000 times. The obtained results are used to compute the rest of the parameters ($\overline{B}$, P, ΔB,ΔP) in accordance with equations (2.19)–(2.22). Specifically, in our simulation, three groups of

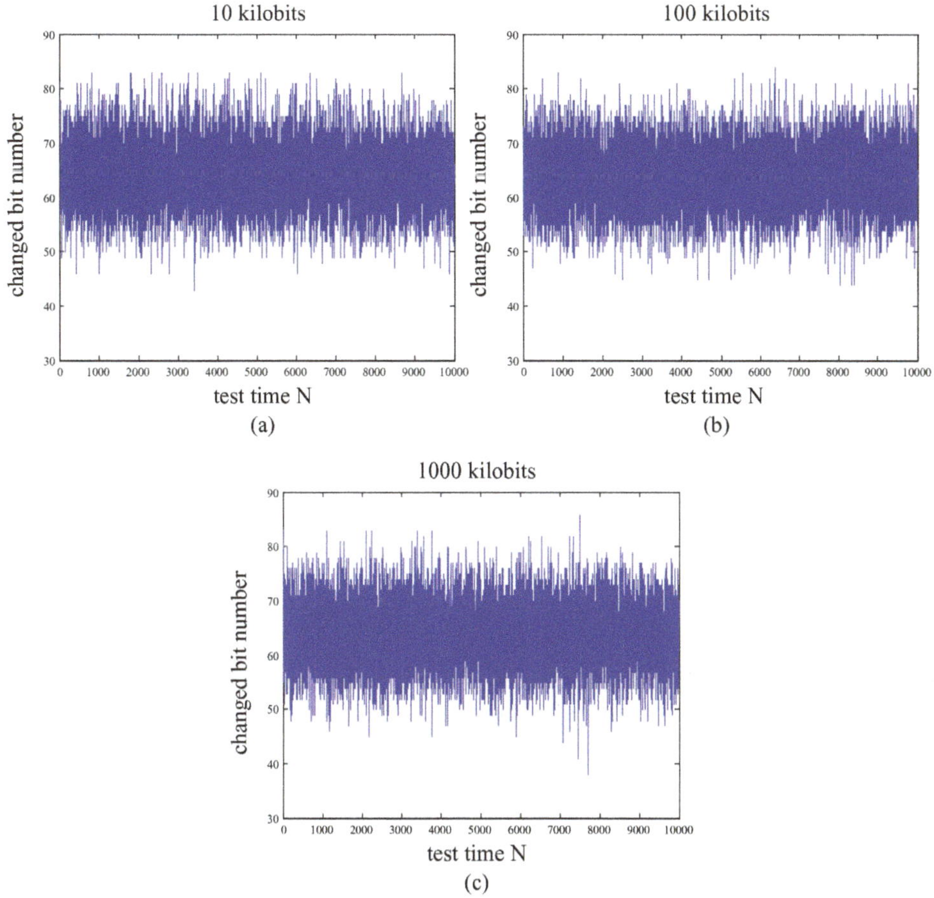

Figure 2.19. Distribution of values of B_i in tests for messages with: (a) 10 kilobits, (b) 100 kilobits, (c) 1000 kilobits, respectively. Reproduced from [12].

messages with ten, a hundred, and a thousand kilobits are tested separately and the results are shown in figure 2.19 and table 2.2. Furthermore, we have made a comparison of the traditional famous OWHF (MD5, SHA-1) and the proposed optical OWHF; the result is given in table 2.3.

From the test results, we can see that a bit change to the original message will on average cause almost a half bit's difference in the hash values in 10 000 tests for all three groups of messages. Besides, ΔB of them are 5.70, 5.69 and 5.70, respectively. This implies that the sensitivity of the hash value to the different scale of the messages, as well as the stability of the sensitivity, is good. Meanwhile, we can also find out that the change probability P of the hash values for the three scales of messages are 50.26, 49.99 and 50.01, respectively. Besides, ΔP of them are 4.46, 4.44 and 4.45, respectively. This means that the change probability of any bit of the hash value is pretty close to the ideal value (50%) and is stationary. Furthermore, the

Table 2.2. The result of avalanche effect testing.

	10 kilobits	100 kilobits	1000 kilobits	Mean
$\overline{B}$	64.34	63.99	64.01	64.11
$P(\%)$	50.26	49.99	50.01	50.08
ΔB	5.70	5.69	5.70	5.70
$\Delta P(\%)$	4.46	4.44	4.45	4.45

Table 2.3. Comparison with MD5 and SHA-1.

	Optical OWHF	MD5	SHA-1
$\overline{B}$	64.11	64.05	79.99
$P(\%)$	50.08	50.04	49.99
ΔB	5.70	5.68	6.27
$\Delta P(\%)$	4.45	4.44	3.92

results in table 2.3 prove that the comprehensive performance of the optical OWHF based on two-beam interference is almost as perfect as traditional OWHFs (MD5 and SHA-1).

Conclusion

In this chapter, we mainly introduced two kinds of optical OWHFs which are, to the best of our knowledge, the only two attempts to solve the integrity checking of a message in the research area of optical information security. The two methods have taken advantage of the techniques of phase-truncated Fourier transform (PTFT) and two-beam interference, respectively. The simulation results imply that both of them have good and equivalent performance compared with traditional OWHFs.

Lastly, we would like to emphasize that there should be many ways in which to introduce a secret key in the construction process of two optical OWHFs described in part B, and therefore obtain two types of optical K-OWHFs (MACs), which have the ability to check the integrity and authenticity of a message simultaneously.

Acknowledgements

We are grateful for the support of the Science and Technology Innovation Commission of Shenzhen under Grant JCYJ20160520164642478, Natural Science Foundation of SZU under Grant 2016028, and National Natural Science Foundation of China under Grant 61377017, Grant 61307003, and Grant 61171073.

References

[1] Bahram J *et al* 2016 Roadmap on optical security *J. Opt.* **18** 083001

[2] Javidi B, Zhang G and Li J 1996 Experimental demonstration of the random phase encoding technique for image encryption and security verification *Opt. Eng.* **35** 2506–12

[3] Chen W, Javidi B and Chen X 2014 Advances in optical security systems *Adv. Opt. Photon.* **6** 120–55

[4] Lu D, He W, Liao M and Peng X 2017 Discussion and a new method of optical cryptosystem based on interference *Opt. Lasers Eng.* **89** 13–21

[5] Liao M, He W, Lu D and Peng X 2017 Ciphertext-only attack on optical cryptosystem with spatially incoherent illumination: from the view of imaging through scattering medium *Sci. Rep.* **7** 41789

[6] He W, Peng X, Meng X and Liu X 2012 Optical hierarchical authentication based on interference and hash function *Appl. Opt.* **51** 7750–7

[7] Qin W and Peng X 2010 Asymmetric cryptosystem based on phase-truncated Fourier transforms *Opt. Lett.* **35** 118–20

[8] Situ G and Zhang J 2004 Double random-phase encoding in the Fresnel domain *Opt. Lett.* **29** 1584–6

[9] Peng X, Zhang P, Wei H Z and Yu B 2006 Known-plaintext attack on optical encryption based on double random phase keys *Opt. Lett.* **31** 1044–6

[10] Peng X, Wei H and Zhang P 2006 Chosen-plaintext attack on lensless double-random phase encoding in the Fresnel domain *Opt. Lett.* **31** 3261–3

[11] He W, Peng X, Qin W and Meng X 2010 The keyed optical hash function based on cascaded phase-truncated Fourier transforms *Opt. Commun.* **283** 2328–32

[12] Lai H, He W and Peng X 2013 Optical hash function based on two-beam interference *Appl. Opt.* **52** 6213–9

[13] Schneier B 1996 *Applied Cryptography: Protocols, Algorithms, and Source Code in C* 2nd edn (Hoboken, NJ: John Wiley), pp 359–83

[14] Rivest R 1992 *The MD4 Message Digest Algorithm* (Lecture Notes Computer Science vol 537) (Springer) pp 304–11

[15] Rivest R L 1992 *The MD5 Message Digest Algorithm* (RFC) p 1321

[16] Wang X *et al* 2005 Cryptanalysis of the hash functions MD4 and RIPEMD *EUROCRYPT 2005: Advances in Cryptology—EUROCRYPT 2005: Annual Int. Conf. on Theory and Applications of Cryptographic Techniques* (Springer) pp 1–18

[17] Wang X, Yu H and Yin Y L 2005 Efficient collision search attacks on SHA-0 *Int. Cryptology Conf.* (Berlin: Springer), pp 1–16

[18] Wang X and Yu H 2005 How to break MD5 and other hash functions *Advances in Cryptology—EUROCRYPT 2005* (DBLP) pp 19–35

[19] Xiao D, Liao X F and Wang Y 2009 Parallel keyed hash function construction based on chaotic neural network *Neural Comput.* **72** 2288–96

[20] Goodman Joseph W 2005 *Introduction to Fourier Optics* 3rd edn (Roberts & Company, Inc.)

[21] Wang Y, Liao X F, Xiao D and Wong K W 2008 One-way hash function construction based on 2D coupled map lattices *Inf. Sci.* **178** 1391–406

[22] Kanso A, Yahyaoui H and Almull M 2012 Keyed hash function based on a chaotic map *Inf. Sci* **186** 249–64

IOP Publishing

Advanced Secure Optical Image Processing for Communications

Ayman Al Falou

Chapter 3

Cooperative MIMO and multi-hop relaying techniques for free-space optical communications: a survey

Mohamed Abaza, Raed Mesleh, Ali Mansour and el-Hadi M Aggoune

In this survey, a historical background of new challenges in optical communication systems is given. Then, approaches to address recently raised problems by the wireless communication community are introduced, challenges encountered in real-life applications are described, and state-of-the-art advances in free space optical communication (FSO) techniques are explained.

3.1 Introduction

FSO communication has gained research and commercial attention as it provides a high data rate, license-free operation, and cheap and eco-friendly means of alternative communication to congested radio frequency communications [1, 2]. The achievable data rate of FSO systems is determined by the physical components of the transmitter, receiver and channel environment. Theoretical and experimental results based on several research studies are reported. Cooperative relays can be used as a virtual multiple-input multiple-output (MIMO) configuration and can be utilized to enhance the diversity gain or the multiplexing gain. The performance analysis of multi-hop decode and forward relays for an FSO communication system using intensity-modulation direct-detection (IM/DD) is addressed in this chapter. Enhancement through multi-hop techniques when considering atmospheric turbulence and different weather attenuation effects with geometric losses in FSO communication systems is also studied. Finally, the use of a space shift keying (SSK) MIMO scheme in FSO systems is also reported.

doi:10.1088/978-0-7503-1457-2ch3

3.2 Spatial diversity

Spatial diversity is considered to be a promising solution to combat atmospheric turbulence and maximize the data rate of systems [3, 4]. By using multiple apertures at the transmitter and/or the receiver sides, spatial diversity has the potential to mitigate atmospheric turbulence effects, enhance the performance of the FSO links and overcome the limitations on the transmitted optical power. Also, the possibility of laser beam blockage by obstacles (e.g. birds or moving vehicles) is further reduced and longer distances can be covered in harsh weather conditions [4–7].

3.2.1 Space time codes

For single-input multiple-output (SIMO) systems, usually equal gain combining (EGC) is performed at the receiver, which provides a performance close to the optimal maximal ratio combining (MRC), while having the advantage of lower implementation complexity [8]. It is shown in [8] that the average bit error rate (ABER) expression of a MIMO FSO system using EGC in the LN channel is similar to that of a single-input single-output (SISO) link with the difference of dividing the variance of the normal channel by the number of sub-channels.

Conventional orthogonal space-time block codes (OSTBCs), as used in radio frequency (RF) systems, should be modified to deal with IM/DD techniques because the output of the transmitter must be unipolar [9, 10]. To achieve this, the negative part of a modulated signal is represented by the 1's complement (i.e. bitwise NOT) of a positive signal [9]. However, it has been shown in [11] that repetition codes (RCs) outperform their counterpart OSTBCs using IM/DD in uncorrelated LN channels. In an RCs system, the same signal is transmitted simultaneously from the available transmitters. MIMO systems can also be applied to coherent polarization as introduced in [12]. Moreover, a transmit laser selection (TLS) system, which requires channel knowledge at the transmitter and at the receiver, outperforms the ABER performance of RCs and OSTBCs at the expense of increasing system complexity [13].

In the case of uncorrelated channels, the spacing between transmitters is assumed to be larger than the fading correlation length. In addition, the total length of the transmitters must not exceed the capture zone of the receiver. The capture zone of the receiver is equal to $\theta_f \ell$ m, where θ_f is the field of view (FOV) of the receiver (in rad) and ℓ is the link distance (in m) [9]. However, the last assumption may be difficult in practice as the available space for the transmitters may not be sufficient for this requirement [14]. A correlation model called the exponential model [15], corresponding to the scenario of multi channel transmission from linearly equi-spaced transmitters, can be considered to calculate the correlation [16, 17]. A simple and general method to approximate the PDF of correlated and uncorrelated LN channels using the moment generating function (MGF) is derived in [18].

In [19], for a SIMO system with two 'receive' apertures, the sum of correlated gamma-gamma (G-G) random variables are approximated by an $\alpha - \mu$ distribution [20] in order to evaluate the BER performance at the receiver. This idea is later generalized to the case of multiple diversity in [21]. Also, the Pad approximation method [22] is used in [23] to obtain the probability density function (PDF) of the sum of correlated G-G random variables from their MGF, which is then used to evaluate the system performance analytically. However, due to the limitation of the Pad approximation, this method cannot be used for very low ABERs.

3.2.2 Spatial modulation

A spatial modulation (SM) MIMO system for indoor optical wireless communication has been studied in [4]. In SM, a group of data bits modulate a symbol from certain constellation diagrams and another group activates one of the transmitter beams. A line-of-sight (LOS) indoor wireless optical communication system using SM is proposed [24]. Results revealed that SM outperforms its counterpart RC system with M-ary pulse amplitude modulation (M-PAM) for spectral efficiencies larger than 4 bits/s/Hz. In [25], spatial pulse position amplitude modulation (SPPAM) is proposed for FSO links. It is shown that SPPAM outperforms SSK for spectral efficiencies of 2 bits/s/Hz or less.

Another system is SSK, which is a special case of SM where no data symbol is transmitted and transmitted information is conveyed through the spatial position of the active transmitter. A study conducted in [26] shows that on-off keying (OOK) RC outperforms OSTBCs, spatial multiplexing (SMux) and SSK for FSO communications using IM/DD over G-G channels. However, these results are given only for a spectral efficiency of 1bit/s/Hz. A detailed survey of SM and SSK systems are discussed in [27].

Tight upper bounds are derived for the error probability of coherent SM FSO systems over generalized H–K distribution [28]. Aperture averaging is another simple solution to reduce the fading effect, which uses a relatively large lens at the receiver to average over intensity fluctuations. However, the data rate of the system decreases as large parasitic capacitance is required [29]. The effect of the synchronization error of optical spatial modulation is studied in [30]. Results show that higher channel gain leads to a higher impact of the synchronization error on the system performance.

3.2.3 Recent developments

The impact of the correlation effect among multiple transmitters at the receiver is calculated in [31]. The PDF of the correlated LN channels is expressed using the MGF approach. Based on the MGF approach, approximated ABER expressions for both OSTBCs and RCs in correlated LN channels are obtained. Additionally, RCs are shown to always have higher signal-to-noise ratio (SNR) gain than OSTBCs by approximately 3 dB, as illustrated in figure 3.1. Finally, the severe effects of different weather conditions on the FSO links are studied in [32], where it is shown that fog degrades the FSO performance significantly, as shown in figure 3.2, where ρ is the correlation coefficient, the fading log-amplitude is modeled by normal distribution with a standard deviation σ and a mean μ, N_t is the number of transmitters, N_r is the number of receivers and α is the attenuation coefficient.

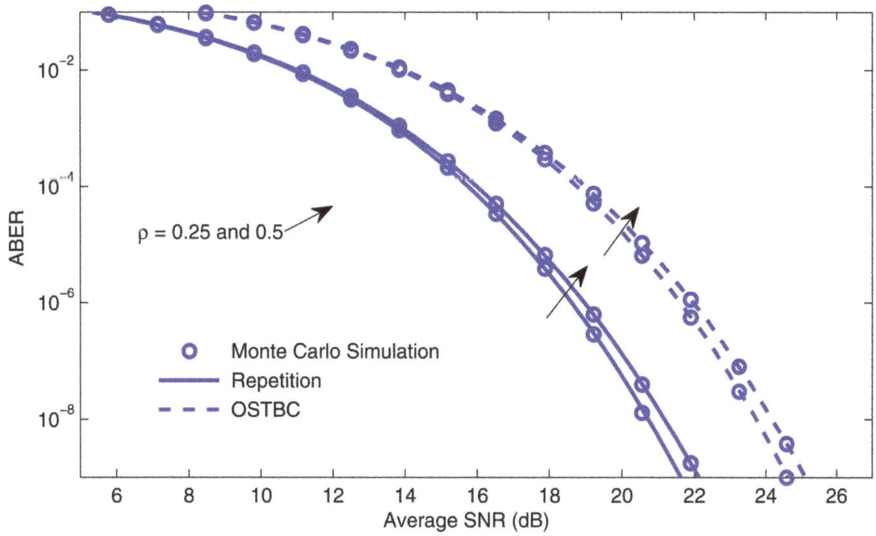

Figure 3.1. Correlated log-normal with $\sigma = 0.1$, $\mu = -\sigma^2$ and $N_t = 2$ [31]. Reproduced from [31].

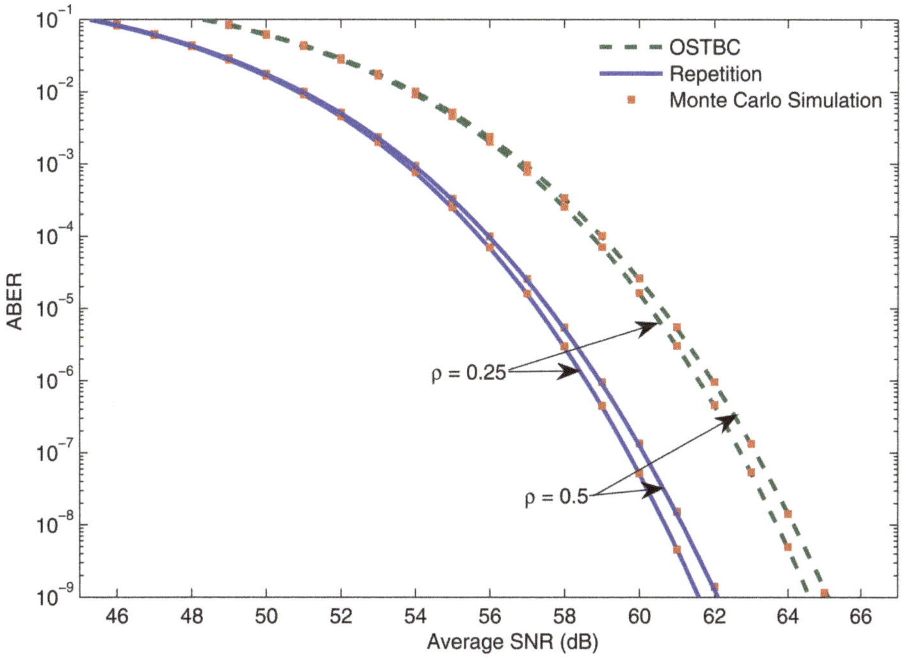

Figure 3.2. Correlated log-normal with $\sigma = 0.1$, $\mu = -\sigma^2$, and $N_t = 2$ and light fog ($\alpha = 20$ dB km^{-1}). © 2014 IEEE. Reprinted, with permission from [32].

Tight upper and lower bounds on the ABER expressions corresponding to negative exponential (NE) channels and LN channels FSO links with SSK are obtained in [33]. In comparison to SISO and RC *M*-PAM techniques, the obtained results are in favor of SSK FSO links for high spectral efficiencies and/or channel with high turbulence effects. Moreover, increasing the number of receivers is shown to yield lower SNR in SSK systems as compared to RC M-PAM systems (shown in figure 3.3).

3.3 Multi-hop relaying systems

To further mitigate turbulence and path loss effects, a multi-hop system is used due to the advantage of shorter hops that yield significant performance improvements.

3.3.1 Conventional and all-optical multi-hop relay systems

Optical-to-electrical and electrical-to-optical conversions are required in the relay for conventional multi-hop relay systems. All-optical FSO relaying, without the need for optical-to-electrical and electrical-to-optical conversions required in conventional FSO relays, is proposed in [34]. Results show that decode-and-forward (DF) systems outperform amplify-and-forward (AF) systems. In [17], results show that FSO systems with all-optical relays achieve a better performance compared with systems having conventional FSO relays. Multi-hop coherent FSO systems are employed over turbulent channels. Moreover, the diversity and multiplexing gains are derived in [35].

All-optical AF FSO systems using an erbium doped fiber amplifier (EDFA) combined with an optical hard-limiter (OHL) over G-G channels are employed in [36]. OHL is used to remove accumulated background noise. Then, a closed-form ABER expression is derived for the system. Moreover, results show that using OHL leads to the same performance of many relays but with fewer relays. In [37], closed-form and asymptotic expressions for the ABER and the ergodic capacity are derived

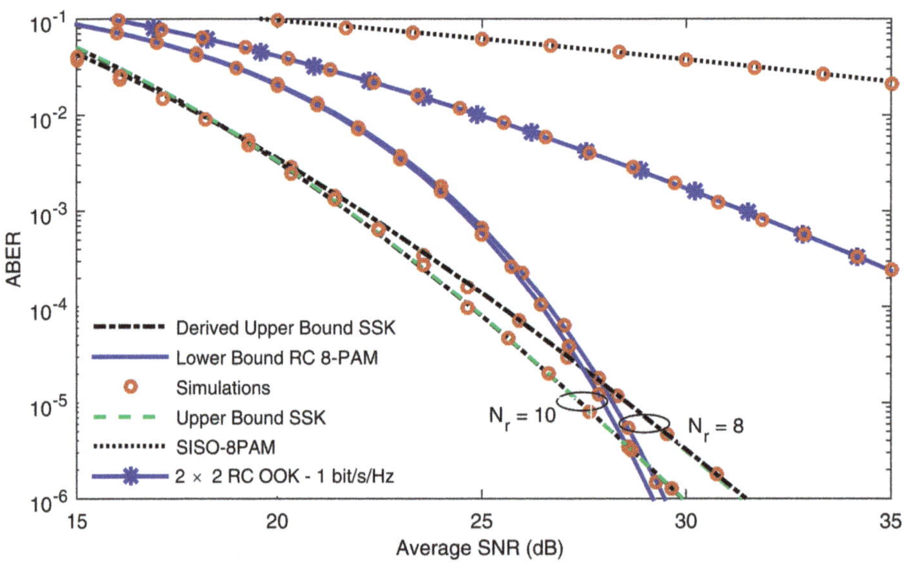

Figure 3.3. Comparison of space shift keying and repetition codes with spectral efficiency = 3 bits/s/Hz in a negative exponential channel with N_t = 8 and different number of receivers. Reproduced from [33] with permission from Elsevier B.V.

for a multi-hop AF heterodyne FSO system over atmospheric turbulence with pointing errors. Experimental results show that an all-optical dual-hop FSO system can achieve 10 Gbps for a link of 500 m in the presence of atmospheric turbulence in [38].

3.3.2 Coherent multi-hop relaying systems

Coherent FSO systems are shown to outperform their non coherent counterpart FSO systems at the expense of laborious receivers in [39]. The lower bound outage probability (OP) for dual-hop AF coherent FSO systems is derived in [40]. Moreover, the results show that the outage performance of AF is comparable with that of DF for coherent FSO systems. The ABER and the diversity order of multi-hop AF fixed-gain relaying systems using heterodyne detection over general Malaga turbulence channels are derived in [41].

3.3.3 Subcarrier intensity modulation multi-hop relaying systems

In [42], a hybrid subcarrier intensity modulation (SIM) and SM systems are proposed. The results show that the proposed system outperforms the SIM system at the spectral efficiency of 4 bits/s/Hz using fewer receivers. The ergodic capacity of SIM dual-hop FSO systems is derived over G-G channels with pointing errors for DF systems and for variable and fixed gains of AF systems in [43]. Mixed dual-hop AF systems using an RF link in the first hop represented by the Nakagami-m channel and using the FSO link represented by the G-G channel with pointing errors is proposed in [44], where SIM is used for FSO transmission. Moreover, the cumulative density function (CDF), the PDF, the MGF, the OP, the ABER and the capacity are derived for the proposed dual-hop AF RF-FSO communication system. The same system of dual-hop AF RF-FSO is employed in [45] with an extra direct RF link, where both IM/DD and coherent modulation techniques are employed, and MRC and selection combining (SC) are employed at the destination side.

3.3.4 Optimum placement for multi-hop relaying systems

Optimum placement for multi-hop relays is generally achieved by placing them at equidistance between the source and the destination nodes, while for cooperative relays (CRs) it is found that the optimum placement for the relays differs according to the link distances, number of relays and the SNRs [46]. The optimum placement for multi-hop relays is determined in the presence of obstacles blocking the direct link in [47].

3.3.5 Photon counting receiver for multi-hop relaying systems

A multi-hop FSO system using a count-and-forward (CF) relay with a photon counting receiver (PCR) that counts the number of received photons is proposed in [48]. The relay employs OOK and uses discrete power levels corresponding to the number of received photons. Results show that using CF with PCR outperforms the hard-decision forwarding scheme.

3.3.6 Beam size variation at the receiver for multi-hop relaying systems

The effect of beam size variation at the receiver due to turbulence is calculated for multi-hop FSO systems using a spatially coherent Gaussian beam model [49].

3.3.7 Power allocation for multi-hop relaying systems

By neglecting background radiation and dark currents, the power allocation for a multi-hop FSO system is discussed in [50].

3.3.8 Recent developments

MISO multi-hop DF FSO systems are considered in [51]. Results demonstrate that a MISO multi-hop system using *M*-PAM shows superior improvement to other existing systems such as a SISO multi-hop system using *M*-PAM, SISO and MISO using OOK at the same spectral efficiency. The impact of weak turbulence with light fog and moderate turbulence with clear air conditions on the performance of FSO systems has been analyzed. Multi-hop systems have the ability to mitigate both turbulence and path loss effects such as weather and geometric loss attenuations while MISO systems can only mitigate turbulence effect. Consequently, increasing the number of relays is shown to significantly enhance the overall system performance. This is why increasing the number of transmitters in MISO systems is shown to lead to small enhancements in SNRs, as can be seen in figures 3.4 and 3.5.

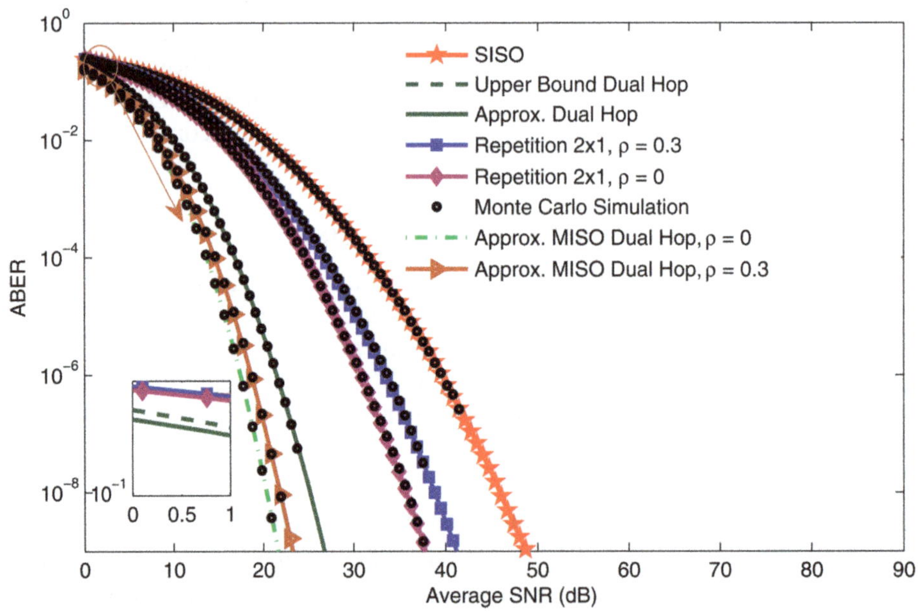

Figure 3.4. ABER of a single-relay multi-hop using one or two transmitters with 4-PAM modulation, two transmitters MISO using RCs and OOK, and SISO using OOK modulation FSO systems in a LN fading channel and clear air conditions [51]. Reproduced from [33] with permission from Elsevier B.V.

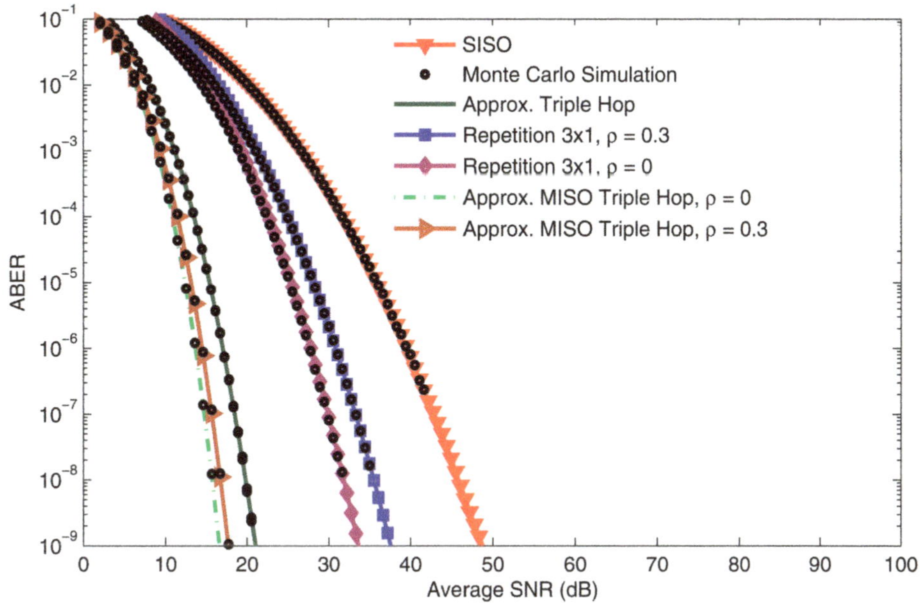

Figure 3.5. ABER of a two-relays multi-hop using one or three transmitters with 8-PAM modulation, three transmitters MISO using RCs and OOK, and SISO using OOK modulation FSO systems in a LN fading channel and clear air [51]. Reproduced from [33] with permission from Elsevier B.V.

3.4 Cooperative relaying systems

Cooperative relaying is a promising technique of wireless communication systems. The relay is used to create a virtual node between the source and destination in order to mitigate turbulence and path losses effects [52].

3.4.1 Transmission relays

Transmission relays are mainly divided to decode-and-forward (DF) and amplify-and-forward (AF). A DF relay decodes, re-modulates and retransmits the received signal. While an AF relay amplifies and retransmits the signal without decoding [53].

3.4.2 DF cooperative relaying systems

The OP of the physical layer of CRs FSO systems using DF and AF systems are derived in [54]. Atmospheric turbulence and path loss attenuations are considered. Results show that multi-hop systems outperform CRs. However, CRs can create artificial broadcasting. The asymptotical relative diversity order (ARDO) of DF FSO for multi-hop systems and CRs is derived in [55]. Results show that ARDO of multi-hop systems is greater than CRs. CRs FSO communication systems with an extra direct link are proposed in [56]. Three CRs are discussed as a bit detect and forward (BDF), adaptive bit detect and forward (ABDF) and adaptive decode and forward (ADF). At BDF, the relay forwards the received bit regardless of its correctness. However, at ABDF both the relay and the direct link re-forward the received bit with low SNR. While in ADF, the relay corrects the false bit, then re-forwards it. If the relay fails to correct the bit, ADF will behave like ABDF. Results show that the proposed systems outperform both the direct link and MISO systems. A novel CRs FSO scheme, called 'improved adaptive decode-and-forward' (IADF) is proposed in [57]. The proposed system transmits two frames in four time slots. In the first time slot, the source transmits the message to the relay. In the second time slot, the relay transmits the message to the destination. The same thing happens for the second frame in the third and fourth time slots. However, if the relay failed to receive the message correctly, in any case, the source performs an XOR operation on the first frame with the second frame and sends the results directly to the destination. In addition, the diversity order is derived for the proposed system as a function of the propagation distance.

In [58], the author showed that the diversity order for an N_{re} relay system using BDF and ABDF is equal to $\lceil \frac{N_{re}}{2} \rceil + 1$ and $N_{re} + 1$, respectively, where the function $\lceil a \rceil$, rounds a to the smallest integer greater than a.

CRs for a DF FSO communication system using a single relay IM/DD in the absence of background noise are employed. The considered system achieves the same ABER compared with direct and MISO systems, which suffer from correlation at low SNR [59]. In the absence of background noise, the BDF scheme is shown to have the same performance as that of ABDF. However, in the presence of background noise, ABDF is shown to have more SNR gain as compared to BDF [60]. In addition, it is shown that using a single feedback bit using RF will have a significant SNR gain, especially in low SNRs.

Asymptotic expressions for OP are derived for a DF IM/DD CRs FSO system using repetition coding over G-G channels with pointing errors and weather attenuation effects [61], where the destination receives two copies of the message over three time slots. Results show that the considered system outperforms the OP of both MISO and conventional CRs systems.

3.4.3 AF cooperative relaying systems

All-optical CRs systems employing AF relays using PCR over LN channels with path loss effects are considered in [62]. Results demonstrate that the proposed system requires less SNR for the same ABER as compared to a multiple-transmitter system, especially for long distances and/or high attenuation factors.

3.4.4 All-optical cooperative relaying systems

All-optical CRs systems using PCR over LN channels with path loss effects are considered for full channel state information (CSI) and semi-blind CSI in [63]. Results show that both cases have comparable performance. However, semi-blind CSI has lower complexity. The OP of all-optical relay selection FSO systems using PCR over G-G turbulent channels is derived for both full CSI and semi-blind relays in [64]. Results demonstrate an SNR gain up to 11.5 dB compared with all-active relays for a system using four relays.

3.4.5 Subcarrier intensity modulation cooperative relaying systems

CRs for SIM FSO communication systems using DF relays and differential modulation are employed over G-G channels in [65]. The ABER, the diversity order and the coding gain are all derived. The OP of CRs binary phase shift keying (BPSK)-SIM for FSO systems over G-G turbulent channels with pointing errors is derived in [66]. A closed-form ABER expression of selective CRs for a DF FSO system using SIM for BPSK over a generalized $\mathcal{M}$-distributed turbulent channel is derived in [67].

3.4.6 Selected relay for cooperative relaying systems

The OP of CRs FSO communication systems over G-G turbulence channels with pointing errors using DF relays are derived in [68]. CRs can be classified according to their transmission protocol into an all-active transmission and selective transmission systems. In all-active transmission, relays forward all received signals to the destination [54]. However, in selective transmission, a single relay is selected and it will only forward the received message based on the max–min SNR criterion [69]. The selective transmission is considered an asynchronous low complexity system and outperforms the OP of all-active transmission. In [70], the OP of the relay selection for FSO systems using DF relays over LN channels with path loss effects is discussed. Results demonstrate that selecting the participated relay depending on the SNR of all links enhances the performance by 6 dB as compared to selecting the participated relay based on the SNR of the relay to destination (R–D) at OP 10^{-5}. DF CRs in a FSO IM/DD communications system are employed over G-G channels and MRC is used at the receiver in [71]. The OP, diversity and coding gains are derived.

In [72], opportunistic CRs for FSO links are proposed, where the relay participates only if the source to relay (S–R) link is strong. In the case that S–R and R–D links have the same performance, opportunistic CRs has the same OP of conventional CRs. While if the S–R link is weak, opportunistic CRs has better OP compared with conventional CRs.

The partial relay section according to the best S–R path, based on the max–min SNR criterion, is employed and an approximate ABER over G-G channels is derived in [73]. In [74], OP and diversity gain of multi-hop parallel relaying over LN channels are derived. Results show that multi-hop parallel relaying outperforms both multi-hop and parallel relaying.

Multi-hop parallel relaying, where only one participating path (i.e. the strongest channel gain) is selected is discussed in [75]. In addition, the second and the third strongest path is considered in a case where the first strongest path is not available due to scheduling and load balancing issues. The results show that the OP of the second and the third strongest path outperforms the multi-hop system using the same number of relays per path. A study in [76] proposes and analyzes the OP of asynchronous low complexity CRs with the best relay selection based on the max–min SNR criterion.

The ABER performance of a single relay selection according to the max-min criterion of SNR for dual-hop parallel DF FSO over G-G channels employing adaptive subcarrier quadrature amplitude modulation is obtained using the power series approximation method in [77]. The ABER for CRs for FSO communication systems using selective transmission is obtained for LN and Rayleigh channels in [78].

3.4.7 Power allocation for cooperative relaying systems

Power allocation for quantum-limited CRs FSO systems is discussed in [79]. In the absence of CSI, the results demonstrate that it is better to distribute the power among the links. While in the presence of CSI, the ABER performance can be significantly enhanced by allocating all the power to the strongest link. In [80], a relay sends its information simultaneously with the information coming from the source to the destination. Optimal power allocation is calculated to avoid interference between the information of the source with the information of the relay.

3.4.8 Multiple-source cooperation relaying

Multiple-source cooperation relaying for FSO systems over strong turbulence is proposed in [81], where many users send their information to a single relay. The relay will then forward all of the messages to the final destination. Compared with conventional CRs, the proposed system requires lower equipment cost. In [82], the performance of the n-th best user point-to-multipoint selection scheme and the performance of the multipoint-to-multipoint selection scheme are evaluated under the LN and G-G channels.

3.4.9 Optimum placement for cooperative relaying systems

The optimum placement for CRs is found by placing the relays close to the source node and it is the function of link distance, number of relays and SNRs as discussed in [46]. The ABER and the diversity order of relay placement in ADF CRs for FSO systems over G-G channels with pointing errors are derived in [83]. Unlike [46], which concludes that placing the relays near the source achieves the optimum OP in the absence of pointing errors, the authors in [83] show that the diversity order is better when the relay nodes are closer to the destination node if pointing errors are present. Moreover, the authors prove that the ADF system based on S–R selection always outperforms the ADF system based on R–D regardless of the position and number of relays.

Asymptotic ABER of CR for a FSO communication system using EGC over G-G channels with pointing errors is derived [84]. This also results in the diversity order being independent of the pointing error in some cases. However, the diversity order strongly depends on the placement of the relay.

3.4.10 Multi-hop parallel relaying

The diversity order for various distributions, as in a special case of a generalized $\mathcal{M}$-distributed turbulent channel, is derived. The ABER and the OP of selective transmission for multi-hop parallel DF relaying using BPSK is derived for exponentiated Weibull (EW) fading channels using the Gauss–Laguerre quadrature rule [85].

3.4.11 Mean time between failures for cooperative relaying systems

In [86], results show that the mean time between failures (MTBF) is 12 years for a SISO FSO system using a low data rate RF back up, while MTBF is 20 years for a CRs FSO system using a low data rate RF back up.

3.4.12 Inter-relay cooperation of cooperative relaying systems

The OP and the diversity of inter-relay cooperation for all-active and selection CRs FSO systems are introduced and derived in [87]. Results show that the inter-relay cooperation outperforms conventional CRs for certain relay positions. Moreover, power allocation and ABER of inter-relay cooperation are derived in [88].

3.4.13 Two-way relay systems

Two-way relays (TWRs) using DF for FSO systems is employed in [89] over G-G channels, where two users communicate using three time slots, thus achieving a higher data rate than conventional relaying. TWR using two time slots is investigated in [90, 91]. In the first time slot, the two source nodes send their information to the relay. The relay decodes the received signals and performs an XOR operation on the decoded signals. The resultant signal is then forwarded to each user in the second time slot. The user can obtain the intended received signal by performing an XOR operation with the received message. The symbol error rate (SER) and the OP for TWR are derived over an $\mathcal{M}$-channel in [90] and LN channel in [91].

In [92], TWR DF FSO is employed over $\mathcal{M}$-distribution FSO channels and the ABER and the OP are derived taking the impact of pointing error effects into consideration. A multi-user TWR is proposed in [93], where the received signal is affected by atmospheric turbulence, path loss and building sway. Partial relay selection for TWR coherent FSO using AF is employed over G-G channels and affected by path losses and pointing errors in [94].

The results show that the TWR scheme almost doubles the ergodic capacity compared to that of the half-duplex (HD) scheme with the same outage performance as reported in [95].

3.4.14 Recent developments

The selection of a single relay based on the max–min SNR criterion for FSO communication systems is investigated in [96]. Dual-hop DF relaying systems over different atmospheric turbulence channels affected by path losses and misalignment errors are employed. OP is obtained for LN and G-G channels under the considered challenges. Moreover, approximated ABER expressions for full-duplex (FD) relays, HD relays and direct link are derived using the Gauss–Laguerre quadrature rule assuming full CSI. Simulation results show the superiority of FD relays systems compared to their counterparts in terms of ABER and OP, especially for strong turbulence, as shown in figures 3.6 and 3.7, with ξ denoting the severity of the pointing error. The lower value of ξ represents a larger pointing error severity.

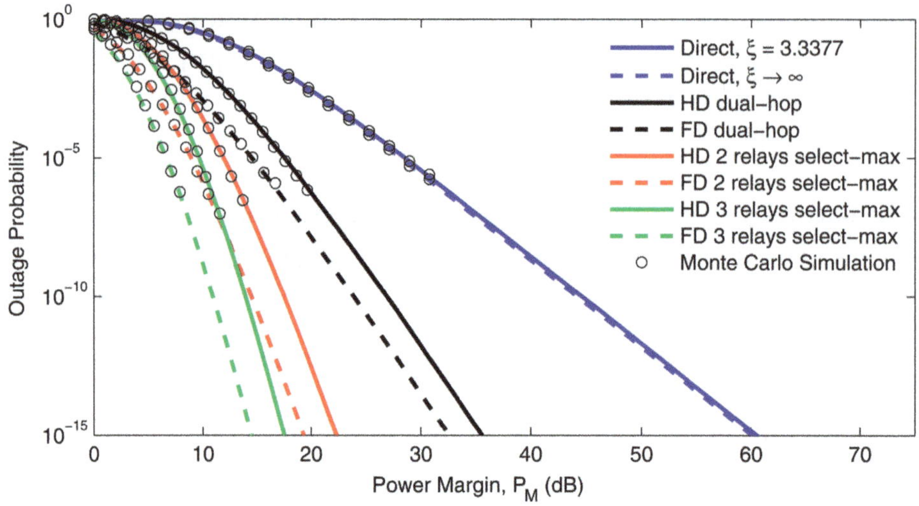

Figure 3.6. OP for FSO over a strong fading channel [96]. Copyright 2016 IEEE. Reprinted, with permission from [64].

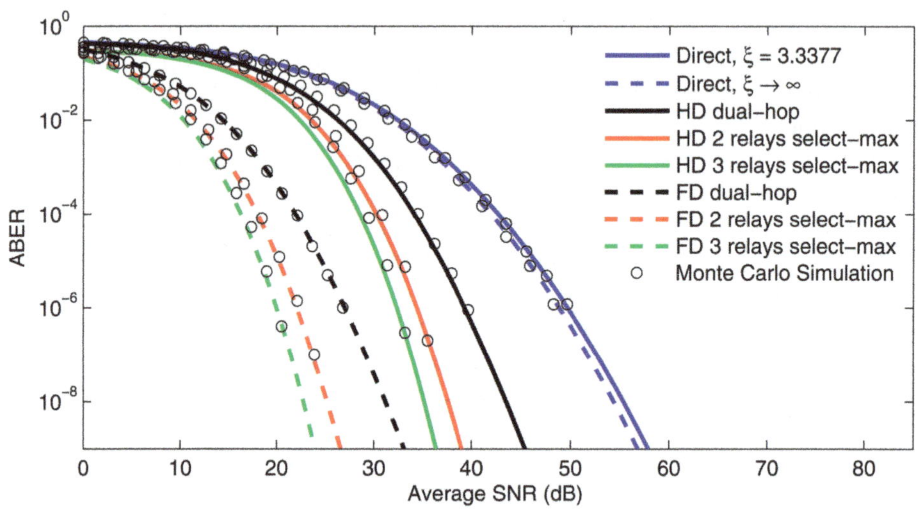

Figure 3.7. ABER of FSO over a moderate fading channel [96]. Copyright 2016 IEEE. Reprinted, with permission from [64].

3.5 Summary

In this chapter, a review of major state-of-the-art developments for FSO communication systems has been discussed. We show that spatial diversity and relay-assisted FSO communication systems proved to be potential candidates for next generation communication networks (5G) and the Internet of Things (IoT). Spatial diversity combats atmospheric turbulence and maximizes the data rate of the systems. In addition, relay-assisted systems mitigate turbulence and path loss effects due to the advantage of shorter hops that yield significant performance improvements.

References

[1] Khalighi M A and Uysal M 2014 Survey on free space optical communication: a communication theory perspective *IEEE Commun. Surveys and Tutorials* **16** 2231–58
[2] Mansour A, Mesleh R and Abaza M 2017 New challenges in wireless & free space optical communications *Opt. Lasers Eng.* **89** 95–108
[3] Abbas G, Ahmed E, Aziz W, Saleem S and ul Islam Q 2012 Performance enhancement of multi-input multi-output (MIMO) system with diversity *Int. J. Multidisciplinary Sci. Eng.* **3** 8–12 www.ijmse.org/Volume3/Issue5/paper2.pdf
[4] Mesleh R, Mehmood R, Elgala H and Haas H 2010 Indoor MIMO optical wireless communication using spatial modulation, in *Proc. of IEEE Int. Conf. on Communications, (Cape Town, South Africa)* pp. 1–5
[5] Tsiftsis T A, Sandalidis H G, Karagiannidis G K and Uysal M 2009 Optical wireless links with spatial diversity over strong atmospheric turbulence channels *IEEE Trans. Wireless Commun.* **8** 951–7
[6] Safari M and Hranilovic S 2012 Diversity gain for near-field MISO atmospheric optical communications, in *Proc. of Int. Conf. on Communications, (Ottawa, Canada)* pp. 3128–3132
[7] Abaza M R, Mesleh R, Mansour A and Alfalou A 2013 MIMO techniques for high data rate free space optical communication system in log-normal channel, in *Proc. of Int. Conf. on Technological Advances in Electrical, Electronics and Computer Engineering, (Konya, Turkey)* pp. 1–5
[8] Navidpour S M, Uysal M and Kavehrad M 2007 BER performance of free-space optical transmission with spatial diversity *IEEE Trans. Wireless Commun.* **6** 2813–9
[9] Simon M K and Vilnrotter V 2005 Alamouti-type space-time coding for free-space optical communication with direct detection *IEEE Trans. Wireless Commun.* **4** 35–9
[10] Premaratne M and Zheng F-C 2007 Orthogonal space-time block codes for free-space IM/DD optical links *Electron. Lett.* **43** 822–3
[11] Safari M and Uysal M 2008 Do we really need OSTBC for free-space optical communication with direct detection? *IEEE Trans. Wireless Commun.* **7** 4445–8
[12] Han Y and Li G 2005 Coherent optical communication using polarization multiple-input-multiple-output *Opt. Express* **13** 7527–34
[13] Garcia-Zambrana A, Castillo-Vazquez C, Castillo-Vazquez B and Hiniesta-Gomez A 2009 Selection transmit diversity for FSO links over strong atmospheric turbulence channels *IEEE Photonics Tech. Lett.* **21** 1017–9
[14] Zhu X and Kahn J M 2000 Maximum-likelihood spatial-diversity reception on correlated turbulent free-space optical channels, *Proc. of IEEE Global Communications Conf. (San Francisco)* vol 2 pp 1237–41.
[15] Karagiannidis G K, Zogas D A and Kotsopoulos S A 2003 An efficient approach to multivariate Nakagami-m distribution using Green's matrix approximation *IEEE Trans. Wireless Commun.* **2** 883–9
[16] Moradi H, Refai H H and LoPresti P G 2011 A switched diversity approach for multi-receiving optical wireless systems *Appl. Opt.* **50** 5606–14

[17] Bayaki E, Michalopoulos D S and Schober R 2012 EDFA-based all-optical relaying in free-space optical systems *IEEE Trans. Commun.* **60** 3797–807

[18] Mehta N B, Molisch A F, Wu J and Zhang J 2007 Approximating the sum of correlated lognormal or lognormal-Rice random variables *IEEE Trans. Wireless Commun.* **6** 2690–9

[19] Yang G, Khalighi M-A, Bourennane S and Ghassemlooy Z 2012 Approximation to the sum of two correlated gamma-gamma variates and its applications in free-space optical communications *IEEE Wireless Commun. Lett.* **1** 621–4

[20] Yacoub M D 2007 The $\alpha - \mu$ distribution: a physical fading model for the Stacy distribution *IEEE Trans. Vehicular Tech.* **56** 27–34

[21] Yang G, Khalighi M A, Bourennane S and Ghassemlooy Z 2014 Fading correlation and analytical performance evaluation of the space-diversity free-space optical communications system *J. Opt.* **16** 1–10

[22] Karagiannidis G K 2004 Moments-based approach to the performance analysis of equal gain diversity in Nakagami-m fading *IEEE Trans. Commun.* **52** 685–90

[23] Yang G, Khalighi M A, Ghassemlooy Z and Bourennane S 2014 Performance analysis of space-diversity FSO systems over the correlated Gamma-Gamma fading channel using Pade approximation method *IET Commun.* **8** 2246–55

[24] Fath T, Haas H, Di Renzo Marco and Mesleh R 2011 Spatial modulation applied to optical wireless communications in indoor LOS environments *Proc. of IEEE Global Communications Conf. (Houston)* pp. 1–5.

[25] Ozbilgin T and Koca M 2014 Optical spatial pulse position amplitude modulation over atmospheric turbulence channels, in *Proc. of Int. Conf. on Communications (Sydney)* pp. 3412–7.

[26] Yang G, Khalighi M A, Virieu T, Bourennane S and Ghassemlooy Z 2012 Contrasting space-time schemes for MIMO FSO systems with non-coherent modulation, in *Proc. of Int. Workshop on Optical Wireless Communications (Pisa)* pp. 1–3

[27] Renzo M D, Haas H, Ghrayeb A, Sugiura S and Hanzo L 2014 Spatial modulation for generalized mimo: challenges, opportunities, and implementation *Proc. IEEE* **102** 56–103

[28] Peppas K P and Mathiopoulos P T Free space optical communication with spatial modulation and coherent detection over H-K atmospheric turbulence channels, *IEEE/OSA J. Lightwave Technology*, accepted in 2015

[29] Khalighi M-A, Schwartz N, Aitamer N and Bourennane S 2009 Fading reduction by aperture averaging and spatial diversity in optical wireless systems *IEEE/OSA J. Opt. Commun. Networking* **1** 580–93

[30] Olanrewaju H G and Popoola W O 2017 Effect of synchronisation error on optical spatial modulation, *IEEE Trans. Commun.* **65** 12

[31] Abaza M, Mesleh R, Mansour A and M Aggoune E-H 2014 Diversity techniques for a free-space optical communication system in correlated log-normal channels *SPIE Opt. Eng.* **53** 1–6

[32] Abaza M, Mesleh R, Mansour A and Aggoune E H M 2014 Spatial diversity for FSO communication systems over atmospheric turbulence channels, in *Proc. IEEE Wireless Commun. Network. Conf. (Istanbul)*

[33] Abaza M, Mesleh R, Mansour A and Aggoune e H M 2015 Performance analysis of space-shift keying over negative-exponential and log-normal FSO channels *Chin. Opt. Lett.* **13** 051001

[34] Kazemlou S, Hranilovic S and Kumar S 2011 All-optical multihop free-space optical communication systems *IEEE/OSA J. Lightwave Tech.* **29** 2663–9
[35] Aghajanzadeh S M and Uysal M 2011 Multi-hop coherent free-space optical communications over atmospheric turbulence channels *IEEE Trans. Commun.* **59** 1657–63
[36] Trinh P V, Dang N T and Pham A T 2015 All-optical AF relaying FSO systems using EDFA combined with OHL over gamma-gamma channels *Proc. of Int. Conf. on Communications (London)*
[37] Zedini E and Alouini M-S 2015 On the performance of multihop heterodyne FSO systems with pointing errors *IEEE Photonics J.* **7**
[38] Libich J, Komanec M, Zvanovec S, Pesek P, Popoola W O and Ghassemlooy Z 2015 Experimental verification of an all-optical dual-hop 10 Gbit/s free-space optics link under turbulence regimes *Opt. Lett.* **40** 391–4
[39] Bayaki E and Schober R 2012 Performance and design of coherent and differential space-time coded FSO systems *IEEE/OSA J. Lightwave Tech.* **30** 1569–77
[40] Chae C-B and Yoon G 2015 Outage probability analysis of a coherent FSO amplify-and-forward relaying system *IEEE Photonics Tech. Lett.* **27** 1204–7
[41] Alheadary W G, Park K H and Alouini M S 2017 BER analysis of multi-hop heterodyne FSO systems with fixed gain relays over general Malaga turbulence channels *Proc. of Int. Wireless Communications and Mobile Computing Conference (Valencia)* pp. 1172–1177
[42] Lee J-H and Hwang S-H 2015 Selection diversity-aided subcarrier intensity modulation/spatial modulation for free-space optical communication *IET Optoelectron.* **9** 116–24
[43] Aggarwal M, Garg P and Puri P 2015 Exact capacity of amplify-and-forward relayed optical wireless communication systems *IEEE Photonics Tech. Lett.* **27** 903–6
[44] Anees S and Bhatnagar M R 2015 Performance of an amplify-and-forward dual-hop asymmetric RF-FSO communication system *IEEE/OSA J. Opt. Commun. Networking* **7** 124–35
[45] Zedini E, Ansari I S and Alouini M S 2014 On the performance of hybrid line of sight RF and RF-FSO fixed gain dual-hop transmission systems *Proc. of IEEE Global Communications Conf. (Austin)* pp. 2119–24
[46] Kashani M A, Safari M and Uysal M 2013 Optimal relay placement and diversity analysis of relay-assisted free-space optical communication systems *IEEE/OSA J. Opt. Commun. Networking* **5** 37–47
[47] Zhu B, Cheng J and Wu L 2014 Optimal FSO relay nodes placement with link obstacles and infeasible regions *Proc. of IEEE Global Communications Conf. (Austin)* pp. 2137–42
[48] Gong C and Xu Z 2015 Non-line of sight optical wireless relaying with the photon counting receiver: a count-and-forward protocol *IEEE Trans. Wireless Commun.* **14** 376–88
[49] Pham H T T, Dang N T and Pham A T 2014 Effects of atmospheric turbulence and misalignment fading on performance of serial-relaying M-ary pulse-position modulation free-space optical systems with partially coherent Gaussian beam *IET Commun.* **8** 1762–8
[50] Abou-Rjeily C 2012 Power allocation for quantum-limited multihop free-space optical communication systems *IEEE Commun. Lett.* **16** 2068–71
[51] Abaza M, Mesleh R, Mansour A and M Aggoune E-H 2015 Performance analysis of MISO multi-hop FSO links over log-normal channels with fog and beam divergence attenuations *Opt. Commun.* **334** 247–52
[52] Elia P, Vinodh K, Anand M and Kumar P V 2009 D-MG tradeoff and optimal codes for a Class of AF and DF cooperative communication protocols *IEEE Trans. Inf. Theory* **55** 3161–85

[53] Levin G and Loyka S 2012 Amplify-and-forward versus decode-and forward relaying: which is better? *Proc. of Int. Zurich Seminar on Communications (Zurich)* pp. 123–6
[54] Safari M and Uysal M 2008 Relay-assisted free-space optical communication *IEEE Trans. Wireless Commun.* **7** 5441–9
[55] Safari M and Uysal M 2008 Diversity gain analysis of free-space optical communication systems *Proc. of Canadian Conf. on Electrical and Computer Engineering (Niagara Falls)* pp. 1239–44
[56] Karimi M and Nasiri-Kenari M 2009 BER analysis of cooperative systems in free-space optical networks *IEEE/OSA J. Lightwave Tech.* **27** 5639–47
[57] Zhu B, Cheng J and Wu L 2015 A distance-dependent free-space optical cooperative communication system *IEEE Commun. Lett.* **19** 969–72
[58] Abou-Rjeily C 2013 Simple-DF versus selective-DF relaying over Rayleigh turbulence-induced FSO fading channels *Proc. of Int. Symp. on Personal, Indoor and Mobile Radio Communications (London)* pp. 1082–6
[59] Abou-Rjeily C and Slim A 2011 Cooperative diversity for free-space optical communications: transceiver design and performance analysis *IEEE Trans. Commun.* **59** 658–64
[60] Abou-Rjeily C and Haddad S 2011 Novel cooperation strategies for free-space optical communication systems in the absence and presence of feedback *Communications in Computer and Information Science* (Springer).
[61] Castillo-Vazquez C, Boluda-Ruiz R, Castillo-Vazquez B and Garcia-Zambrana A 2015 Outage performance of DF relay-assisted FSO communications using time-diversity *IEEE Photonics Tech. Lett.* **27** 1149–52
[62] Karimi M and Nasiri-Kenari M 2011 Free space optical communications via optical amplify-and-forward relaying *IEEE/OSA J. Lightwave Tech.* **29** 242–8
[63] Wang J Y, Wang J B, Chen M, Hu Q S, Huang N, Guan R and Jia L 2013 Free-space optical communications using all-optical relays over weak turbulence channels with pointing errors *Proc. of Int. Conf. on Wireless Communications and Signal Processing (Hangzhou)* pp. 1–6
[64] Mostaani Z and Uysal M 2014 Relay selection in FSO systems with all-optical relaying over Gamma-Gamma turbulence channels, in *Proceedings of IEEE Signal Processing and Communications Applications Conference, (Trabzon, Turkey)* pp. 710–713
[65] Bhatnagar M R 2012 Average BER analysis of differential modulation in DF cooperative communication system over gamma-gamma fading FSO links *IEEE Commun. Lett.* **16** 1228–31
[66] Prabu K and Kumar D S 2014 Outage analysis of relay-assisted BPSK-SIM based FSO systems over strong atmospheric turbulence with pointing errors *Int. J. Com. Comm. Eng.* **3** 317–20
[67] Sharma N, Bansal A and Garg P 2014 On the performance of selective transmission based DF FSO system with generalized $\mathcal{M}$-distributed turbulence, in *Proceedings of Annual IEEE India Conference, (Pune, India)* pp. 1–5
[68] Feng M, Wang J B, Sheng M, Cao L L, Xie X X and Chen M 2011 Outage performance for parallel relay-assisted free-space optical communications in strong turbulence with pointing errors, in *Proceedings of International Conference on Wireless Communications and Signal Processing, (Nanjing, China)* pp. 1–5
[69] Chatzidiamantis N D, Michalopoulos D S, Kriezis E E, Karagiannidis G K and Schober R 2013 Relay selection protocols for relay-assisted free-space optical systems *IEEE/OSA J. Opt. Commun. Networking* **5** 92–103

[70] Celik Y and Odabasioglu N 2012 On relay selection for cooperative free-space optical communication *Proc. of European Conf. on Networks and Optical Communications (Vilanova i la Geltru)* pp. 1–5
[71] Aghajanzadeh S M and Uysal M 2012 Outage performance and DMT analysis of DF parallel relaying in FSO IM/DD communications *Proc. of IEEE Vehicular Technology Conf. (Quebec City)* pp. 1–5
[72] He Q and Xu Z 2012 Opportunistic cooperation for FSO links aided by decode and forward relay *Proc. of IEEE Workshop on Optical Wireless Communications (Anaheim)* pp. 1193–7
[73] Bhatnagar M R 2013 Average BER analysis of relay selection based decode-and-forward cooperative communication over gamma-gamma fading FSO links *Proc. of Int. Conf. on Communications (Budapest)* pp. 3142–7
[74] Kashani M A and Uysal M 2013 Outage performance and diversity gain analysis of free-space optical multi-hop parallel relaying *IEEE/OSA J. Opt. Commun. Networking* **5** 901–9
[75] Zhalehpour S and Uysal M 2014 Outage performance analysis of mth best path selection protocol in FSO communications *Proc. of Int. Conf. on Transparent Optical Networks (Graz)* pp. 1–4
[76] Chatzidiamantis N D, Michalopoulos D S, Kriezis E E, Karagiannidis G K and Schober R 2011 Relay selection in relay-assisted free space optical systems *Proc. of IEEE Global Communications Conf. (Houston)* pp. 1–6
[77] Vu B T, Thang T C and Pham A T 2014 Selection decode-and-forward cooperative FSO systems with adaptive rate strategy over gamma-gamma fading channels *Proc. of IEEE Global Communications Conf. (Austin)* pp. 3585–90
[78] Abou-Rjeily C 2013 Performance analysis of selective relaying in cooperative free-space optical systems *IEEE/OSA J. Lightwave Tech.* **31** 2965–73
[79] Abou-Rjeily C and Haddad S 2011 Cooperative FSO systems: performance analysis and optimal power allocation *IEEE/OSA J. Lightwave Tech.* **29** 1058–65
[80] Li Y, Guo S, Geng T, Ma S, Gao S and Gaoa H 2017 Evaluation on the capacity and outage performance of the free space optical system impaired by timing jitters over an aggregate channel *Opt. Eng.* **56** 076108
[81] Zhou X, Zhang D, Yang Y and Obaidat M S 2012 Network-coded multiple-source cooperation aided relaying for free-space optical transmission *Int. J. Comm. Sys.* **25** 1465–78
[82] Abou-Rjeily C and Chami M 2017 Approximate BER expressions for FSO communication systems with multiuser diversity *Proc. of Int. Wireless Communications and Mobile Computing Conf. (Valencia)* pp. 1683–8
[83] Boluda-Ruiz R, Garcia-Zambrana A, Castillo-Vazquez B and Castillo-Vazquez C 2015 Impact of relay placement on diversity order in adaptive selective DF relay-assisted FSO communications *Opt. Exp.* **23** 2600–17
[84] Garcia-Zambrana A, Castillo-Vazquez C, Castillo-Vazquez B and Boluda-Ruiz R 2012 Bit detect and forward relaying for FSO links using equal gain combining over gamma-gamma atmospheric turbulence channels with pointing errors *Opt. Express* **20** 16394–409
[85] Wang P, Cao T, Guo L, Wang R and Yang Y 2015 Performance analysis of multi-hop parallel free-space optical systems over exponentiated Weibull fading channels *IEEE Photonics J.* **7** 7900715
[86] Dubey V, Chadha D and Chandra V 2014 Bit error rate and reliability analysis of cooperative communication in free-space optical systems *Photonic Network Commun.* **28** 92–101

[87] Abou-Rjeily C 2014 All-active and selective FSO relaying: do we need inter-relay cooperation? *IEEE/OSA J. Lightwave Tech.* **32** 1899–906
[88] Abou-Rjeily C and Haddad S 2014 Inter-relay cooperation: a new paradigm for enhanced relay-assisted FSO communications *IEEE Trans. Commun.* **62** 1970–82
[89] Tang Y, Zhou X, Zhang Z and Tian Q 2011 Performance analysis of a two-way network-coded free space optical relay scheme over strong turbulence channels *Proc. of IEEE Vehicular Technology Conf. (San Francisco)* pp. 1–5
[90] Sharma P K and Garg P 2014 Bi-directional decode-XOR-forward relaying over $\mathcal{M}$-distributed free space optical links *IEEE Photonics Tech. Lett.* **26** 1916–9
[91] Puri P, Garg P and Aggarwal M 2014 Analysis of spectrally efficient two-way relay assisted free space optical systems in atmospheric turbulence with path loss *Int. J. Commun. Syst.* **29** 99–112
[92] Sharma P K, Bansal A and Garg P 2015 Relay assisted bi-directional communication in generalized turbulence fading *IEEE/OSA J. Lightwave Tech.* **33** 133–9
[93] Puri P, Garg P, Aggarwal M and Sharma P K 2014 Multiple user pair scheduling in TWR-FSO systems in presence of building sway *Proc. of Int. Conf. on Signal Processing and Communications (Bangalore)* pp. 1–5
[94] Puri P, Garg P and Aggarwal M 2015 Asymptotic analysis of TWR assisted FSO links with partial dual-relay selection *IEEE Commun. Lett.* **19** 879–82
[95] Al-Eryani Y F, Salhab A M, Zummo S A and Alouini M S 2017 On the performance of two-way multiuser mixed RF/FSO relay networks with opportunistic scheduling & asymmetric channel gains *Proc. of Int. Wireless Communications and Mobile Computing Conf. (Valencia)* pp. 1178–83
[96] Abaza M, Mesleh R, Mansour A and Aggoune e H M 2016 Relay selection for full-duplex FSO relays over turbulent channels *Proc. of IEEE Int. Symp. On Signal Processing and Information Technology (Limassol)* pp. 103–8

IOP Publishing

Advanced Secure Optical Image Processing for Communications

Ayman Al Falou

Chapter 4

Coded aperture correlation holography system for recording secured digital holograms of incoherently illuminated 3D scenes

A Vijayakumar, Manoj Kumar, Mani Ratnam Rai and Joseph Rosen

4.1 Introduction

Holography methods are attractive candidates for security and encryption applications [1–4]. Digital holography techniques using coherent illumination have been demonstrated for encryption and security [5–7]. Recently, imaging and digital holography systems using three-color coherent light [8] and incoherent light [9] have proven useful for security applications. Encryption systems which use incoherent light are simpler and cheaper as they do not require lasers. In this line of research, we propose incoherent digital holography systems for securing 3D scenes. Incoherent digital holography systems have been developed for various reasons such as exploiting the higher spatial bandwidth of incoherent light sources and to achieve super resolution [10]. Incoherent light sources are also used to avoid edge ringing effects and speckle noise during imaging and sometimes even to avoid the use of lasers [11, 12]. In this chapter, we present incoherent digital holography techniques which can also be applied for security applications.

4.2 Coded aperture correlation holography for 3D encryption

Coded aperture correlation holography (COACH) is a recently developed incoherent digital holography technique [13]. The optical configuration of COACH is shown in figure 4.1. An incoherent light source such as a light emitting diode (LED) is used to critically illuminate a point object such as a pinhole using a refractive lens L_1 [14]. The light diffracted from the point object is collected by a refractive lens L_2 and then passes through a polarizer P_1 and through a spatial light modulator (SLM). The polarizer P_1 is oriented at 45° with respect to the active axis of the SLM such

doi:10.1088/978-0-7503-1457-2ch4

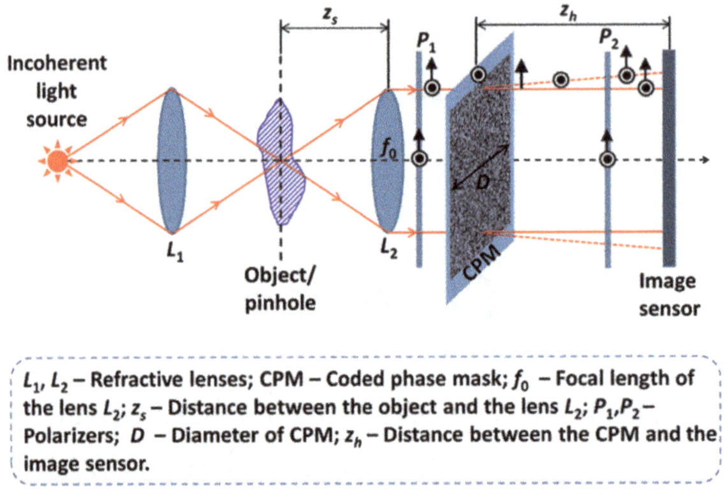

Figure 4.1. Optical configuration of coded aperture correlation holography.

that only about half of the intensity of the light is modulated by the SLM and the remaining part remains unaffected. On the SLM, a random phase mask is displayed to encrypt the incident light. A second polarizer P_2 oriented at 45° with respect to the active axis of the SLM is mounted after the SLM in order to enable interference at the sensor plane between the light modulated by the random phase mask and the unaffected light.

The resulting encrypted intensity pattern, which is the impulse response, is recorded by the image sensor. As the digital holography configuration is an in-line optical configuration, three intensity patterns are recorded by introducing three different phase shifts ($\theta = 0°$, 120° and 240°) to the random phase mask and the intensity patterns are superposed [10] in order to remove the twin image and the bias terms. The resulting complex function is considered as the key hologram H_K of the system. Following the training stage with the point object, a complicated object is then placed exactly at the same axial location of the point object and with identical conditions; again three intensity patterns are recorded corresponding to the three phase shifts and superposed to obtain the object hologram H_{obj}. The image of the object is decrypted by a cross-correlation between H_{obj} and H_K. During the decryption, there is a background noise due to the randomness of the mask. Assuming that the point object is placed at the front focal plane of the lens L_2, the light entering the SLM is assumed to be a pure phase function. If in the sensor plane the light distribution is also a pure phase, the decrypted image can have minimum side lobes. Therefore, instead of using an arbitrary random mask, the SLM mask was synthesized using the Gerchberg–Saxton algorithm (GSA) [15], such that in the sensor plane, the response is a pure phase function, i.e. the function has a uniform intensity. The synthesized random phase mask using GSA is called a coded phase mask (CPM), and since the CPM has a random-like profile it has the same security strength as any other random phase mask. The object information can be retrieved only by using the key hologram.

COACH belongs to a special category of interferometers called self-informative reference interferometers as the information about the object is present in both interfering waves [16]. In most digital holography techniques for security applications, the reference wave is not derived from the object. However, in the case of COACH, both the interfering waves carry information about the object. It must be noted that along with the CPM information, the axial location of the object is also required because the axial correlation length of COACH is the same as the axial resolution of a conventional imaging system with the same numerical aperture [13]. The specifications of optical components and the SLM are also necessary to simulate the key hologram and to decrypt the object information from H_{obj}. This is because the CPM displayed on the SLM is different from the synthesized CPM in the computer due to the interpixel gap of the available SLMs. Therefore, even though the CPM information is known, the key hologram cannot be simulated without the axial location of the object and specifications of the SLM and optical components.

4.2.1 Methodology

The mathematical formulation of the encryption and decryption processes based on figure 4.1 is as follows. For a point object at the front focal plane of L_2, two beams, namely a plane wave and a random-like complex function $G(u,v)$ generated by the CPM modulation, are interfered on the sensor plane. The intensity profile recorded by the image sensor for one of the three phase shifts is given by

$$I_k(u, v) = |A + G(u, v)\exp(\mathrm{i}\theta_k)|^2, \qquad k = 1, 2, 3, \tag{4.1}$$

where θ_k is the kth phase value of the three phase shifts. The phase-shifting procedure produces a complex hologram $H_K = A^* G(u,v)$, the impulse response which is the key that can be used for decryption. A 2D object is placed exactly at the same axial location of the point object and a complex hologram is synthesized from three intensity recordings given by (4.1). Any 2D object illuminated by a spatially incoherent light can be represented as a collection of uncorrelated point objects as follows,

$$t(x, y) = \sum_j a_j \delta(x - x_j, y - y_j). \tag{4.2}$$

The central wavelength of the incoherent source is λ. In the case of the 2D object at a distance f_o from L_2, each object point j in the object plane generates two mutually coherent beams on the sensor plane: $A_j \exp[\mathrm{i}2\pi(ux_j + vy_j)/\lambda f_o]$ which is a tilted plane wave; the second wave is a shifted version of $G(u,v)$, due to the linear phase, and multiplied by the same plane wave represented as $\exp[\mathrm{i}2\pi(ux_j + vy_j)/\lambda f_o]G(u - u_j, v - v_j)$, where $(u_j, v_j) = (x_j, y_j)z_h/f_o$. The intensity profile of the 2D object at the sensor plane is given by

$$I_k(u, v) = \sum_j \left| A_j \exp\left[\frac{\mathrm{i}2\pi\left(x_j u + y_j v\right)}{\lambda f_o}\right] + \exp(\mathrm{i}\theta_k) B_j \exp\left[\frac{\mathrm{i}2\pi\left(x_j u + y_j v\right)}{\lambda f_o}\right] G(u - u_j, v - v_j) \right|^2. \tag{4.3}$$

The complex hologram is synthesized from the three intensity recordings as similar to [10]

$$H_{obj}(u, v) = I_1(u, v)[\exp(-i\theta_3) - \exp(-i\theta_2)] \\ +I_2(u, v)[\exp(-i\theta_1) - \exp(-i\theta_3)] + I_3(u, v)[\exp(-i\theta_2) - \exp(-i\theta_1)]. \quad (4.4)$$

From (4.4), the expression can be simplified as

$$H_{obj}(u, v) = \sum_j A_j^* B_j G(u - u_j, v - v_j). \quad (4.5)$$

The image of the object can be decrypted by the cross-correlation between the key hologram $H_K = G(u,v)$ and H_{obj} as follows:

$$P(u', v') = \iint \left\{ \sum_j A_j^* B_j G(u - u_j, v - v_j) \right\} G^*(u - u', v - v')\mathrm{d}u\mathrm{d}v \\ \approx \sum_j A_j^* B_j \delta(u' - u_j, v' - v_j) \propto t(u'/M_T, v'/M_T). \quad (4.6)$$

The decryption of the object hologram is carried out using the cross-correlation and therefore, the lateral and axial resolutions will be similar to that of regular imaging with the same NA. The magnification of the system is given by $M_T = z_h/f_o$.

4.2.2 Experiments and results

The experimental two-channel digital holography setup of COACH is shown in figure 4.2. The setup consists of two identical LEDs (Thorlabs LED631E, 4 mW, λ = 635 nm, $\Delta\lambda$ = 10 nm) to critically illuminate the objects in the two-channels using two identical objective lenses L_{1A} and L_{1B}. Two irises were mounted after the objective lenses to control the NA of the two-channels. The objects were mounted at

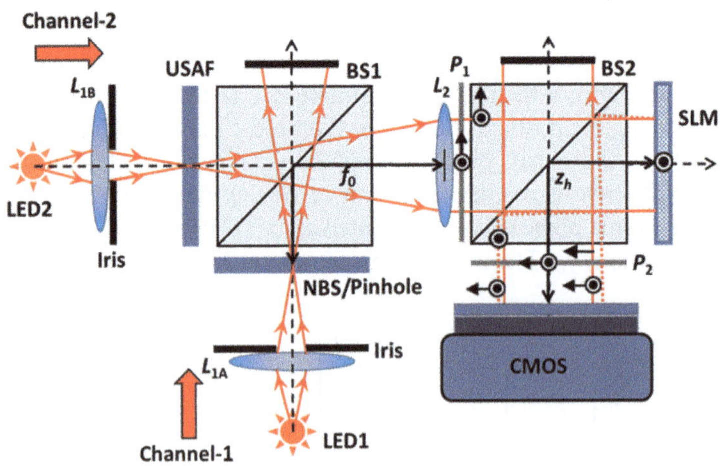

Figure 4.2. Experimental setup of two-channel COACH for 3D encryption.

a distance of 15 cm from the irises each with a diameter of 0.6 cm, resulting in an NA of 0.04 and lateral and axial resolutions of roughly 10 μm (0.61λ/NA) and about 0.8 mm [$2\lambda/(\text{NA})^2$], respectively. At first, a pinhole with a diameter of 100 μm was mounted in channel 1 at a distance of 20 cm from the lens L_2 with a focal length f_0 = 20 cm and channel 2 was blocked. The lens L_2 collects the light diffracted from the pinhole and collimates it. The collimated light is polarized by a polarizer P_1 oriented at an angle of 45° with respect to the active axis of the SLM (Holoeye PLUTO, 1920 × 1080 pixels, 8 μm pixel pitch, phase-only modulation).

A CPM calculated using the GSA was displayed on the SLM in the central 1080 × 1080 pixels. Due to the polarizer P_1, only about half of the intensity of light is modulated by the CPM while the remaining part propagates without any modulation. Three phase masks were generated from the CPM by applying three different phase shifts $\theta = 0°$, 120° and 240° and the corresponding intensity patterns were superimposed using (4.4) to obtain the complex key hologram H_K without twin image and bias components.

In the first experiment, the axial tolerance of the decryption by the key hologram was analyzed. The location of the pinhole was moved from −6 cm to 6 cm with respect to the front focal plane of the lens L_2 and the corresponding key holograms were recorded. The key holograms at different axial locations $H_K(z)$ were correlated with the key hologram recorded at the front focal plane of the lens $H_K(z = 0)$ and the image intensity at $(x, y) = (0,0)$ was calculated. The plot of the intensity values at $(x, y) = (0, 0)$ is shown in figure 4.3, which shows how rapidly the information deteriorates with an error in the axial location.

In the second experiment, a negative National Bureau of Standards (NBS 1963A Thorlabs) resolution chart was mounted at the front focal plane of the lens L_2 and the element 7.1 lp mm^{-1} was illuminated. Three intensity patterns corresponding to three phase shifts of the same CPM were recorded and composed into H_{obj}. The image of the object is decrypted by a cross-correlation between the H_{obj} and H_K. The three intensity patterns corresponding to phase shifts $\theta = 0°$, 120° and 240° recorded for the pinhole are shown in figures 4.4(a)–(c), respectively. The amplitude and

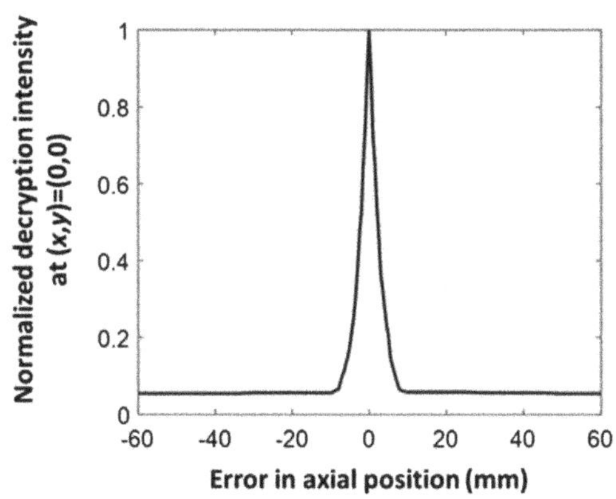

Figure 4.3. Normalized intensity of decryption at (x = 0, y = 0) versus the error in the axial distance with respect to the front focal plane of the lens L_2.

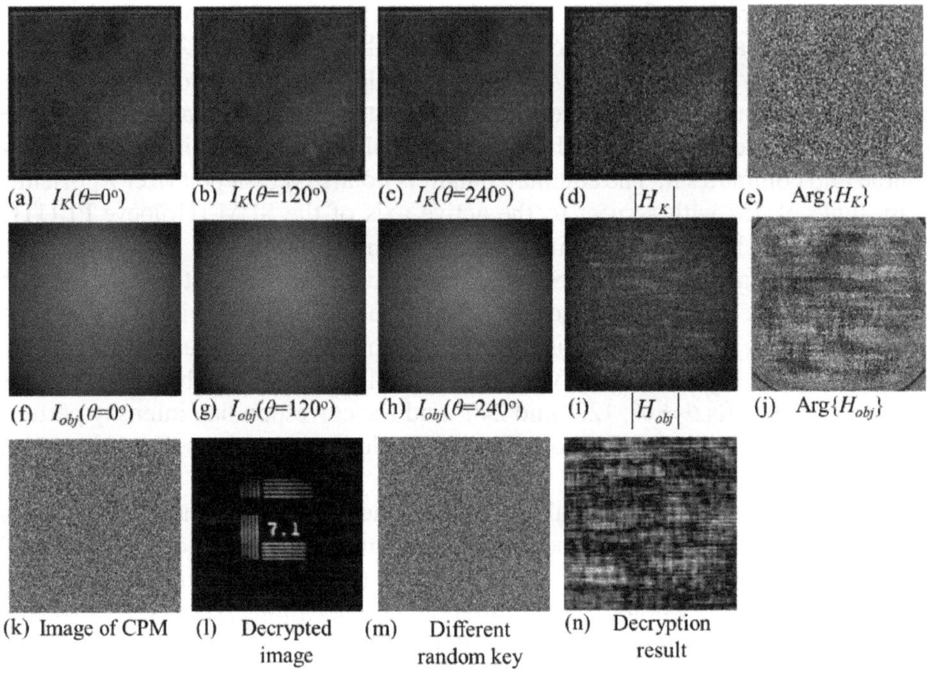

Figure 4.4. (a)–(c) Intensity patterns for the pinhole for phase shifts $\theta = 0°$, 120° and 240°, (d) amplitude and (e) phase, of the complex H_K. (f)–(h) Intensity patterns for the object for phase shifts $\theta = 0°$, 120° and 240°, (i) amplitude and (j) phase of the complex H_{obj}. (k) Image of the CPM, (l) decrypted image, (m) image of a different random CPM, and (n) decryption result.

phase of the complex H_K are shown in figures 4.4(d) and (e), respectively. The intensity patterns corresponding to three phase-shifts, the amplitude and the phase of the encrypted H_{obj} are shown in figures 4.4(f)–(j), respectively. The image of the CPM is shown in figure 4.4(k). The decrypted image is shown in figure 4.4(l). The image of a different random CPM is shown in figure 4.4(m) and the decryption result is shown in figure 4.4(n). The results show that the hologram can be decrypted only with the key hologram recorded with the same CPM.

In the third experiment, 3D encryption is carried out using a single CPM for a multiplane object. A negative NBS resolution chart was mounted in channel 1 and the element 7.1 lp mm^{-1} was illuminated, while, a United States Air Force chart (USAF 1951 1X Edmund Optics) was mounted in channel 2 and elements 2 (4.49 lp mm^{-1}) and 3 (5.04 lp mm^{-1}) of group 2 were illuminated. The location of the NBS object was kept constant while the USAF object was shifted in steps of ± 1 cm away from the NBS object up to a maximum separation of ± 3 cm and the corresponding encrypted intensity patterns were recorded and H_{obj}s was calculated for each case.

The encrypted H_{obj}s were decrypted with the key holograms recorded at different axial locations in the first experiment. The decryption results are shown in figure 4.5. These results reveal that it is necessary to have the complete key hologram library corresponding to the different axial locations in order to completely decrypt a

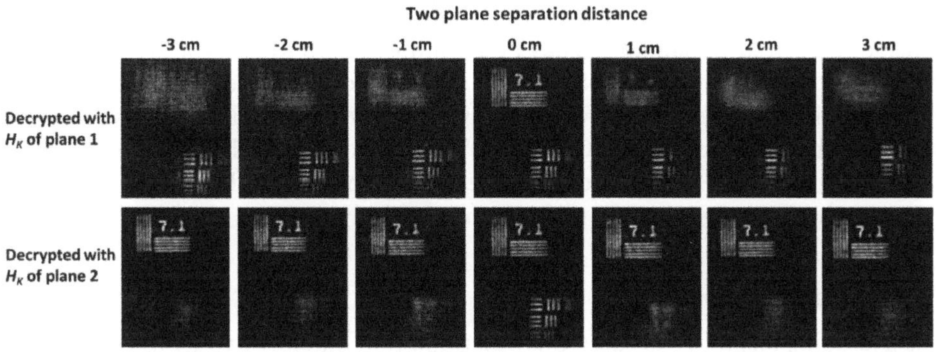

Figure 4.5. Decryption results of the two-plane objects with different thicknesses using the key holograms corresponding to the two planes.

hologram of a double plane or 3D object. Therefore, 3D encryption is possible with COACH but by using only one CPM, as the axial resolution of COACH is relatively high. Once again, if a different random key is employed then the information cannot be decrypted.

4.2.3 Conclusion

COACH is an incoherent digital holography encryption technique which has 3D encryption capability with only one random-like CPM due to the high axial resolution obtained because of the narrow axial correlation length. A key hologram library is created corresponding to the different axial locations of the pinhole using the CPM. An object hologram is recorded using the same CPM. The information of the object at different planes can be decrypted using the key hologram corresponding to the axial locations. The random nature of CPM which enables the possibility of 3D encryption is also the reason for the background noise during decryption. In the next section, different noise reduction techniques are presented to improve the signal-to-noise ratio (SNR) in the decryption results.

4.3 Noise reduction techniques for COACH

In COACH, instead of a random key as used in many optical security systems [1, 3], the random mask is synthesized using GSA, in order to reduce the background noise arising during decryption. However, in our experiments, there was still substantial background noise in the decryption results. The GSA cannot simultaneously completely satisfy two constraints. Since the constraint in the CPM domain needs to be satisfied by the use of a phase-only SLM, the other constraint in the spectrum domain can be only approximated [17]. Therefore, additional noise reduction techniques are required in order to obtain a higher SNR.

4.3.1 Phase-only filtering technique

The first technique employed to reduce the background noise is dubbed the phase-only filtering (POF) technique, while in the previous decryption technique (section 4.2) the matched filtering technique was applied. In the POF method, a phase-only filtered version of the key hologram was used to decrypt the information from the object hologram. An NBS object is mounted in channel 1 and element 10 lp mm^{-1} is critically illuminated. The H_{obj} is synthesized from three intensity recordings with phase shifts $\theta = 0°$, $120°$ and $240°$ and correlated by the phase-only version of the key hologram $\tilde{H}_K$. The decryption results for the matched filter and phase-only filter are shown in figures 4.6(a) and (b), respectively.

4.3.2 Averaging technique

The averaging technique is based on the assumption that any two CPMs synthesized using GSA from different initial random profiles are different from one another, i.e. their cross correlation is close to zero. In the averaging technique, the noise reduction in the decryption is achieved at the expense of the time resolution. Therefore, if N independent complex key holograms and object holograms are recorded using non-correlating CPMs and their complex decryptions are averaged, the noise level will decrease as a function of $\sqrt{N}$. However, the time resolution decreases, since instead of 3 intensity patterns, $3N$ intensity patterns are required to reduce the background noise.

The experiment was carried out and intensity patterns for the pinhole and the NBS object with element 10 lp mm^{-1} corresponding to phase shifts $\theta = 0°$, $120°$ and $240°$ were recorded but with 20 independent CPMs. The key holograms were created for the 20 CPMs and their phase-only filtered versions were extracted and correlated with the object holograms created for the same 20 CPMs. The complex decryptions were averaged. The decryption results after averaging for 1, 5, 10, 15 and 20 CPMs are shown in figures 4.7(a)–(e) respectively, and the corresponding normalized SNR values were calculated as 1, 2, 2.8, 3.2 and 3.3, respectively.

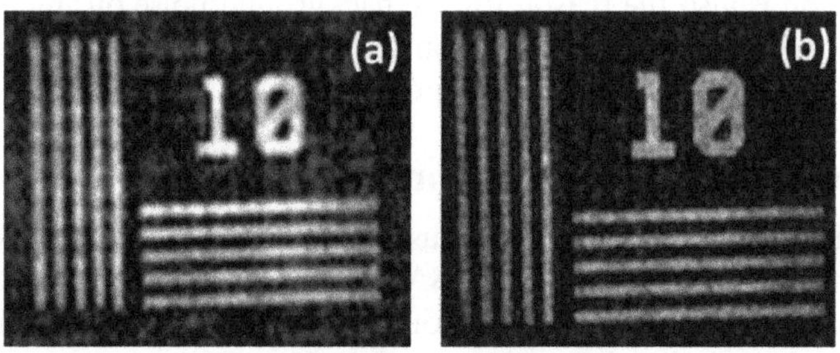

Figure 4.6. Decryption result of COACH with (a) matched filter and (b) phase-only filter.

Figure 4.7. Decryption results after averaging the complex decryptions using (a) 1 CPM, (b) 5 CPMs, (c) 10 CPMs, (d) 15 CPMs and (e) 20 CPMs.

The averaging technique has shown to improve the SNR of the decryption at the expense of the time resolution of the system.

4.4 Spectral encryption in COACH

In the regular COACH configuration, the encryption has two parameters, namely the CPM and the axial location of the object. Key holograms at different axial locations are necessary in order to decrypt the object information in the respective axial planes. In this section, we propose and demonstrate an encryption technique in which the wavelength of the light source is also a parameter that should be *a priori* known for the decryption. In other words, the library of the key hologram is expanded to another dimension which is the wavelength. Recalling that the 3D encryption capability of COACH arose from the narrow axial correlation length, the goal, in this case, is to convert the variation in the spectrum into a variation in the focal distance such that the shift is larger than the axial correlation length of COACH. In this case, the information can be decrypted with a key hologram of not only a particular axial plane but also a particular wavelength. Without the information about the wavelength, the information cannot be decrypted.

4.4.1 Methodology

The optical configuration of COACH is modified as shown in figure 4.8. The refractive objective lens is replaced by a diffractive objective lens such that there is a different focusing response for different wavelengths. The encryption in wavelength is discussed using two wavelengths. In the first step, the key hologram library is created using a CPM for the two wavelengths, and for the different axial locations $H_K(\lambda, z)$, by recording the intensity patterns for the phase shifts $\theta = 0°$, 120° and 240°. In the next step, an object is placed within the axial limits of the key hologram library and it is encrypted using the same CPM but using two wavelengths. The wavelength-specific information can be decrypted by using the appropriate element of the key hologram library.

The spectral response of the modified COACH with a diffractive objective lens was compared with the previous COACH with a refractive objective lens. The key hologram for 635 nm was simulated and the wavelength was varied by ±100 nm in steps of 1 nm and the key holograms were simulated for all wavelengths. The

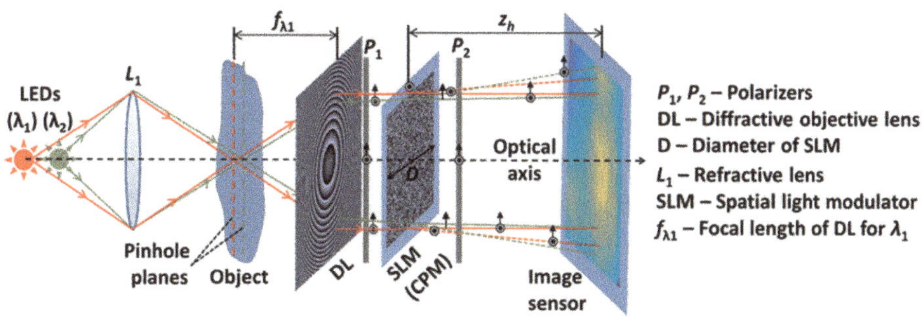

Figure 4.8. Optical configuration of modified COACH for 3D and spectral encryption.

wavelength response of the decryption was studied by a cross-correlation of the $H_K(\lambda)$ with $H_K(\lambda = 635\text{ nm})$ and the decrypted intensity at $(x, y) = (0, 0)$ was calculated. The plot of the normalized decryption intensity at $(x, y) = (0, 0)$ for different wavelengths is shown in figure 4.9 for COACH with a refractive objective lens and with a diffractive objective lens. It can be seen that with a diffractive objective lens, the spectral response of decryption is sharper. Therefore, besides the axial plane specification, the wavelength information is also required in order to decrypt the information of the object from the hologram [18].

4.4.2 Experiments and results

The experimental demonstration of the modified COACH was carried out using an experimental setup, as shown in figure 4.10. The two optical channels were illuminated by two incoherent light sources with wavelengths, namely 635 nm (Thorlabs LED631E, 4 mW, $\lambda = 635$ nm, $\Delta\lambda = 10$ nm) and 587 nm (Thorlabs M565L3, 880 mW, $\lambda = 565$ nm, $\Delta\lambda = 104$ nm), and their spectrum was constricted using band pass filters (BPFs) ($\lambda = 633$ nm, $\Delta\lambda = 10$ nm) and ($\lambda = 587$ nm, $\Delta\lambda = 16$ nm), respectively. Two identical lenses L_{1A} and L_{1B} were used to critically illuminate the objects 8 lp mm^{-1} of NBS resolution chart in channel 1 and elements 10.1 lp mm^{-1} and 11.3 lp mm^{-1} of group 3 of USAF resolution chart in channel 2, respectively. The light diffracted by the two objects was collimated by the diffractive lens (*DL*) displayed on SLM_1 and a polarizer P_1 oriented along the active axis of SLM_1 was installed to enable full modulation by SLM_1.

The *DL* induces different effects for different wavelengths and directs the light towards SLM_2 on which the CPM is displayed. The polarizers P_2 and P_3 are used to create two beams at SLM_2 and to form interference between them in the sensor plane. The resulting intensity patterns have information encrypted by the CPM for the two wavelengths. Three intensity patterns were recorded corresponding to phase shifts $\theta = 0°$, $120°$ and $240°$ and were composed into a complex object hologram

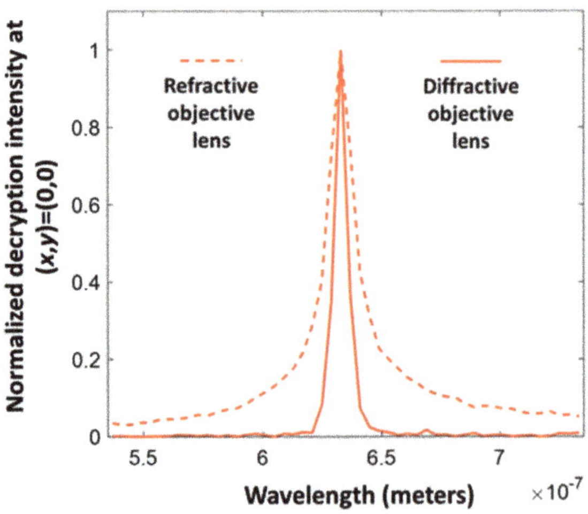

Figure 4.9. Plot of the normalized decryption intensity at $(x, y) = (0, 0)$ for variation in the wavelength.

H_{obj}. The object information corresponding to different axial planes and wavelengths can be decrypted using the prerecorded key holograms $H_K(z, \lambda)$.

The decryption results for the two objects in the same axial location using $H_K(z = 0, \lambda = 635$ nm), $H_K(z = 0, \lambda = 587$ nm), $H_K(z = +1$ cm, $\lambda = 635$ nm) and $H_K(z = -1$ cm, $\lambda = 587$ nm) are shown in figures 4.11(a)–(d), respectively. From the results, it can be concluded that the object information can be decrypted only when the axial plane and the wavelength are known accurately. The results are obtained

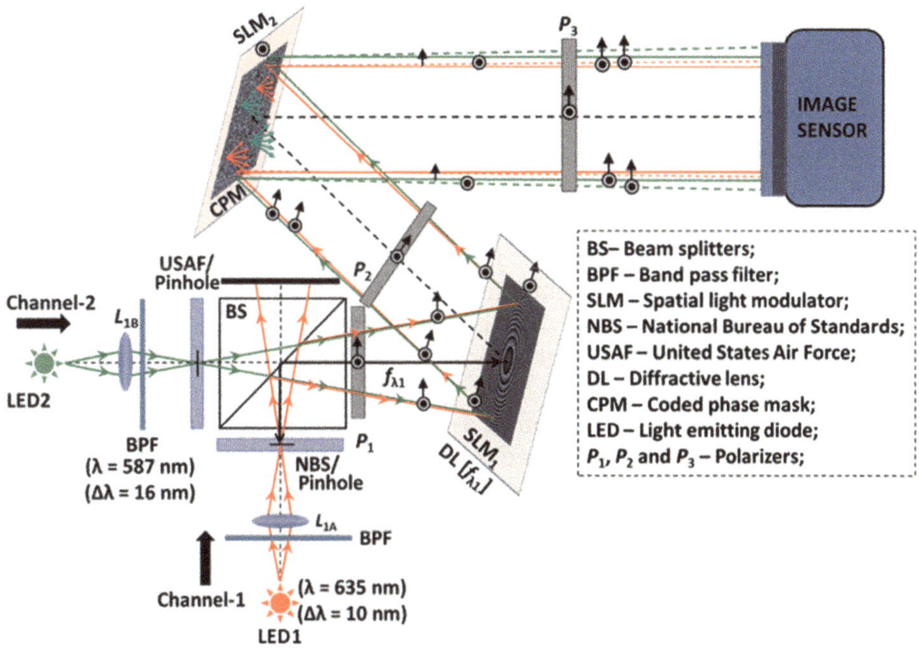

Figure 4.10. Experimental setup of COACH with a diffractive objective lens for encrypting 3D and spectral information.

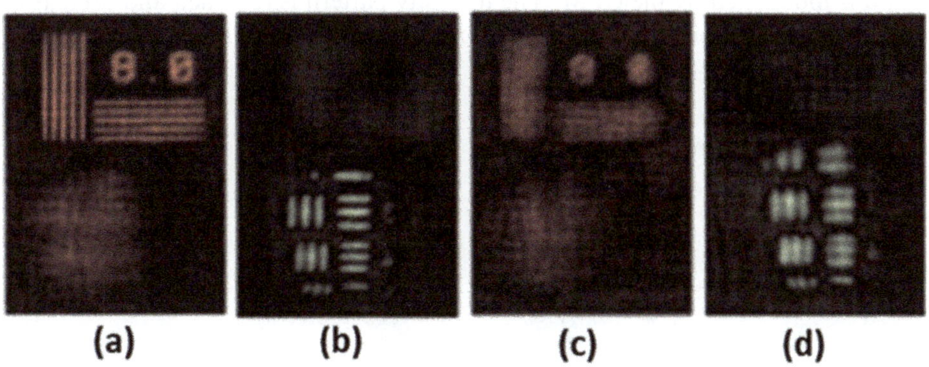

Figure 4.11. Decryption results of H_{obj} using (a) $H_K(z = 0, \lambda = 635$ nm), (b) $H_K(z = 0, \lambda = 587$ nm), (c) $H_K(z = +1$ cm, $\lambda = 635$ nm) and (d) $H_K(z = -1$ cm, $\lambda = 587$ nm) (Colors are given artificially for better understanding of the results).

after averaging over the complex decryptions of 20 different CPMs for both the object as well as for the pinhole.

4.4.3 Conclusion

A modified version of COACH using a diffractive objective lens has been proposed and demonstrated for encryption of information not only in the different axial planes but also for different wavelengths. The strength of the encryption can be improved using this technique. The preliminary demonstration using two wavelengths has proved that the specific visual information can be decrypted only when the wavelength information is known.

4.5 Referenceless 3D encryption in COACH

In the COACH configuration and in most of the digital holography encryption techniques [5–7], it is necessary to retrieve the phase information by an interference with a reference wave. The prerequisites of interferometers are numerous, ranging from the need of vibration isolation, power matching between the two interfering beams, optical alignment, high temporal coherence, etc.

Recently, we have discovered that the 3D location information of the object is present not only in the phase but also in the amplitude of the intensity pattern recorded by COACH. Therefore, interference is not necessary in the case of COACH to retrieve the information of a 3D location of the object and hence the optical configuration can be simplified to the level of a regular imaging system, but with a capability to encrypt and decrypt 3D information [19]. In this section, the COACH technique without two-wave interference termed interferenceless COACH (I-COACH) is discussed for security applications.

4.5.1 Methodology

The optical configuration of I-COACH is shown in figure 4.12. The light from an incoherent source is used to critically illuminate a pinhole and the light diffracted from it is modulated by the CPM. The intensity pattern is recorded and used as the key hologram. The pinhole location is shifted to various axial locations and a key

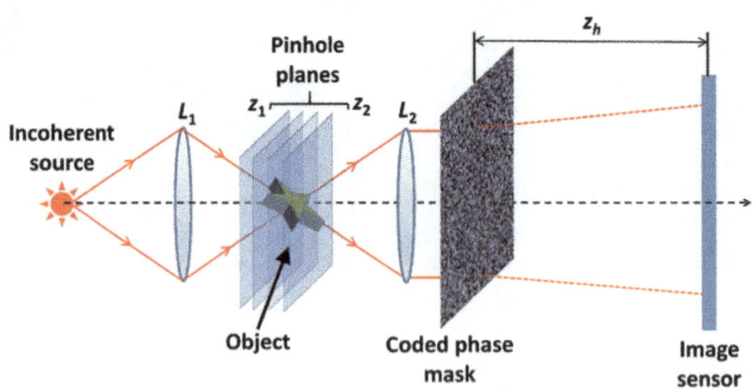

Figure 4.12. Optical configuration for encrypting 3D information with I-COACH.

hologram library is created. An object is then placed within the axial boundaries of the key holograms and the object hologram is recorded using the same CPM. The image of the object at a particular plane is decrypted by a cross-correlation between the object hologram and the key hologram corresponding to the plane. However, the decryption results showed a poor SNR. Therefore, in order to improve the SNR, intensity patterns are recorded using three different CPMs synthesized from three different initial random phases using GSA and they are projected on the complex space for both the pinhole and the object.

4.5.2 Experiment and results

Encryption and decryption using I-COACH was experimentally demonstrated using a setup of figure 4.13. A pinhole with a diameter of ~25 μm was mounted in channel 1 and a complex key hologram library was generated for different axial locations using three different CPMs. An LED was mounted in channel 2 and critically illuminated the reflective 3D object (LED) in the reflection mode. The light diffracted by the LED passed through a polarizer, P_1, oriented along the active axis of the SLM to enable full modulation. Then the beam was collected and collimated by a lens L_2. The collimated light was modulated by the three different CPMs displayed one after the other on the SLM, and the corresponding intensity patterns were recorded by the image sensor. The recorded intensity patterns were projected into the complex space with phase shifts $\theta = 0°$, 120° and 240° and composed into a complex hologram. The image of the object at any plane can be decrypted by a cross-correlation between the object hologram and the key hologram corresponding to that particular plane. The decryption results using key holograms of three different axial locations after averaging using 20 different CPMs are shown in the same figure 4.13.

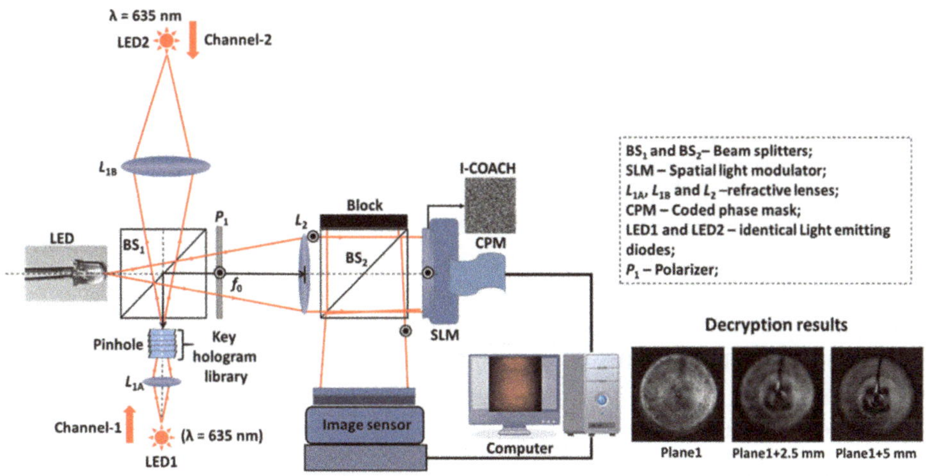

Figure 4.13. Experimental setup for encrypting 3D information in I-COACH.

4.5.3 Conclusion

In I-COACH, the optical configuration is simplified to the level of a regular imaging system without any interference, nonetheless, the 3D encryption capability is retained. The phase shifts are used only to improve the SNR and are not really necessary to perform the encryption and decryption. However, using three uncorrelated random-like CPMs improves the SNR and simultaneously strengthens the encryption. To break the encryption in I-COACH, it is necessary to synthesize the hologram key using three random-like independent CPMs.

4.6 Single shot 3D encryption using I-COACH

In I-COACH the optical configuration is as simple as regular imaging. However, the background noise is still higher and needs the averaging technique in order to improve the SNR. In this section, we present an advanced version of I-COACH, capable of encryption in just one camera shot and with relatively high SNR. This method is called single camera shot I-COACH (SCS-I-COACH).

4.6.1 Methodology

In SCS-I-COACH, the need for averaging is eliminated by modifying the optical configuration to have better background noise suppression. In the previous optical configuration, the GSA was constrained to have pure phase function in the SLM and in the sensor plane. However, the condition of the GSA which requires a Fourier transform relation between the SLM and sensor plane is not fulfilled. Nevertheless, in SCS-I-COACH, the condition is satisfied by implementing a lens function at the SLM plane such that the lens Fourier transforms the CPM distribution on the sensor plane.

Furthermore, it was noted that three different CPMs are not necessary to suppress the noise, but instead, two CPMs are sufficient to achieve a reasonable SNR. In the case of two CPMs, the recorded two intensity patterns should be subtracted from each other. The two CPMs are simultaneously displayed on the SLM such that the entire recording process can be simplified to one camera shot. In order to fit the two intensity patterns within the sensor area, an additional constraint on the area of the pure phase function must be imposed by the GSA. Since both the CPMs share the SLM space, the penalty is paid in the form of some reduction in the field of view (FOV). The optical configuration of SCS-I-COACH is shown in figure 4.14.

The two CPMs are displayed side by side and a quadratic phase function (QPM) is spatially multiplexed over the entire area of the SLM. The distance between the SLM and sensor is selected to match the focal length of the QPM. When the QPM is multiplexed with the two CPMs, the two intensity patterns are superimposed in the sensor plane. In order to separate the two intensity patterns I_1 and I_2 in the sensor plane, equal and opposite-in-sign linear phases are introduced into the two CPMs.

4.6.2 Experiment and results

The experimental setup is similar to that of a regular I-COACH. In the first step, a library of key holograms was generated corresponding to different axial locations.

The point object was moved in steps of 2 mm and the intensity patterns I_1 and I_2 were recorded, subtracted from each other and the resulting bi-polar matrices were stored as the key holograms.

Following the training stage, an object was placed within the axial limits of the key hologram library and the intensity patterns were recorded using the same CPM pairs and composed into the object hologram. The object information at any plane was decrypted by a cross-correlation between the object hologram and the key hologram recorded at the corresponding axial location. The experiment was carried out by illuminating the number '0' near the element 8 lp mm^{-1} of the NBS object and the number '6' near the element 14.25 lp mm^{-1} of group 3 of USAF in channel 1 and channel 2, respectively. The images of the intensity patterns for the pinhole and for the two plane object, when the two objects were in the same axial plane, are shown in figure 4.15(a) and (b), respectively.

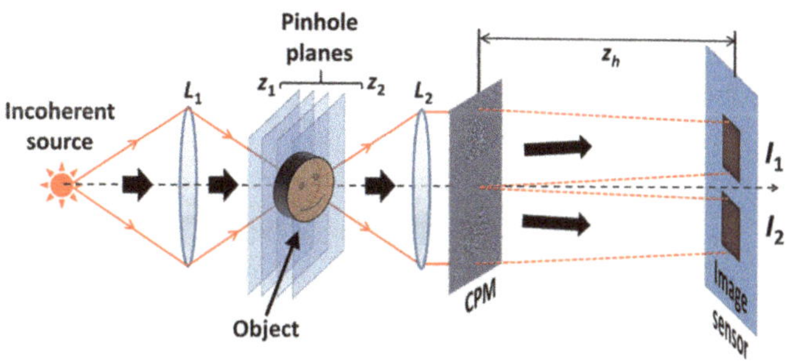

Figure 4.14. Optical configuration of SCS-I-COACH for 3D encryption.

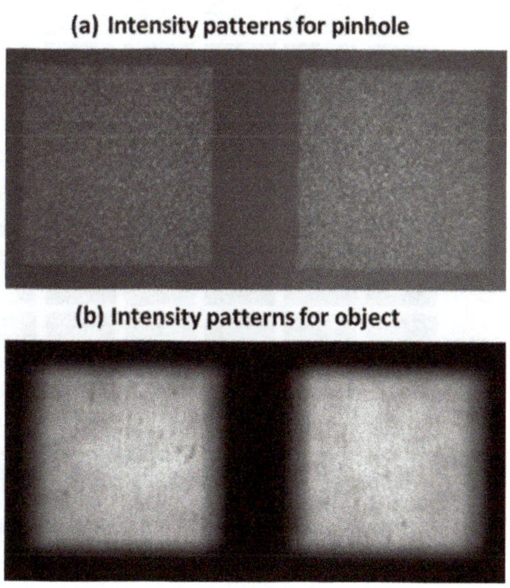

Figure 4.15. Intensity patterns recorded in the sensor plane for (a) pinhole and (b) two plane object made up of numbers '0' and '6'.

The distance between the two plane objects was increased in steps of 1 mm up to a maximum of 10 mm, and the intensity patterns were recorded and composed into the object hologram for every case. The object holograms were decrypted using the key hologram library. The decryption results for different thicknesses of the two-plane object, and for different elements of the key hologram library, are shown in figure 4.16. From the decryption results, it can be observed that the background noise is not completely removed but the SNR was reasonable for a single shot scenario without any time consuming averaging.

4.6.3 Conclusion

In SCS-I-COACH, the single shot capability enables the possibility of video encryption and decryption. Unlike the previous versions of COACH which require many recordings with different CPMs, in SCS-I-COACH, dynamic scenes can be encrypted by recording the video of the intensity patterns I_1 and I_2. This can be a powerful tool for transmitting and receiving encrypted videos of 3D information. Additional noise reduction is needed either by improving the GSA or by alternative techniques without affecting the time resolution of the system.

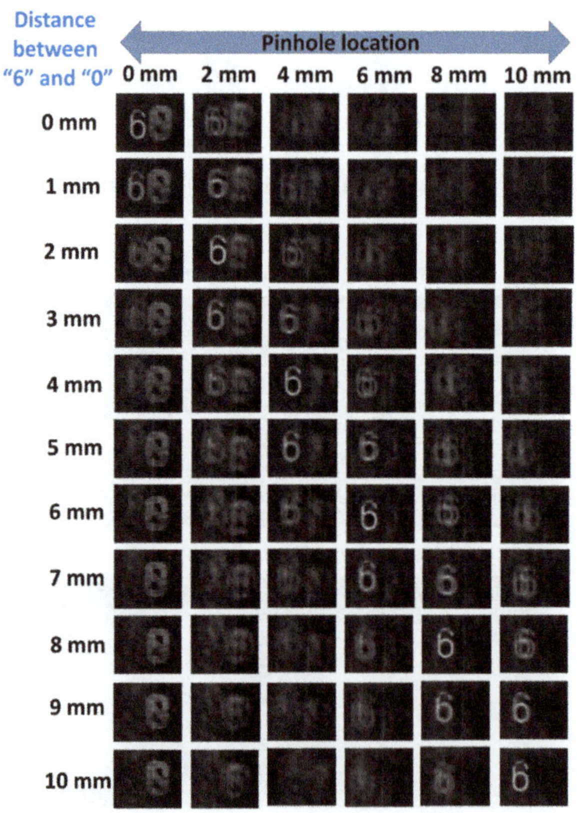

Figure 4.16. Decryption results of the two plane objects made of '0' and '6' using key holograms corresponding to different axial locations.

4.7 Lensless and interferenceless COACH for 3D encryption

In this section, an even simpler version of COACH without interference and without any refractive lens called lensless, interferenceless COACH (LI-COACH) is discussed for the application of 3D encryption. Lensless systems can offer many advantages such as low cost, large field of view and fewer aberrations. Moreover, since manufacturing of lenses for microscopy, telescopic systems and for non-visible wavelengths is a challenging task, an imaging system without lenses can be a practical solution for these applications.

4.7.1 Methodology

The optical configuration of LI-COACH is shown in figure 4.17. An incoherent light source critically illuminates a pinhole and the light diffracted by the pinhole is directly modulated using two CPMs displayed one after the other on the SLM. The resulting intensity patterns are recorded by the image sensor and subtracted from each other to obtain the key hologram. In this case, the GSA is modified into a Fresnel propagator GSA and the constant phase at the input plane is replaced by the quadratic phase produced by the pinhole. The pinhole is moved to different axial locations and a library of key holograms is created. An object is then placed within the axial limits of the key hologram library and two intensity patterns are recorded and composed into the object hologram. The image of the object is decrypted by a cross-correlation of the object hologram with the key hologram.

4.7.2 Experiment and results

The demonstration of LI-COACH for 3D encryption is shown in figure 4.18. The key hologram library was created in the first step. A 3D object in the form of two LEDs (LED-A and LED-B), axially separated by a distance of 15 mm, was illuminated in reflection mode and the object holograms were recorded. The decryption results for two LEDs using the key holograms recorded at these two planes are shown in figure 4.19. It can be noted that the focal depth of the system is larger and therefore even the information that is not decrypted is dominantly present.

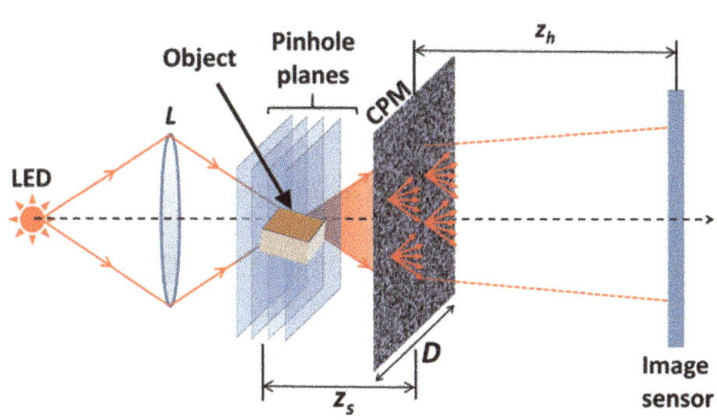

Figure 4.17. Optical configuration of LI-COACH for 3D encryption.

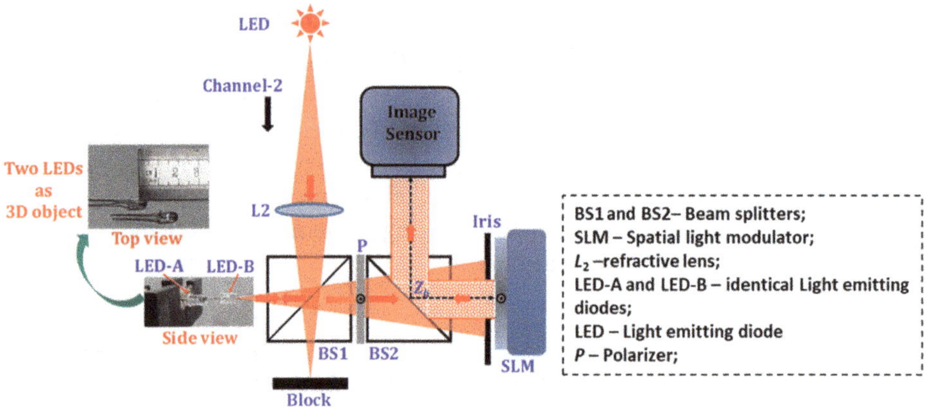

Figure 4.18. Experimental setup of LI-COACH for 3D encryption.

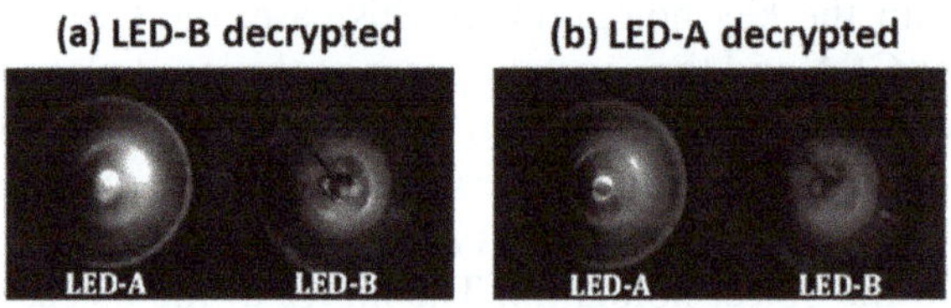

Figure 4.19. Decryption results of (a) LED-B and (b) LED-A.

4.8 Conclusion

We have proposed and demonstrated incoherent 3D encryption techniques based on coded aperture correlation holography. This chapter reinstates the fact that the choice of the optical system plays a crucial role in security [20]. The chapter describes the evolution of COACH from the more complicated regular COACH to the simplest I-COACH and LI-COACH configurations. The advanced techniques such as spectral encryption of I-COACH and video encryption techniques can be useful for many applications. Different noise reduction techniques have also been proposed and demonstrated to improve the SNR during decryption of the information. Further research is necessary to study the strength of encryption in the above techniques.

References

[1] Javidi B 2006 *Optical and Digital Techniques for Information Security* vol 1 (Springer Science & Business Media) pp 240–69
[2] Tajahuerce E and Javidi B 2000 *Appl. Opt.* **39** 6595–601
[3] Javidi B and Nomura T 2000 *Opt. Lett.* **25** 28–30
[4] Nishchal N K, Joseph J and Singh K 2004 *Opt. Commun.* **235** 253–9
[5] Nishchal N, Joseph J and Singh K 2004 *Opt. Eng.* **43** 2959–66
[6] Naughton T J and Javidi B 2004 *Opt. Eng.* **43** 2233–8
[7] Kim H, Kim D H and Lee Y H 2004 *Opt. Express* **12** 4912–21

[8] Chen L and Zhao D 2006 *Opt. Express* **14** 8552–60
[9] Zang J, Xie Z and Zhang Y 2013 *Opt. Lett.* **38** 1289–91
[10] Rosen J, Siegel N and Brooker G 2011 *Opt. Express* **19** 26249–68
[11] Kim M K 2012 *Opt. Lett.* **37** 2694–6
[12] Yanagawa T, Abe R and Hayasaki Y 2015 *Opt. Lett.* **40** 3312–5
[13] Vijayakumar A, Kashter Y, Kelner R and Rosen J 2016 *Opt. Express* **24** 12430–41
[14] Pawley J B 1990 *Handbook of Biological and Confocal Microscopy* (New York: Plenum), pp 73–5
[15] Gerchberg R W and Saxton W O 1972 *Optik* **35** 227–46
[16] Rosen J and Kelner R 2014 *Opt. Express* **22** 29048–66
[17] Vijayakumar A, Kashter Y, Kelner R and Rosen J 2017 *Appl. Opt.* **56** F67–77
[18] Vijayakumar A and Rosen J 2017 *Opt. Lett.* **42** 947–50
[19] Vijayakumar A and Rosen J 2017 *Opt. Express* **25** 13883–96
[20] Muniraj I, Guo C, Malallah R, Cassidy D, Zhao L, Ryle J P, Healy J J and Sheridan J T 2017 *Opt. Eng.* **56** 063103

IOP Publishing

Advanced Secure Optical Image Processing for Communications

Ayman Al Falou

Chapter 5

Equal modulus decomposition based asymmetric optical cryptosystems

Areeba Fatima and Naveen K Nishchal

5.1 Introduction

Image encryption using the method of double random phase encoding (DRPE) has emerged as one of the most prominent methods in optical security applications [1–6]. The technique required the input image to be bonded with a random phase mask (RPM) and use this as input to a 4*f* (*f* being the focal length of a converging lens) system. The output of this procedure is a complex-amplitude stationary white noise. The encoded image thus obtained was quite robust to blind convolutions. The 4*f* system practically meant subjecting the input to double Fourier transform, along with RPMs placed in the optical path. The RPMs served as the keys. The decryption involved back tracing the optical path, using either conjugate of the RPMs or conjugate of the encrypted image.

With the success of the DRPE system, various other encryption schemes using different optical transformation were reported [7–13]. Extending DRPE to other transformations was done to exploit their features in making the system more robust. For example, DRPE in the fractional Fourier transform (FRT) domain introduced the quadratic phase structure in the encryption path [7]. The quadratic phase parameters were in addition to the RPMs required in the original DRPE scheme. Thus, introducing a new optical transform domain enhanced the security of the system. Similarly, the Fresnel transform, Hartley transform, gyrator transform, Mellin transform, and wavelet transform were among other transforms to be used for optical encryption [3–5, 8–13].

Over a period of time, different optical principles and image processing techniques have been employed to set-up encryption systems based on free space optics. These involved the polarization principle, holographic principle, ghost imaging, compressive sensing to name a few [14–17]. Among these methods,

doi:10.1088/978-0-7503-1457-2ch5

interference has found wide application. The basic idea behind this method involves separating the image to be encrypted into two entities [18]. These entities can either be phase-only or complex-valued functions. The original image is retrieved by superposing these entities by means of optical interference. Hence, this method is also known as coherent superposition. The first interference based techniques involved evaluating phase-only masks (POMs) corresponding to the image. The scheme was so designed that it would require knowledge of the generated POMs to obtain the plaintext. It was a basic criteria to make the system secure. However, it was later found out that each POM served as a stand-alone entity that could reconstruct the silhouette of the image [19, 20]. This weakened the interference based cryptosystem. Various modifications were made to strengthen this architecture [21–23].

It should be mentioned here that the interference-based encoding technique, like the DRPE method, falls under the category of symmetric cryptosystems. Such cryptosystems have the same encryption and decryption keys. The linearity in these systems made them vulnerable to various kinds of attacks, such as, the known plaintext and the chosen plaintext attacks [24–26]. To remove this anomaly, asymmetric cryptosystems were introduced that had dissimilar encryption and decryption keys.

In this chapter, an overview of an optical cryptosystem based on phase-truncated Fourier transform (PTFT) is given and equal modulus decomposition (EMD) with a coherent superposition based scheme is elaborated. The following sections discuss its vulnerability to applicable specific attacks. One of the attack algorithms has been presented to give an idea of the formulation of such attacks. This is followed by a discussion on the methods devised to overcome the shortcomings of EMD based cryptosystems.

5.2 Overview of a phase-truncated Fourier transform based cryptosystem

PTFT-based techniques were the first attempt to introduce asymmetric nature in optical cryptosystems [27]. This method involved using RPMs as public keys that were bonded with the input image and then Fourier transformed. The phase of the output was reserved as the private key and the amplitude, again bonded with another RPM, was Fourier transformed. The phase of this second output was again reserved as the second private key while the amplitude served as the ciphertext. Mathematically, this can be expressed as:

$$G(u, v) = \iint (f(x, y) \times R1(x, y))\exp[-i2\pi(ux + vy)]dxdy \tag{5.1}$$

$$K1 = AT[G(u, v)] \tag{5.2}$$

$$H(u, v) = PT[G(u, v)] \tag{5.3}$$

$$H'(u, v) = H(u, v) \times R2(u, v) \tag{5.4}$$

$$P(\xi, \eta) = \iint (H'(u, v))\exp[-i2\pi(\xi u + \eta v)]\mathrm{d}u\mathrm{d}v \tag{5.5}$$

$$K2 = AT[P(\xi, \eta)] \tag{5.6}$$

$$E(\xi, \eta) = PT[P(\xi, \eta)]. \tag{5.7}$$

Here AT and PT denote amplitude- and phase-truncation, respectively. The process is illustrated in figure 5.1. For decryption the inverse path is followed, using the generated keys K1 and K2. The mathematical equations leading to the decryption can be stated as:

$$H'(u, v) = \iint (E(\xi, \eta) \times K2)\exp[i2\pi(um + vn)]\mathrm{d}m\mathrm{d}n \tag{5.8}$$

$$H(u, v) = PT[H'(u, v)] \tag{5.9}$$

$$f(x, y) = PT\left\{\iint (H(u, v) \times K1)\exp[i2\pi(xu + yv)]\mathrm{d}u\mathrm{d}v\right\}. \tag{5.10}$$

It can be seen from the schematic, that the encryption keys R1 and R2 are different from the decryption keys, K1 and K2. This process introduced non-linearity in the cryptosystem. The generated decryption keys are plaintext dependent. Every time a new plaintext is encrypted, it generates corresponding private keys required for decryption. This made the system a one-time pad. Though the non-linearity made the system resistant to some attacks, such as the known-plaintext attack, it was

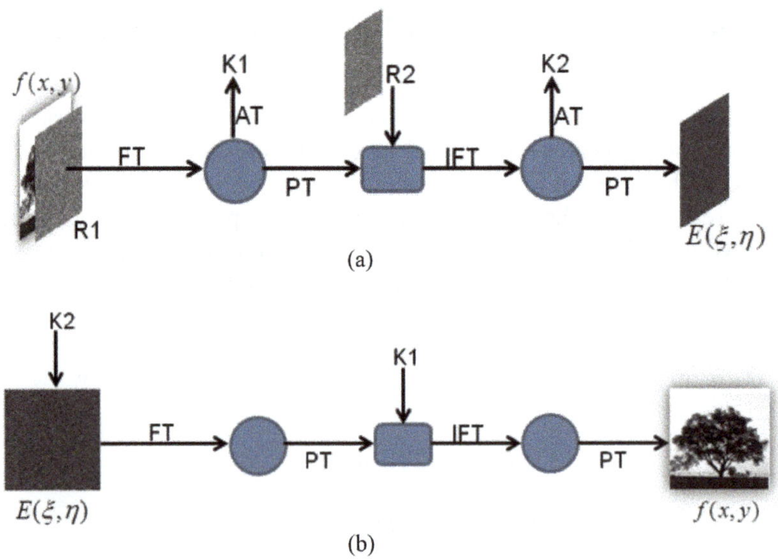

Figure 5.1. Schematic for PTFT-based (a) encryption scheme, (b) decryption scheme. $f(x,y)$ is the plaintext, R1 and R2 are the RPMs (public keys). K1 and K2 are the private keys, FT denotes Fourier transform.

shown later that it was vulnerable to yet another attack known as the specific attack [28–31]. The vulnerability stems from the fact that the decryption keys are not independent of the ciphertext. In the scenario where the public keys of these optical asymmetric cryptosystems are known to the attacker, an iterative phase retrieval algorithm can be devised that uses the appropriate constraints to retrieve the input image even if the private keys are not known. Variants of the PTFT-based system have been formulated for robust encryption [32–40].

Bearing in mind the aforementioned advantages and disadvantages of the interference-based encryption techniques and asymmetric optical encryption, a new method was introduced that aimed at removing the silhouette problem of the interference-based scheme as well as the vulnerability of the asymmetric cryptosystem. This method was called equal modulus decomposition (EMD) [41]. The method involved the decomposition of the Fourier spectrum of the input image into two complex-valued functions with equal amplitudes. The decryption consisted of interference of these two entities. Initially, the EMD-based scheme was shown to be resistant to the specific attack which made it more robust than the PTFT scheme. It provided an alternative method to secure information in the asymmetric encryption protocol with enhanced security.

5.3 Equal modulus decomposition

The basic idea behind EMD is the splitting of a complex-valued function into two complex functions having equal amplitude. Let $f(x,y)$ denote the image to be encrypted. A random phase function distributed uniformly in the interval [0,1] is used as a public key, say, $R(x,y)$. The plaintext $f(x,y)$ is modulated with this random phase and Fourier transformed:

$$I(u, v) = \iint (f(x, y) \times R(x, y))\exp[-i2\pi(ux + vy)]\mathrm{d}x\mathrm{d}y. \tag{5.11}$$

To proceed, a random argument function θ is defined as another public key of the system. Now, splitting of the Fourier spectrum $I(u, v)$ is done such that one of the components $P_1(u,v)$ would have its argument as θ, i.e.

$$\theta = 2\pi rand(u) = \arg\{P_1\}. \tag{5.12}$$

This would mean that when projected on an Argand diagram, the complex function P_1 would make an angle of θ with the x axis. If the argument of the Fourier spectrum is ϕ, then the expression of the functions, $P_1(u,v)$ and $P_2(u,v)$ can be written as [37]:

$$P_1(u, v) = \frac{A/2}{\cos[\varphi - \theta]}.\ \exp[i\theta] \tag{5.13}$$

$$P_2(u, v) = \frac{A/2}{\cos[\varphi - \theta]} \exp\{i[2\varphi - \theta]\}. \tag{5.14}$$

The above mathematical relation can be clearly understood by the projection of the complex quantities $I(u, v)$, $P_1(u,v)$ and $P_2(u,v)$ on the Argand diagram shown in figure 5.2.

One of the components, P_1, is saved as the ciphertext while the other component P_2 serves as the private key, which is required by the authentic user to retrieve the input image. The decryption process, as illustrated in figure 5.3, is done by the

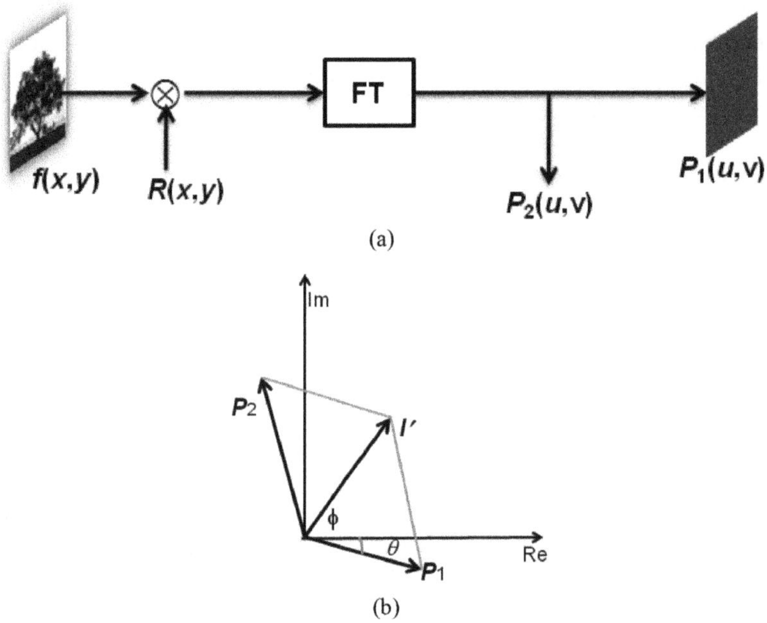

Figure 5.2. (a) Schematic showing the encryption using EMD, and (b) EMD in the Argand plane. (b) Reproduced from [45].

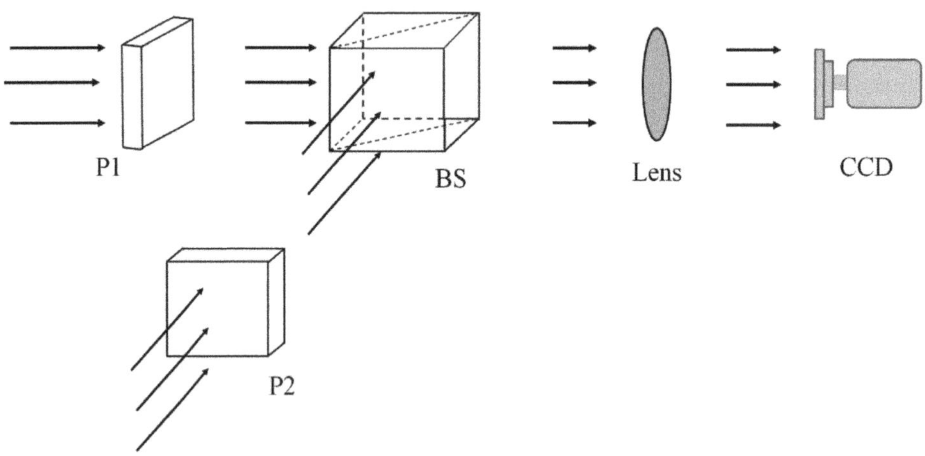

Figure 5.3. Proposed optical set-up for decryption in the EMD scheme. P1 and P2 combine after interference and are then Fourier transformed by the lens to retrieve the input image.

coherent superposition of P_1 and P_2. It is a typical optical set-up for decrypting ciphertexts generated from interference based schemes. P_1 and P_2 can be displayed on a spatial light modulator. Beams of coherent illumination through these modulators carry their information. They are brought together by means of a beam splitter and made to interfere at a point. After interference the beam is subjected to the inverse Fourier transform by means of a converging lens. A CCD placed at the focal plane would record the intensity of the beam, which will give the input image. Mathematically, the decryption process can be explained as:

$$P_1(u, v) + P_2(u, v) = I(u, v) \tag{5.15}$$

$$f(x, y) = |FT^{-1}\{I(u, v)\}|^2. \tag{5.16}$$

Numerical simulation of the EMD-based scheme can be carried out on the MATLAB platform. The results of such a simulation are shown in figures 5.4(a)–(f). The gray-scale image of the size 256 × 256 pixels, shown in figure 5.4(a) is used as the input to the cryptosystem. Figure 5.4(b) shows the RPM, $R(x,y)$, that serves as the first public key, while the argument θ which is the second public key is shown in figure 5.4(c). Figures 5.4(d) and (e) show the private key and the ciphertext generated after the EMD, respectively. The decrypted image using the correct key is shown in figure 5.4(f).

In this cryptosystem, the random phase $R(x,y)$, and the argument are the encryption keys. However, the decryption key is the private key P_2. This makes the system asymmetric. Moreover, the functions, P_1 and P_2, do not carry sufficient information to give the silhouette of the input image upon individual inverse Fourier transform. With all these properties, EMD-based encryption systems promise to be an efficient optical cryptosystem.

5.4 Cryptanalysis

Being an asymmetric cryptosystem, attack algorithms, such as the known plaintext attack on the EMD, would be of no meaning as they aim to retrieve the keys and in asymmetric optical cryptosystems the keys vary with the plaintexts. Asymmetric optical cryptosystems are vulnerable to the specific attack in which the attacker tries to retrieve plaintext with the knowledge of the ciphertext and the public key [28]. In the EMD-based cryptosystems, different iterative Fourier transform algorithms have been proposed that show that the system can be breached [42–44]. Any iterative amplitude phase retrieval algorithm (APRA) involves going back and forth from the Fourier plane to the input plane. In each iteration step, the phase or the amplitude in either (or both) planes is substituted with the available information, which is generally referred to as the constraints. The iterations continue until the Fourier- and object-domain constraints are satisfied and the process converges to the required solution. In the case of the EMD-based system, the RPM, $R(x,y)$ and the random argument θ can serve as the object constraints. In the Fourier domain, the ciphertext is the constraint. Other information about the system, such as the equality of the amplitude of the private key and the ciphertext can be utilized as additional

Figure 5.4. Simulation results for EMD-based encryption scheme: (a) Input image, (b) RPM used as the first public key, (c) random distribution for the second public key θ, (d) private key generated after EMD, (e) ciphertext, (f) decrypted image.

constraints. Different algorithms can be designed depending on the constraints available to the attacker.

For the purpose of illustration, one of the specific attacks reported by Wu *et al* is explained [43]. This algorithm considers that the attacker has obtained knowledge of the ciphertext P_1 and the public key, $R(x,y)$. Trigonometric manipulation of the ciphertext expression helps construct the constraints. As shown in the Argand diagram (figure 5.2), the Fourier spectrum $I(u, v)$ has an argument of ϕ, therefore, equation (5.11) can be written as:

$$I(u, v) = FT\{f(x, y) \times R(x, y)\} = A(u, v)\exp(i\varphi). \qquad (5.17)$$

Here, FT denotes Fourier transform. The ciphertext P_1 (5.3) can be expressed in terms of $A(u,v)$ and $\phi(u,v)$ as:

$$\begin{aligned} P_1(u, v) &= A(u, v)\exp(i\varphi(u, v)) \times \left\{\frac{1}{2} - \frac{i}{2}\tan(\varphi(u, v) - \theta(u, v))\right\} \\ &= X(u, v).\ Y(u, v). \end{aligned} \qquad (5.18)$$

Here

$$X(u, v) = A(u, v)\exp(i\varphi(u, v)) \qquad (5.19a)$$

and

$$Y(u, v) = \left\{\frac{1}{2} - \frac{i}{2}\tan(\varphi(u, v) - \theta(u, v))\right\}. \qquad (5.19b)$$

Hence it can be said that:

$$\mathrm{Re}\,\{P_1/X\} = \mathrm{Re}\,\{Y\} = 1/2 \qquad (5.20)$$

always holds true and can be used as a constraint in each step of iteration. The iterative steps of the attack scheme proceeds as follows:

Step 1: For the kth iteration, the approximated input image $f_k(x,y)$ (taken as 1 for $k = 1$) is multiplied with the public key $R(x,y)$ to get an intermediate function denoted by Z_k. Z_k is Fourier transformed and the Fourier spectrum approximates the value of $X(u,v)$ given in (5.18). For kth iteration let us denote the approximated value of X as X_k.

$$Z_k(x, y) = f_k(x, y)\exp(iR(x, y)) \qquad (5.21)$$

$$X_k = FT\{Z_k\} \qquad (5.22)$$

Step 2: The ciphertext P_1 is divided by the quantity X_k obtained in step 1, to get Y_k (5.18).
Step 3: The constraint is applied, where the real part of Y_k is restricted to assume the value of ½ as evaluated in (5.20).

Step 4: The ciphertext P_1 is divided by Y_k obtained in Step 3 which would give a new value for $X(u,v)$ as shown by (5.18). This new value gives X_{k+1} to be used in the next iteration step.
Step 5: X_{k+1} is inverse Fourier transformed, to get an intermediate function denoted as Z_k'. The phase of Z_k' is substituted with $R(x,y)$. Its amplitude is reserved to form the $(k+1)$th approximation of the input image, i.e. $f_{k+1}(x,y)$. Hence after applying the constraint Z_k' can be written as:

$$Z_{k+1} = |Z_k'|\exp(iR(x, y)) = f_{k+1}(x, y)\exp(iR(x, y)). \tag{5.23}$$

The algorithm proceeds with Z_{k+1} as its input for the $(k+1)$th iteration. The process continues until the normalized mean square error (NMSE) between the phase of Z_k' and $R(x,y)$ for any consecutive iteration step is less than a preset threshold value. Figure 5.5 shows the flowchart for the attack algorithm.

Generally, 200–500 iterations are sufficient to converge the process to get the plaintext. Results of the numerical simulation of the aforementioned algorithm are shown in figures 5.6(a) and (b). The retrieved image through this attack is shown in figure 5.6(a). It can be seen that it is similar to the original image. Figure 5.6(b) shows the plot of NMSE values with respect to the iteration number. It can be seen that the NMSE value saturates to the low value of 0.0556, which indicates that the image is recovered. The correlation coefficient (CC) is evaluated to be 0.8928 in this case.

Similarly, various other algorithms can be developed with different choices of constraints to retrieve the input image. For example, another algorithm can be

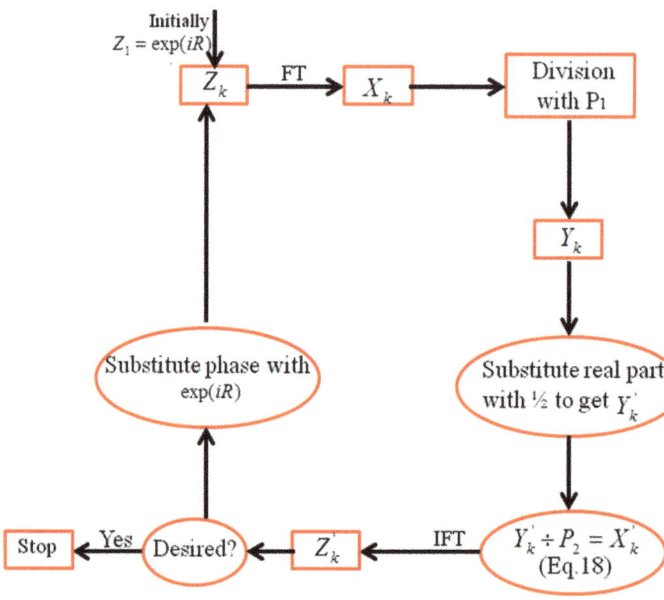

Figure 5.5. Flowchart showing the attack algorithm based on APRA.

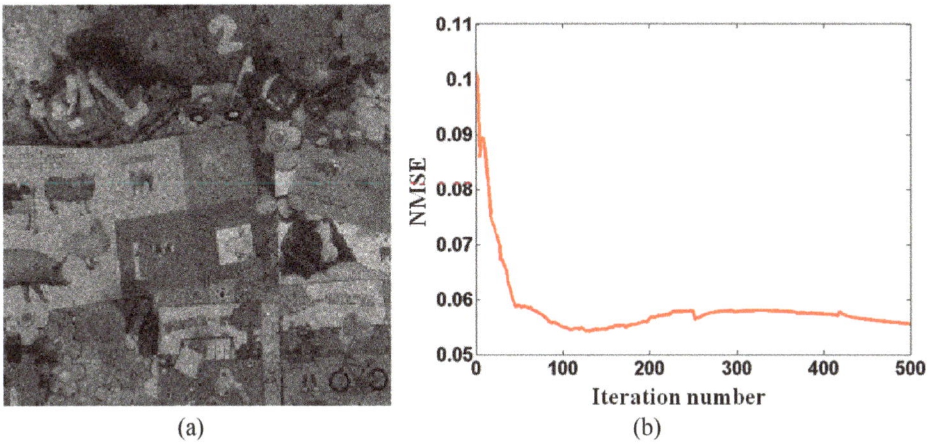

Figure 5.6. (a) The retrieved image using the attack algorithm based on APRA, and (b) relationship between the number of iterations and NMSE for the specific attack.

constructed using the constraint of equal amplitude of the ciphertext and the private key in the Fourier plane.

The above-mentioned optical cryptosystem based on coherent superposition and EMD show vulnerability to the specific attack because the private key is dependent on the ciphertext [42–44]. This allows the construction of various constraints that would supply information in the iterative APRA so that the process converges to a solution. One of the ways in which the scheme can be made secure is to suppress the available information in the path so that the iterative process does not converge. This brings us to the next topic that discusses the modifications made in the EMD scheme that make it resistant to the specific attack.

5.5 Modifications in EMD-based optical cryptosystems

The modifications in EMD aim at making fewer numbers of constraints available to the attacker. This involves multiple diffractive imaging based EMD or simply cascading two EMDs consecutively [45–48]. As seen in the previous section, the public key, $R(x,y)$ was used as one of the constraints. In this section, an example of an optical cryptosystem that suppressed this information to make the system secure is discussed. Multiple diffraction intensity patterns of the plaintext modulated with a random phase ($f(x,y)\exp[iR(x,y)]$) can be used as the input to the EMD-based system [45]. To start with, the Fourier transform (or the gyrator transform) of the random phase modulated plaintext is obtained, which is further modulated with a random phase and then is free space propagated for three different distances. The corresponding diffraction patterns are recorded. The diffraction patterns undergo the gyrator transform and subsequently the output is subjected to the EMD. As a result, three private keys and three ciphertexts are generated. The schematic to this cryptosystem is shown in figure 5.7.

The key point here is that not using the RPM just before the EMD removes a prospective constraint from the path of iterative APRA, as opposed to the available RPM $R(x,y)$ in the APRA path of the classic EMD scheme.

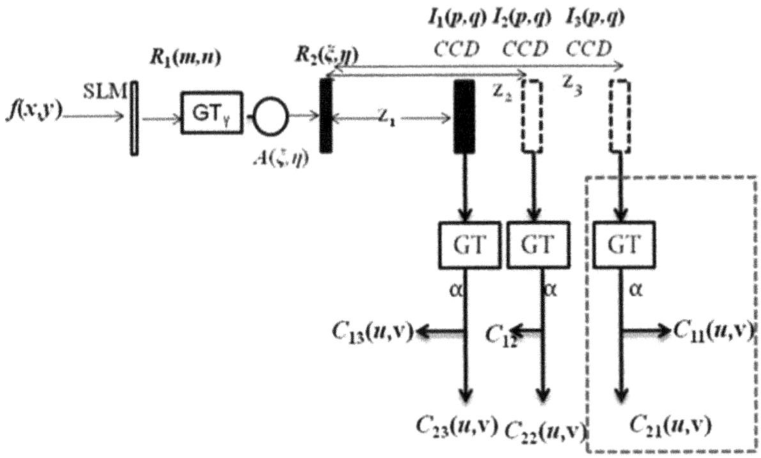

Figure 5.7. Schematic of a diffractive imaging based EMD cryptosystem. SLM: spatial light modulator, GT: gyrator transform, CCD: charge-coupled device camera. Reproduced from [45].

The specific attack on the above-mentioned system would consist of two steps: in the first stage, the attacker tries to retrieve the private keys. This step involves the boxed part of the schematic shown in figure 5.7. Using the superposition of the private and public key, the diffraction patterns can be obtained after the inverse gyrator transform. The second stage involves the use of the diffraction intensity pattern and the public keys to run an APRA to retrieve the input image. But in the given scenario, in the absence of proper constraints, the first stage will not converge to the proper solution and hence, the second stage of the attack cannot be initiated. The first stage would proceed in the following steps. For clarity, the boxed part of figure 5.7 can be referred to.

Step 1: The private key is (C_{21} in the schematic diagram) approximated by a random function, say, P_k. The guessed key is superposed with the ciphertext and inverse gyrator transformed:

$$H_k = GT^{-\alpha}\{P_k + C_{11}\} \tag{5.24}$$

Step 2: Since there is no constraint available in this domain to update the function H_k, hence it is simply gyrator transformed and from the output, an intermediate function is obtained by subtracting the ciphertext from the GT spectrum:

$$P_k'' = GT^{\alpha}\{H_k\} - C_{11} \tag{5.25}$$

Step 3: Since the ciphertext and the private key have equal moduli, hence the amplitude of P_k'' is substituted with that of the ciphertext and the phase is reserved to construct the new private key for the next iteration.

$$P_{k+1} = |C_{11}|\exp\{\mathrm{i}\arg(P_k'')\}. \tag{5.26}$$

At the end of each iteration step in the above-mentioned attack the retrieved keys are used to obtain the input to the EMD. The iterative phase retrieval process fails to

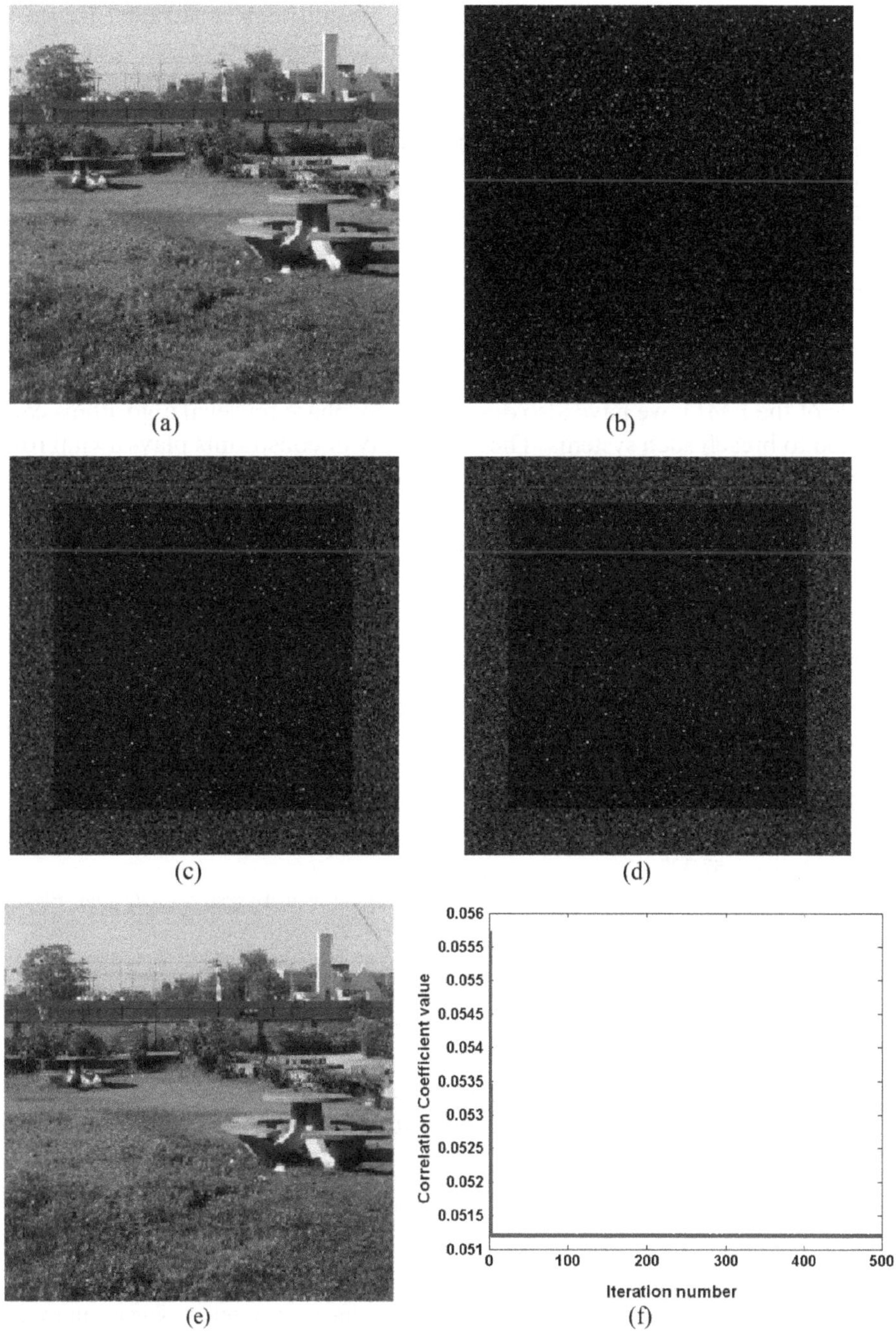

Figure 5.8. Simulation results for a diffractive imaging based EMD encryption scheme: (a) Input image, (b) one of the diffractive images, (c) one of the three private keys (d) one of the three ciphertexts, (e) decrypted image, and (f) plot of number of iterations versus CC for specific attack.

converge at a proper solution in the absence of sufficient constraint. Hence, the second stage of the attack cannot be initiated. This makes the cryptosystem secure against the specific attack. Figure 5.8 shows the simulation results of the above-mentioned scheme. For analyzing the specific attack, the CC value between the obtained EMD input and the original EMD input used during encryption is evaluated. Figure 5.8(f) shows that the CC value saturates to a small value of 0.05. Hence, the original image cannot be retrieved using this method.

5.6 Conclusion

The EMD-based optical cryptosystem is one of the asymmetric cryptosystems that aims at removing the silhouette problem of interference based encoding. Besides the principle of the EMD, we have also discussed how phase retrieval algorithms can be developed to breach such systems. The availability of constraints plays a vital role in the successful formulation of such attack algorithms. Hence, modification of the classic scheme to suppress constraints can make the system secure. Further variants may include using a classic EMD-based scheme in different optical transform domains, such as Fresnel, fractional Fourier, wavelet, Hartley etc.

References

[1] Javidi B (ed) 2005 *Optical and Digital Techniques for Information Security* (New York: Springer)

[2] Javidi B (ed) 2006 *Optical Imaging Sensors and Systems for Homeland Security Applications* (New York: Springer)

[3] Liu S, Guo C and Sheridan J T 2014 A review of optical image encryption techniques *Opt. Laser Technol.* **57** 327–42

[4] Chen W, Javidi B and Chen X 2014 Advances in optical security systems *Adv. Opt. Photon.* **6** 120–55

[5] Javidi B *et al* 2016 Roadmap on Optical Security *J. Opt.* **18** 083001

[6] Alfalou A and Mansour A 2009 Double random phase encryption scheme to multiplex and simultaneous encode multiple images *Appl. Opt.* **48** 5933–47

[7] Unnikrishnan G, Joseph J and Singh K 2000 Optical encryption by double-random phase encoding in the fractional Fourier domain *Opt. Lett.* **25** 887–89

[8] Zhu B, Liu S and Ran Q 2000 Optical image encryption based on multifractional Fourier transforms *Opt. Lett.* **25** 1159–61

[9] Hennelly B and Sheridan J T 2003 Optical image encryption by random shifting in fractional Fourier domains *Opt. Lett.* **28** 269–71

[10] Situ G and Zhang J 2004 Double random phase encoding in the Fresnel domain *Opt. Lett.* **29** 1584–6

[11] Mehra I and Nishchal N K 2015 Optical asymmetric image encryption using gyrator wavelet transform *Opt. Commun.* **354** 344–52

[12] Chen L and Zhao D 2006 Optical image encryption with Hartley transforms *Opt. Lett.* **31** 3438–40

[13] Zhou N, Li H, Wang D, Pan S and Zhou Z 2015 Image compression and encryption scheme based on 2D compressive sensing and fractional Mellin transform *Opt. Commun.* **343** 10–21

[14] Alfalou A and Brosseau C 2009 Optical image compression and encryption methods *Adv. Opt. Photon.* **1** 589–636
[15] Alfalou A and Brosseau C 2010 Dual encryption scheme of images using polarized light *Opt. Lett.* **35** 2185–7
[16] Nishchal N K, Joseph J and Singh K 2004 Securing information using fractional Fourier transform and digital holography *Opt. Commun.* **235** 253–9
[17] Nishchal N K and Naughton T J 2011 Flexible optical encryption with multiple users and multiple security levels *Opt. Commun.* **284** 735–9
[18] Zhang Y and Wang B 2008 Optical image encryption based on interference *Opt. Lett.* **33** 2443–5
[19] Zhang Y, Wang B and Dong Z 2009 Enhancement of image hiding by exchanging two phase masks *J. Opt.* A **11** 125406
[20] Wang X and Zhao D 2012 Optical image hiding with silhouette removal based on the optical interference principle *Appl. Opt.* **51** 686–91
[21] Kumar P, Joseph J and Singh K 2010 Optical image encryption based on interference under convergent random illumination *J. Opt.* **12** 095402
[22] Kumar P, Joseph J and Singh K 2011 Optical image encryption using a jigsaw transform for silhouette removal in interference-based methods and decryption with a single spatial light modulator *Appl. Opt.* **50** 1805–11
[23] Chen W, Quan C and Tay C J 2009 Optical color image encryption based on Arnold transform and interference method *Opt. Commun.* **282** 3680–5
[24] Carnicer A, Montes-Usategui M, Arcos S and Juvells I 2005 Vulnerability to chosen-cyphertext attacks of optical encryption schemes based on double random phase keys *Opt. Lett.* **30** 1644–6
[25] Peng X, Zhang P, Wei H and Yu B 2006 Known-plaintext attack on optical encryption based on double random phase keys *Opt. Lett.* **31** 1044–6
[26] Frauel Y, Castro A, Naughton T J and Javidi B 2007 Resistance of the double random phase encryption against various attacks *Opt. Express* **15** 10253–65
[27] Qin W and Peng X 2010 Asymmetric cryptosystem based on phase-truncated Fourier transforms *Opt. Lett.* **35** 118–20
[28] Wang X and Zhao D 2012 A special attack on the asymmetric cryptosystem based on phase-truncated fractional Fourier transforms *Opt. Commun.* **285** 1078–81
[29] Wang Y, Quan C and Tay C J 2015 Improved method of attack on an asymmetric cryptosystem based on phase-truncated Fourier transform *Appl. Opt.* **54** 6874–81
[30] Wang X, Chen Y, Dai C and Zhao D 2014 Discussion and a new attack of the optical asymmetric cryptosystem based on phase-truncated Fourier transform *Appl. Opt.* **53** 208–13
[31] Fatima A and Nishchal N K 2016 Discussion on comparative analysis and a new attack on optical asymmetric cryptosystem *J. Opt. Soc. Am.* A **33** 2034–40
[32] Rajput S K and Nishchal N K 2012 Image encryption based on interference that uses fractional Fourier domains asymmetric keys *Appl. Opt.* **51** 1446–52
[33] Wang X and Zhao D 2011 Multiple image encryption based on non-linear amplitude-truncation and phase-truncation in Fourier domain *Opt. Commun.* **284** 148–52
[34] Wang X, Zhao D and Chen Y 2014 Double-image encryption without information disclosure using phase-truncation Fourier transforms and a random amplitude mask *Appl. Opt.* **53** 5100–8

[35] Rajput S K and Nishchal N K 2013 Image encryption using polarized light encoding and amplitude and phase truncation in the Fresnel domain *Appl. Opt.* **52** 4343–52
[36] Rajput S K and Nishchal N K 2013 Optical asymmetric cryptosystem using polarized light *Proc. SPIE* **8855** 885503
[37] Rajput S K and Nishchal N K 2013 Known plaintext attack-based optical cryptosystem using phase-truncated Fresnel transform *Appl. Opt.* **52** 871–8
[38] Mehra I and Nishchal N K 2014 Image fusion using wavelet transform and its application to asymmetric cryptosystem and hiding *Opt. Express* **22** 5474–82
[39] Wang X and Zhao D 2011 Security enhancement of a phase-truncation based image encryption algorithm *Appl. Opt.* **50** 6645–51
[40] Ding X, Deng X, Song K and Chen G 2013 Security improvement for asymmetric cryptosystem based on spherical wave illumination *Appl. Opt.* **52** 467–73
[41] Cai J, Shen X, Lei M, Lin C and Dou S 2015 Asymmetric optical cryptosystem based on coherent superposition and equal modulus decomposition *Opt. Lett.* **40** 475–8
[42] Deng X 2015 Asymmetric optical cryptosystem based on coherent superposition and equal modulus decomposition: comment *Opt. Lett.* **40** 3913
[43] Wu J, Liu W, Liu Z and Liu S 2015 Cryptanalysis of an asymmetric optical cryptosystem based on coherent superposition and equal modulus decomposition *Appl. Opt.* **54** 8921–4
[44] Wang Y, Quan C and Tay C J 2016 New method of attack and security enhancement on an asymmetric cryptosystem based on equal modulus decomposition *Appl. Opt.* **55** 679–86
[45] Fatima A, Mehra I and Nishchal N K 2016 Optical asymmetric cryptosystem using equal modulus decomposition and multiple diffractive imaging *J. Opt.* **18** 085701
[46] Chen H, Tanougast C, Liu Z and Sieler L 2017 Asymmetric optical cryptosystem for color image based on equal modulus decomposition in gyrator transform domains *Opt. Lasers Eng.* **93** 1–8
[47] Chen L, Gao X, Chen X, He B, Liu J and Li D 2016 A new optical image cryptosystem based on two-beam coherent superposition and unequal modulus decomposition *Opt. Laser Tech.* **78** 167–74
[48] Xu H, Xu W, Wang S and Wu S 2017 Asymmetric optical cryptosystem based on modulus decomposition in Fresnel domain *Opt. Commun.* **402** 302–10

IOP Publishing

Advanced Secure Optical Image Processing for Communications

Ayman Al Falou

Chapter 6

Information security using fractional transforms

Zhengjun Liu, Hang Chen, Camel Tanougast and Shutian Liu

Fractional transform is a class of expanded linear canonical transform [1], such as fractional Fourier transform (FrFT) [2] and gyrator transform (GT) [3]. When the parameter in the transforms is changed, the output spectrum has different distributions, which includes amplitude and phase. In mathematics, the FrFT and GT can become Fourier transform, if their transformation angle is equal to $\pi/2$. FrFT can be implemented with a single lens system [4]. A system having six thin cylinder lenses [3] can achieve GT optically. In recent years, two optical fractional transforms have been applied widely with the aim of optical encryption. This parameter is regarded as an additional key in some optical information security methods.

6.1 Optical fractional transforms

In this section, the definition, optical implementation and discrete algorithm of the transforms will be represented. This is necessary to design a new optical hiding scheme.

6.1.1 Definition and optical implementation

The integral definition of FrFT [2] of the one-dimensional function $f(x)$ is expressed as follows

$$\mathcal{F}^{\alpha}\{f(x)\}(u) = \int f(x)K^{\alpha}(x, u)\mathrm{d}x, \tag{6.1}$$

where $\mathcal{F}^{\alpha}$ is the fractional transform operator. The parameter α is the order of the transform. The function $K^{\alpha}(x, u)$ is the kernel of this transform and is defined as

doi:10.1088/978-0-7503-1457-2ch6

$$K^{\alpha}(x, u) = \begin{cases} A_\alpha \exp\left[i\pi(u^2 \cot\theta_\alpha - 2xu \csc\theta_\alpha + x^2 \cot\theta_\alpha)\right], \text{ if } \alpha \neq 2n, \\ \delta(x + u), \text{ if } \alpha = 4n + 2, \\ \delta(x - u), \text{ if } \alpha = 4n, \end{cases} \tag{6.2}$$

$$A_\alpha = \sqrt{1 - i \cdot \cot\theta_\alpha}, \qquad \theta_\alpha = \alpha\pi/2. \tag{6.3}$$

Here θ_α is the transform angle. The two-dimensional case can be reduced directly according to the property of the separable variables. FrFT has some properties that are similar to the Fourier transform, but they are more complex [5]. Differing from the Fourier transform, FrFT has addditivity properties:

$$\mathcal{F}^{\alpha}\mathcal{F}^{\beta} = \mathcal{F}^{\beta}\mathcal{F}^{\alpha} = \mathcal{F}^{\alpha+\beta}. \tag{6.4}$$

The optical implementation of FrFT with a single thin lens has been proposed by Lohmann [6]. The corresponding system is illustrated in figure 6.1. The input pattern is illuminated by a plane light wave and the diffracted light field is obtained in the output plane. This structure is extensively used for information processing system and technique. FrFT is also performed by quadratic graded index media (lenses or fiber) [5], which is beneficial for the application of information transmission in fiber communication.

For a true optical system, FrFT of the function $p(x)$ is defined as [6]

$$\mathcal{F}^{\alpha}\{p(x)\}(u) = \frac{\exp(2\mathrm{i}kd)}{\mathrm{i}\lambda f \sin\alpha} \int p(x) \exp\left[\frac{\mathrm{i}\pi(x^2 + u^2)}{\lambda f \sin\alpha \tan\alpha} - \frac{2\mathrm{i}\pi x u}{\lambda f \sin\alpha}\right] \mathrm{d}x, \tag{6.5}$$

where λ is the wavelength and k is the wavenumber ($k = 2\pi/\lambda$).

Gyrator transform is another fractional transform. GT, however, has no one-dimensional expression. GT of the function $f(x, y)$ is written as follows [3]

$$\mathcal{G}^{\alpha}\{f(x, y)\}(u, v) = \frac{1}{|\sin\alpha|} \iint f(x, y) \exp\left[\mathrm{i}2\pi \frac{(xy + uv)\cos\alpha - xv - yu}{\sin\alpha}\right] \mathrm{d}x\mathrm{d}y. \tag{6.6}$$

When α is equal to $\pi/2$, $G^{\pi/2}\{f(x, y)\} = \mathcal{F}\{f(y, x)\}$, in which the symbol $\mathcal{F}$ is the Fourier operator. If the transform angle $\alpha = 0$, the transform kernel is an impulse function $\delta(x - u, y - v)$. The additivity property is also satisfied in GT. Due to equation (6.6), GT can be used for image processing by replacing the Fourier transform.

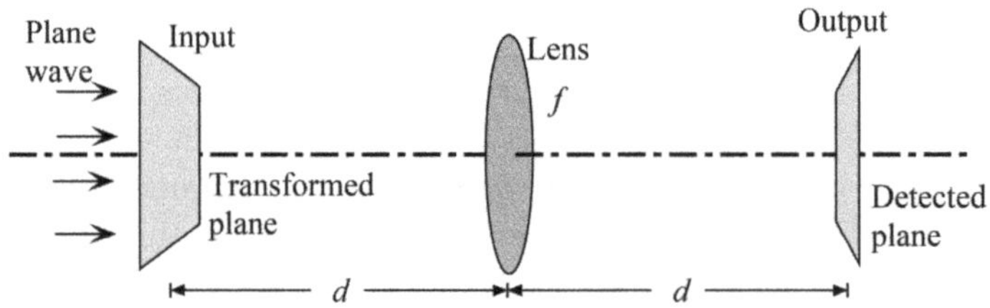

Figure 6.1. Optical system achieving FrFT.

An optical gyrator transform is shown in figure 6.2. This system includes six thin cylinder lenses. Each pair of lenses is arranged and displayed in figure 6.2(b). The transform angle α is determined by the intersection angles ϕ_2 and ϕ_1 as

$$\sin 2\phi_1 = \cot \frac{\alpha}{2}, \qquad \sin 2\phi_2 = \frac{1}{2} \sin \alpha, \tag{6.7}$$

and the lenses (L_1 and L_3) have the focal length f_1. The focal length of Lens L_2 is f_2. Between the focal length and distance z, a relation is requested as

$$f_1 = 2f_2 = z. \tag{6.8}$$

6.1.2 Optical random transform

Except for the integral format, FrFT has an expression with series, which is addressed as follows

$$\mathcal{F}^{\alpha}\{f(x)\}(u) = \sum_{n=0}^{\infty} c_n \exp(in\pi\alpha/2)\phi_n(u), \tag{6.9}$$

where the eigenfunction ϕ_n is calculated as

$$\phi_n(u) = \frac{1}{\sqrt{n!2^n\sqrt{\pi}}} H_n(u) \exp(-u^2/2). \tag{6.10}$$

The function ϕ_n is called the normalized Hermite–Gaussian and is the eigenfunction of both the Fourier transform and FrFT. The function H_n is an n-order Hermite polynomial. The coefficient c_n is computed as

$$c_n = \int_{-\infty}^{+\infty} f(x)\phi_n(x)\mathrm{d}x. \tag{6.11}$$

The eigenvalue of FrFT is $\exp(-in\pi\alpha/2)$, which are the points in the unit circle of two-dimensional coordinates.

When the order α is an irrational number [7], the eigenvalue of FrFT will be randomized. The transform will become a random Fourier transform. The random transform has a random output and can be implemented with an optical 4F system

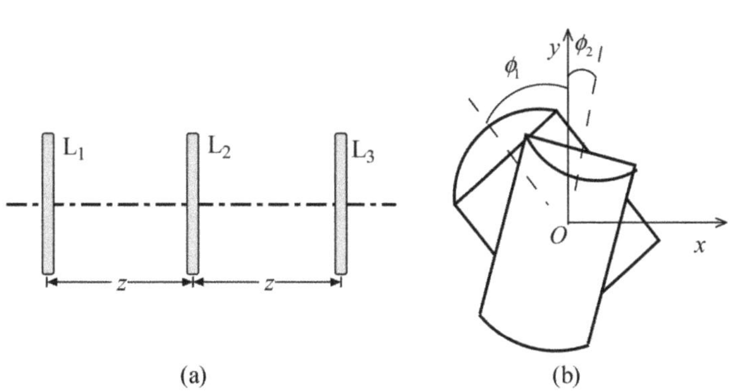

Figure 6.2. Optical gyrator transform system: (a) lens structure, (b) thin cylinder unit.

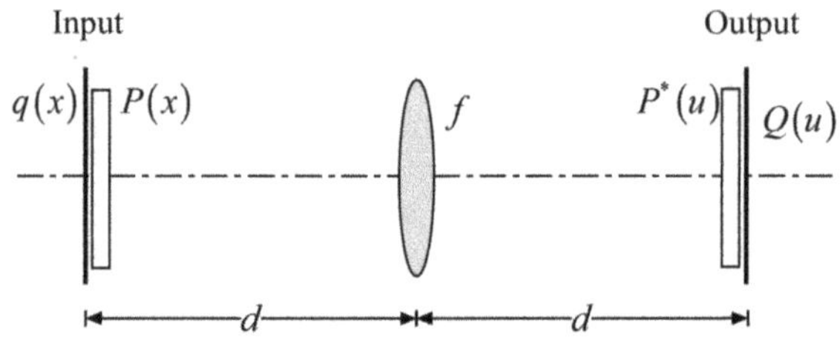

Figure 6.3. Optical system of random FrFT.

approximately. It is also used for hiding a gray-level image. The discrete algorithm of this transform is related to discrete FrFT.

A random FrFT was also proposed by changing the kernel function of the FrFT [8]. An optical system implementing random FrFT is displayed in figure 6.3. Here the functions P and P^* are used to express a pair of phase masks having a conjugate relation, which are placed at the planes of input and output, respectively. In random FrFT, the kernel function of FrFT is randomized.

Mathematically, the input function $q(x)$ and output function $Q(u)$ satisfy the equation as follows

$$Q(u) = P^*(u)\mathcal{F}^\alpha\{q(x)P(x)\}(u), \tag{6.12}$$

which can also be rewritten into an integral expression as

$$Q(u) = S^\alpha\{q(x)\}(u) = \int_{-\infty}^{+\infty} q(x)\underline{P(x)K_\alpha(x, u)P^*(u)}\mathrm{d}x, \tag{6.13}$$

the operator S^α is random FrFT. The term underlined is the kernel function of the random transform. When the function P is constant, the transform becomes FrFT.

The transform defined in equation (6.12) has some mathematical properties, which is from FrFT. The transform is linear as

$$S^\alpha\{c_1q_1(x) + c_2q_2(x)\} = c_1S^\alpha\{q_1(x)\} + c_2S^\alpha\{q_2(x)\}, \tag{6.14}$$

In which c_1 and c_2 are complex constants. For this transform, the additivity is given as

$$S^\alpha S^\beta = S^\beta S^\alpha = S^{\alpha+\beta}. \tag{6.15}$$

When $\alpha = 0$, the transform is an identical transform. Thereby the inverse transform of S^α is $S^{-\alpha}$. Since the phase masks are employed in the system in figure 6.3, the random FrFT has the energy conservation property:

$$\int_{-\infty}^{+\infty} |q(x)|^2 \,\mathrm{d}x = \int_{-\infty}^{+\infty} |S^\alpha\{q(x)\}(u)|^2 \,\mathrm{d}u. \tag{6.16}$$

6.2 The discrete algorithm of fractional transforms

In this section, the discrete procedure calculating fractional transforms will be given for image and signal processing.

The discrete transform matrix of FrFT can be obtained from this matrix [8] as

$$S = \begin{bmatrix} 2 & 1 & 0 & 0 & \cdots & 0 & 1 \\ 1 & 2\cos\omega & 1 & 0 & \cdots & 0 & 0 \\ 0 & 1 & 2\cos 2\omega & 1 & \cdots & 0 & 0 \\ \vdots & \vdots & \vdots & \ddots & \vdots & \vdots & \vdots \\ 1 & 0 & 0 & 0 & \cdots & 1 & 2\cos(N-1)\omega \end{bmatrix}, \tag{6.17}$$

where $\omega = 2\pi/N$. Here N is the length of signal. The matrix S has a relation $SF = FS$, in which F is the transform of the discrete Fourier transform. The commutative property is satisfied for FrFT. According to matrix theory [8], the matrices S and F (or F^α) have the same eigenvectors. All eigenvectors of discrete FrFT can be calculated from the real symmetrical matrix S. The eigenvectors are closed to the Hermite–Gaussian functions, which are the eigenfunctions of the FrFT and Fourier transform.

The kernel matrix of the FrFT is expressed as follows

$$F^\alpha = VD^\alpha V^t, \tag{6.18}$$

where V is a matrix composed of all eigenvectors. The symbol t is to obtain a transpose matrix. The diagonal matrix D^α is determined as

$$D^\alpha = \begin{cases} \text{diag}\left\{\exp\left(-i\frac{\alpha\pi}{2}[0, 1, \ldots, N-1]\right)\right\}, & \text{if } N \text{ is odd} \\ \text{diag}\left\{\exp\left(-i\frac{\alpha\pi}{2}[0, 1, \ldots, N-2, N]\right)\right\}, & \text{if } N \text{ is even,} \end{cases} \tag{6.19}$$

where the function 'diag' is to form a diagonal matrix from a vector. The parameter α is the order of the transform. When N is even, the values have a jumping phenomenon after N-2. The algorithm has the properties of energy conservation and additivity due to the definition of the eigenvalue matrix D^α in equation (6.19).

The discrete of FrFT of image q and signal p can be expressed as

$$Q = F^\alpha qF^\alpha, \tag{6.20}$$

$$P = F^\alpha p, \tag{6.21}$$

where Q and P are the output of the discrete transform.

An example of a discrete FrFT is given in figure 6.4. The one-dimensional (1D) function is defined as

$$f(x) = \begin{cases} 1, & \text{if } |x| < 2, \\ 0, & \text{if } 2 \leqslant |x| \leqslant 6, \end{cases} \tag{6.22}$$

where the total number of sampling points is 129. According to sampling theory, the variable u in the frequency domain is in the range [−16/3, 16/3]. The two-dimensional (2D) function is a squared aperture having 128 × 128 pixels, in which

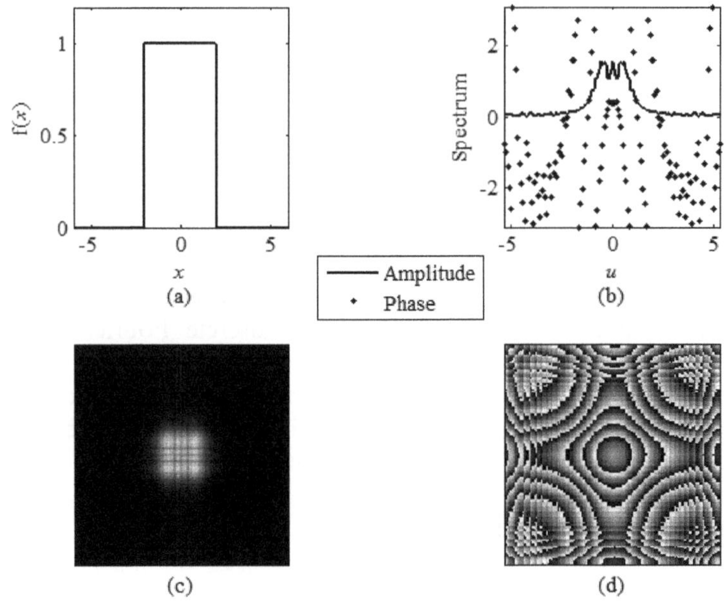

Figure 6.4. The result of discrete FrFT: (a) rectangle function, (b) 1D FrFT, (c) amplitude, (d) phase.

the pixel values in the center 50 × 50 are equal to 1 and the others are 0. The order α is fixed at 0.6. The output patterns are symmetrical, because the input signals are even functions. As an extension, the random Fourier transform and random FrFT can be designed by using the eigenvector matrix V.

The integral expression of FrFT can also be written into a format composed of a convolution operation [9] as

$$\begin{aligned}\mathcal{F}^{\alpha}\{f(x)\}(u) = A_{\alpha}\exp\left[-i\pi\tan\left(\frac{\alpha\pi}{4}\right)u^2\right] \times \int_{-\infty}^{+\infty} f(x) \\ \times \exp\left[-i\pi\tan\left(\frac{\alpha\pi}{4}\right)x^2\right]\exp\left[i\pi\csc\left(\frac{\alpha\pi}{2}\right)(u-x)^2\right]\mathrm{d}x,\end{aligned} \tag{6.23}$$

where the integral is a convolution, which can be calculation by a fast Fourier transform (FFT) algorithm in the discrete case. Equation (6.23) is given as

$$\mathcal{F}^{\alpha}\{f(x)\}(u) = A_{\alpha}\exp\left[-i\pi\tan\left(\frac{\alpha\pi}{4}\right)u^2\right] \times [f'(x)\otimes B_{\alpha}(x)], \tag{6.24}$$

where the symbol '⊗' represents the convolution operator. The functions $f'(x)$ and $B_{\alpha}(x)$ are calculated as

$$f'(x) = f(x)\exp\left[-i\pi\tan\left(\frac{\alpha\pi}{4}\right)x^2\right], \tag{6.25}$$

$$B_{\alpha}(x) = \exp\left[i\pi\csc\left(\frac{\alpha\pi}{2}\right)x^2\right], \tag{6.26}$$

by assuming

$$B'_{\alpha}(x) = \exp\left[i\pi \tan\left(\frac{\alpha\pi}{4}\right)x^2\right], \tag{6.27}$$

the function f' can be re-written as

$$f'(x) = f(x)B'_{\alpha}(x). \tag{6.28}$$

In the Fourier domain, the FrFT can be expressed as

$$\mathcal{F}^{\alpha}\{f(x)\}(u) = A_{\alpha}B_{\alpha}(u)\mathcal{F}^{-1}\{\mathcal{F}\{f(x)\}(u')\mathcal{F}\{B_{\alpha}(x)(u')\}\}(u). \tag{6.29}$$

Here the Fourier transform of the function $B_{\alpha}(x)$ can be calculated as

$$\mathcal{F}\{A_{\alpha}B_{\alpha}(x)\}(u) = \exp\left[-i\pi x^2 \sin\left(\frac{\alpha\pi}{2}\right)\right], \tag{6.30}$$

where the coefficient A_{α} is counteracted. Therefore, two steps of FFT can achieve the computational load in equation (6.29). This method has better speed than the scheme composed of the eigenvector matrix [8].

With the input functions and related parameters in figure 6.4, the example of the convolution algorithm is performed and displayed in figure 6.5. By comparing figure 6.4, it is evident that two algorithms have some differences in the detailed information of the output patterns. The additivity property of FrFT is not satisfied in the convolution method [9], which is also a choice in the field of image and signal

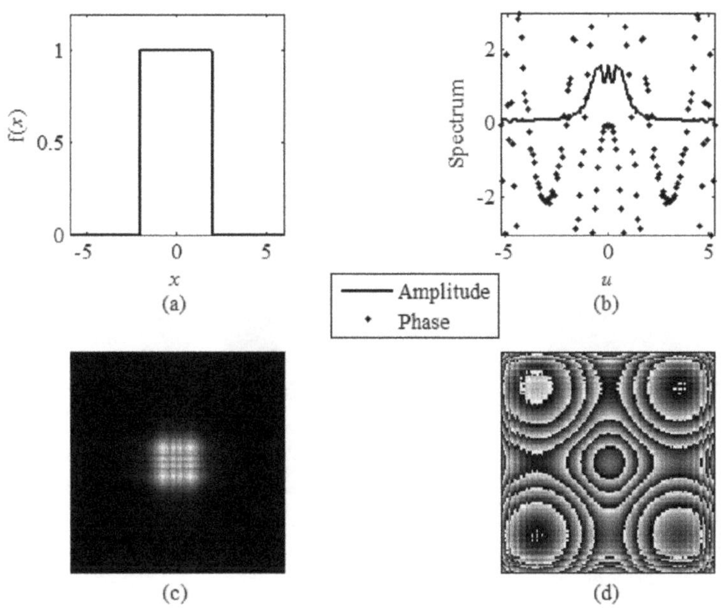

Figure 6.5. The result of discrete FrFT by using the convolution algorithm: (a) rectangle function, (b) 1D FrFT, (c) amplitude, (d) phase.

processing. With the help of random phase modulation, discrete random FrFT can be calculated by using the convolution scheme.

The gyrator transform also has a convolution format [10]:

$$\mathcal{G}^{\alpha}\{f(x, y)\}(u, v) = p_1(u, v)\left\{\left[q(x, y)p_1(x, y)\right] \otimes p_2(x, y)\right\}, \tag{6.31}$$

where the functions p_1 and p_2 are equal to

$$\begin{cases} p_1(x, y) &= \exp\left(-2i\pi xy \tan\dfrac{\alpha}{2}\right), \\ p_2(x, y) &= \dfrac{\exp(2i\pi xy \csc\alpha)}{|\sin\alpha|}, \end{cases} \tag{6.32}$$

and the Fourier transform of the function p_2 is given as follows

$$\begin{aligned} P_2(u, v) &= \mathcal{F}\{p_2(x, y)\}(u, v) \\ &= \exp(-2i\pi uv \sin\alpha) \end{aligned}. \tag{6.33}$$

Thereby the gyrator transform can be rewritten as

$$\mathcal{G}^{\alpha}\{f(x, y)\}(u, v) = p_1(u, v)\mathcal{F}^{-1}\left\{\mathcal{F}\left[f(x, y)p_1(x, y)\right] \otimes P_2(u, v)\right\}. \tag{6.34}$$

The gyrator transform of the input 2D pattern in figure 6.5 is displayed in figure 6.6. Here the transform angle is taken at 0.9. The amplitude distribution is closed to a shrunken square. The phase pattern is symmetrical for the axes ($u = v$ and $u = -v$), which is from the cross term xy and uv in the kernel function of gyrator transform.

6.3 Discrete random fractional transform

From the view of point of the matrix S in equation (6.17), the eigenvectors of FrFT can be explored completely. Here the matrix S is simple and symmetrical in the real number domain. When the matrix S is replaced with a random symmetrical matrix, some new transforms [11] can be defined.

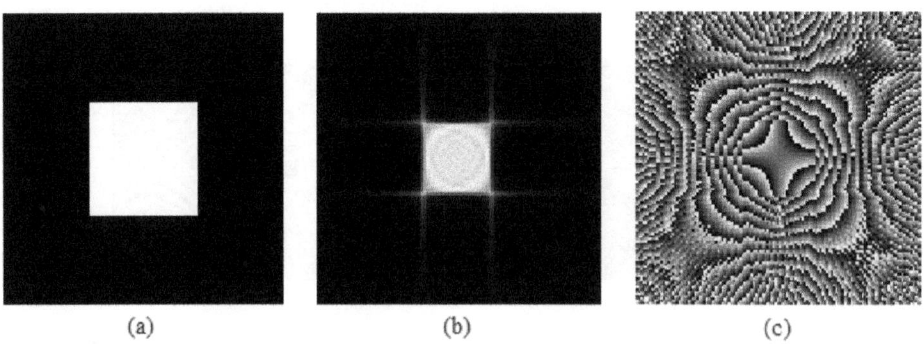

Figure 6.6. The discrete gyrator transforms of the 2D function: (a) input pattern, (b) amplitude, (c) phase.

A real symmetrical matrix Q is generated by using a random matrix P as

$$Q = [P + P^t]/2. \tag{6.35}$$

According to matrix theory, the matrix Q has orthogonal eigenvectors in the real number domain.

The eigenvalue matrix D^α is defined as

$$D^\alpha = \mathrm{diag}\left(1, \exp\left(-\mathrm{i}\frac{2\pi\alpha}{M}\right), \exp\left(-\mathrm{i}\frac{4\pi\alpha}{M}\right), \ldots, \exp\left(-\mathrm{i}\frac{2(N-1)\pi\alpha}{M}\right)\right), \tag{6.36}$$

where α and M are the order and period of this transform, respectively. The transform matrix R^α is determined as

$$R^\alpha = VD^\alpha V^t, \tag{6.37}$$

where the eigenvector matrix V is composed of the eigenvectors of the matrix Q. The matrix R^α can be used for converting a signal or image into its random transform domain.

In particular, the output of the discrete random fractional transform is in the real domain, when the order α is equal to $M/2$. The reason is that all elements of the matrix D^α are of the real matrix. Due to the random characteristic, this transform can be applied for data encryption directly. An example of a gray-level image is shown in figure 6.7. Here the order α and M are fixed at 0.5 and 1, respectively. A random pattern is obtained and is matched with the aim of image encryption.

After the definition of discrete random fractional transform, the discrete fractional angular transform [12], which is generated by using a recursion process, is introduced. Firstly, a coordinate rotation matrix is given as

$$V_2^\beta = \begin{bmatrix} \cos\beta & \sin\beta \\ -\sin\beta & \cos\beta \end{bmatrix}. \tag{6.38}$$

(a) (b)

Figure 6.7. The result of discrete random fractional transform: (a) input image, (b) transformed image.

A 3 × 3 matrix is defined as

$$V_3^\beta = \begin{bmatrix} \cos\beta & \sin\beta & 0 \\ 0 & 0 & 1 \\ -\sin\beta & \cos\beta & 0 \end{bmatrix}, \tag{6.39}$$

which is a special kind of orthogonal matrix. Here the matrices V_2^β and V_3^β are regarded as the initial values of recursion calculation.

If V_N is an orthogonal matrix, then the matrices

$$V_{2N} = \frac{1}{\sqrt{2}}\begin{bmatrix} V_N & V_N \\ -V_N^z & V_N^z \end{bmatrix} \tag{6.40}$$

and

$$V_{2N+1} = \frac{1}{\sqrt{2}}\begin{bmatrix} V_N & V_N & V_0^t \\ V_0 & V_0 & \sqrt{2} \\ -V_N^z & V_N^z & V_0^t \end{bmatrix} \tag{6.41}$$

are orthogonal matrices. The matrix V_N^z is obtained by flipping the matrix V_N in the up–down direction. Here V_0 is a zero vector having $1 \times N$ elements.

By using equations (6.40) and (6.41), we can calculate any orthonormal matrices $V_N^\beta (N > 3)$ using the recurrence method. For instance, the matrices V_{42}^β and V_{53}^β can be determined as follows

$$V_2^\beta \xrightarrow{\text{Eq.(6.41)}} V_5^\beta \xrightarrow{\text{Eq.(6.40)}} V_{10}^\beta \xrightarrow{\text{Eq.(6.41)}} V_{21}^\beta \xrightarrow{\text{Eq.(6.40)}} V_{42}^\beta, \tag{6.42}$$

$$V_3^\beta \xrightarrow{\text{Eq.(6.40)}} V_6^\beta \xrightarrow{\text{Eq.(6.41)}} V_{13}^\beta \xrightarrow{\text{Eq.(6.40)}} V_{26}^\beta \xrightarrow{\text{Eq.(6.41)}} V_{53}^\beta. \tag{6.43}$$

For all cases ($N > 3$), the orthogonal matrix V_N^β can be constructed by a recursion like the description in equations (6.42) and (6.43).

Similar to the discrete random fractional transform, the eigenvalues of the discrete fractional angular transform are generated as

$$\lambda_N^\alpha = [1, \exp(-\mathrm{i}2\pi\alpha), \exp(-\mathrm{i}4\pi\alpha), ..., \exp(-2(N-1)\mathrm{i}\pi\alpha)], \tag{6.44}$$

and the eigenvalue matrix is given as $D_N^\alpha = \mathrm{diag}(\lambda_N^\alpha)$. Here the transform matrix of the discrete fractional angular transform is calculated as

$$A_N^{\beta,\alpha} = V_N^\beta D_N^\alpha (V_N^\beta)^t. \tag{6.45}$$

This transform is also a kind of discrete fractional transform for making a frequency domain.

An example of a discrete fractional angular transform is shown in figure 6.8. The converted images are closed to blocked uniform patterns, in which the energy is located in low frequency range. This characteristic is different from some orthogonal transforms, such as the Fourier transform and discrete cosine transform.

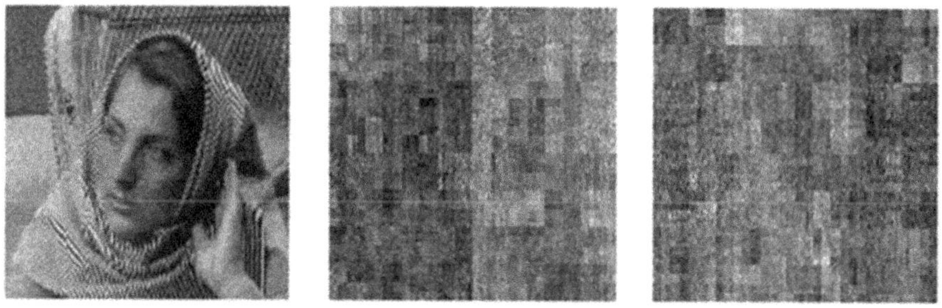

Figure 6.8. The results of discrete fractional angular transform: (a) input image, (b) $\alpha = 0.8$ and $\beta = 0.5\pi$, (c) $\alpha = 0.9$ and $\beta = 0.5\pi$.

Being similar to the discrete Fourier transform, the discrete cosine transform has four different kernel matrices as

$$C_{N+1}^{\mathrm{I}}(m, n) = \sqrt{\frac{2}{N}} k_m k_n \cos\left(\frac{mn\pi}{N}\right), \quad m, n = 0, 1, 2, \ldots, N \tag{6.46}$$

$$C_{N}^{\mathrm{II}}(m, n) = \sqrt{\frac{2}{N}} k_m \cos\left[\frac{m}{N}\left(n + \frac{1}{2}\right)\pi\right], \quad m, n = 0, 1, 2, \ldots, N - 1 \tag{6.47}$$

$$C_{N}^{\mathrm{III}}(m, n) = \sqrt{\frac{2}{N}} k_n \cos\left[\left(m + \frac{1}{2}\right)\frac{n\pi}{N}\right], \quad m, n = 0, 1, 2, \ldots, N - 1 \tag{6.48}$$

$$C_{N}^{\mathrm{IV}}(m, n) = \sqrt{\frac{2}{N}} \cos\left[\left(m + \frac{1}{2}\right)\left(n + \frac{1}{2}\right)\frac{\pi}{N}\right], \quad m, n = 0, 1, 2, \ldots, N - 1. \tag{6.49}$$

For the discrete sine transform, its kernel matrix also has four kinds of definitions as

$$S_{N-1}^{\mathrm{I}}(m, n) = \sqrt{\frac{2}{N}} \sin\left(\frac{mn\pi}{N}\right), \quad m, n = 0, 1, 2, \ldots, N - 1 \tag{6.50}$$

$$S_{N}^{\mathrm{II}}(m, n) = \sqrt{\frac{2}{N}} k_m \sin\left[\left(n - \frac{1}{2}\right)\frac{m\pi}{N}\right], \quad m, n = 1, 2, \ldots, N \tag{6.51}$$

$$S_{N}^{\mathrm{III}}(m, n) = \sqrt{\frac{2}{N}} k_n \sin\left[\left(m - \frac{1}{2}\right)\frac{n\pi}{N}\right], \quad m, n = 1, 2, \ldots, N \tag{6.52}$$

$$S_{N}^{\mathrm{IV}}(m, n) = \sqrt{\frac{2}{N}} \sin\left[\left(m - \frac{1}{2}\right)\left(n - \frac{1}{2}\right)\frac{\pi}{N}\right], \quad m, n = 1, 2, \ldots, N \tag{6.53}$$

where $k_0 = k_N = 1/\sqrt{2}$, and other k_m and k_n are equal to 1. The fractional sine and cosine transforms [13] have been defined by using equations (6.46) and (6.50). When a vector $v = [v_0, v_1, ..., v_{N-2}, v_{N-1}, v_{N-2}, ..., v_1]^t$ is the even eigenvector of the kernel matrix of FrFT of a signal having $(2N - 2)$ elements, namely $\mathcal{F}^{\alpha}_{2N-2}v = \lambda v(\lambda = 1, -1)$, one has

$$v_c = [v_0, \sqrt{2}\, v_1, ..., \sqrt{2}\, v_{N-2}, v_{N-1}]^t, \tag{6.54}$$

which is the eigenvector of the discrete cosine transform of a signal including the N elements. Here the superscript t is used to obtain the transpose of the vector. Similarly, if the vector $v = [v_0, v_1, ..., v_N, 0, -v_N, -v_{N-1}, ..., v_1]^t$ is an odd eigenvector of the kernel matrix of FrFT, namely $\mathcal{F}^{\alpha}_{2N+2}v = \lambda v$ $(\lambda = \mathrm{i}, -\mathrm{i})$, the vector

$$v_s = \sqrt{2}[v_1, v_2, ..., v_N]^t \tag{6.55}$$

is the eigenvector of the discrete sine transform. Therefore, the eigenvector matrices of discrete sine and cosine V_s and V_c can be defined as

$$V_s = [v_{s,1}, v_{s,2}, ..., v_{s,N}], \qquad V_c = [v_{c,1}, v_{c,2}, ..., v_{c,N}]. \tag{6.56}$$

The eigenvalues of the two transforms can be expressed as

$$\lambda_{s,n,\alpha} = \lambda_{c,n,\alpha} = \exp(-in\pi\alpha). \tag{6.57}$$

Thereby the kernel matrices of the transforms are written as

$$C^{\alpha}_N = V_c D^{\alpha} V^t_c,\ S^{\alpha}_N = V_s D^{\alpha} V^t_s \tag{6.58}$$

where α is the fractional order of the transforms. And the diagonal matrix D^{α} is

$$D^{\alpha} = \mathrm{diag}\{\exp(-\mathrm{i}\pi\alpha[0, 1, 2, ..., N - 1]/2)\}. \tag{6.59}$$

The discrete fractional sine and cosine transform have the properties of unitary, additivity and period ($T = 4$).

The eigenvector matrices of the discrete fractional random sine and cosine transforms are computed as follows

$$V_s = [s_1, s_2, ..., s_N],\ V_c = [c_1, c_2, ..., c_N]. \tag{6.60}$$

There the vectors s_n and c_n ($n = 1, ..., N$) are generated by using the same method with the discrete fractional random transfom [11]. And the diagonal matrices composed of their eigenvalues are determined by

$$\begin{aligned} D^{\alpha}_s &= \mathrm{diag}\left\{\exp\left(-\frac{2\mathrm{i}\pi\alpha}{M}[1, 3, 5, ..., 2N - 1]\right)\right\}, \\ D^{\alpha}_c &= \mathrm{diag}\left\{\exp\left(-\frac{2\mathrm{i}\pi\alpha}{M}[0, 2, 4, ..., 2N - 2]\right)\right\}, \end{aligned} \tag{6.61}$$

where M is the period of the discrete fractional random transform. The parameter α is the fractional order of the two transforms. The elements of the diagonal matrices

D_s^α and D_c^α are the eigenvalues of the transforms, in which the orders are even numbers and odd numbers. The discrete fractional random sine and cosine transforms of the one-dimensional signal x and two-dimensional image y are listed using the two transform matrices R_s^α and R_c^α as

$$\begin{cases} X_s^\alpha = R_s^\alpha x, \ X_c^\alpha = R_c^\alpha x, \\ Y_s^\alpha = R_s^\alpha y[R_s^\alpha]^t, \ Y_c^\alpha = R_c^\alpha y[R_c^\alpha]^t. \end{cases} \tag{6.62}$$

For the discrete fractional random transform, the random sine and cosine transforms are expressed as

$$R_s^\alpha = \exp(-2i\pi\alpha/M)R^{2\alpha}, \ R_c^\alpha = R^{2\alpha}, \tag{6.63}$$

where $R^{2\alpha}$ denotes the operator of the discrete fractional random transform. The diagonal eigenvalue matrix is calculated as

$$D_{2N}^\alpha = \mathrm{diag}\left\{\exp\left(-\frac{2i\pi\alpha}{M}[0, 1, 2, \ldots, 2N-1]\right)\right\}, \tag{6.64}$$

or

$$D_{2N+1}^\alpha = \mathrm{diag}\left\{\exp\left(-\frac{2i\pi\alpha}{M}[0, 1, 2, \ldots, 2N]\right)\right\}. \tag{6.65}$$

The eigenvector matrix of the transform $R^{2\alpha}$ is written as

$$V_{2N} = \frac{1}{\sqrt{2}}\begin{bmatrix} c_1, & s_1, & c_2, & s_2, & \ldots, & c_N, & s_N \\ c_1^z, & s_1^z, & c_2^z, & s_2^z, & \ldots, & c_N^z, & s_N^z \end{bmatrix} \tag{6.66}$$

or

$$V_{2N+1} = \frac{1}{\sqrt{2}}\begin{bmatrix} c_1, & s_1, & c_2, & s_2, & \ldots, & c_N, & s_N, & v_0 \\ 0, & 0, & 0, & 0, & \ldots, & 0, & 0, & \sqrt{2} \\ c_1^z, & s_1^z, & c_2^z, & s_2^z, & \ldots, & c_N^z, & s_N^z, & v_0 \end{bmatrix}, \tag{6.67}$$

where the vectors c_n^z and s_n^z ($n = 1, 2, \ldots, N$) are obtained by flipping the vectors c_n and s_n along the up-down direction, respectively. All elements of the vector v_0 are equal to 0.

6.4 Single gray-level image encryption

If an image or two-dimensional data contains secret information, it should be encrypted. There is a reversible random process used to hide the data matrix. The double random phase encoding technique [15] (DRPE) is a classical method of optical data encryption. Figure 6.9 shows a flowchart of image encryption. The mathematical process of DRPE is given as follows

$$I_{\text{out}}(x_1, y_1) = \mathcal{F}\{\mathcal{F}[I_{\text{in}}(x, y)P_1(x, y)]P_2(u, v)\}, \tag{6.68}$$

Figure 6.9. The process of image encryption.

where I_{in} and I_{out} are the input image and output images of this system. The functions P_1 and P_2 denote the two phase masks, which are placed at the input plane and output plane, respectively. In the random process, some optical transforms and random data are employed for obtaining a random pattern, which is regarded as the encrypted image. In DRPE, a two-step Fourier transform is adopted for changing the distribution of intensity and phase in the process of light diffraction. Here, only one image is considered for scrambling, which is the main technique in the early stages of optical information security.

Random phase encoding has been improved for enhancing its security. The DRPE in the fractional Fourier transform domains and Fresnel domain [16, 17] includes an additional key by using the fractional order and diffraction distance (or wavelength), the latter being regarded as an unknown number, respectively. Moreover, the iterative optical structure [18], in which more random phase masks are employed, was designed for enhancing the security of a gray-level image. A greater load of key data in the iterative process is required in the storage and transform of secret information than the rand phase in the DRPE technique. By using chaotic mapping [19], a series of random data was used for encoding the secret image. The iterative encoding process makes the sensitivity enhancement of fractional order. In the following context, several image encryption schemes are presented.

An image encryption method composed of commutation and anti-commutation rules [20] is displayed in figure 6.10. Fourier transform or FrFT is implemented in the two separated optical paths. With the help of the operator and matrix, the transform in figure 6.10 can be expressed as follows

$$I_y = \mathcal{F}^a I_R \tag{6.69}$$

and

$$I_x = I_R \mathcal{F}^a \tag{6.70}$$

where I_R stands for the product of image I and the random phase R. The order a is from the two lenses placed in a vertical and horizontal direction. The functions P_+ and P_- can be obtained with and without HWP in figure 6.10. In the output plane, the transformed patterns are recorded by using a reference light and are given as

$$P_\pm = \mathcal{F}^a I_R \pm I_R \mathcal{F}^a = [\mathcal{F}^a, I_R]_\pm, \tag{6.71}$$

where $[...]_+$ and $[...]_-$ denote the commutator and anto-commutator, respectively. Here the function P_- is the encrypted image and P_- is the key data. If the complex data I_R have $M \times N$ elements, equation (6.71) takes this form

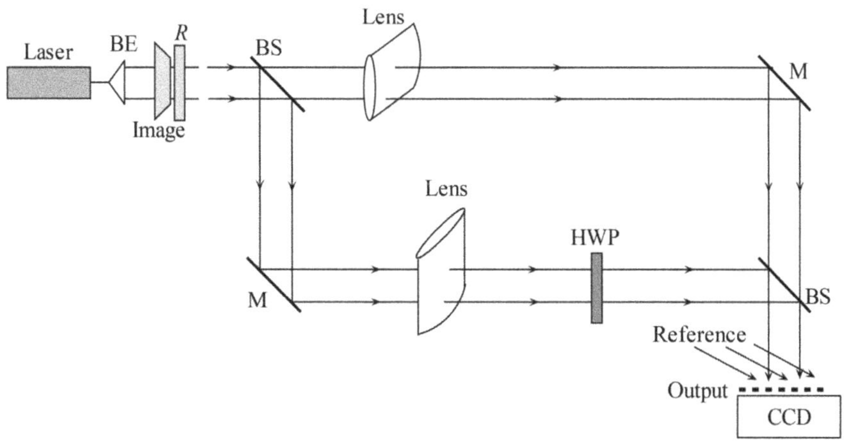

Figure 6.10. The experimental system of image encryption using commutation and anti-commutation operations. Two arms achieving 1D Fourier transform (or FrFT) can implement image encryptions in the y- and x-directions, respectively. The final encrypted image P_- and the key data P_+ can be obtained by superposing them in the computer. BE: beam expander, R: random phase mask, BS: beam splitter, M: mirror, HWP: half wave plate.

$$P_- = \mathcal{F}^{\alpha_y} I_R - I_R \mathcal{F}^{\alpha_x} \tag{6.72a}$$

and

$$P_+ = \mathcal{F}^{\alpha_y} I_R + I_R \mathcal{F}^{\alpha_x}, \tag{6.72b}$$

where α_x and α_y are the orders for the x-direction and y-direction.

In the image decryption process, the functions P_+ and P_- are entered into the commutation and anti-commutation as follows

$$\mathcal{F}^{-\alpha_y} P_+ + P_+ \mathcal{F}^{-\alpha_x} = 2I_R + \mathcal{F}^{-\alpha_y} I_R \mathcal{F}^{\alpha_x} + \mathcal{F}^{\alpha_y} I_R \mathcal{F}^{-\alpha_x} \tag{6.73a}$$

and

$$\mathcal{F}^{-\alpha_y} P_- - P_- \mathcal{F}^{-\alpha_x} = 2I_R - \mathcal{F}^{-\alpha_y} I_R \mathcal{F}^{\alpha_x} - \mathcal{F}^{\alpha_y} I_R \mathcal{F}^{-\alpha_x}. \tag{6.73b}$$

By adding equation (6.73), one can obtain

$$(\mathcal{F}^{-\alpha_y} P_+ + P_- \mathcal{F}^{-\alpha_x}) + (\mathcal{F}^{-\alpha_y} P_- - P_- \mathcal{F}^{-\alpha_x}) = 4I_R. \tag{6.74}$$

Equation (6.74) implies that the image I can be retrieved by using the system in figure 6.10 with the inverse transform.

By using a non-uniform laser beam, an image encryption method [21] is shown in figure 6.11. As a light source, a coherent optical beam $E(x, y)$ at the input plane is expressed as

$$E(x, y) = B(x, y) \exp\left[i\phi_B(x, y)\right]. \tag{6.75}$$

The beam $E(x, y)$ can be selected as a Hermite–Gaussian beam, Laguerre–Gaussian beam or others. At the plane P_1, the amplitude $A(u, v)$ and $\phi_A(u, v)$ is computed by the inverse gyrator transform $\mathcal{G}^{-\alpha}$ with the field at the output plane as

$$A(u,\, v)\exp\left[\mathrm{i}\phi_A(u,\, v)\right] = \mathcal{G}^{-\alpha}\left[c_0\sqrt{h(x',\, y')}\right],$$
$$c_0^2 = \frac{\iint A^2(u,\, v)\mathrm{d}u\mathrm{d}v}{\iint h(x',\, y')\mathrm{d}x'\mathrm{d}y'}, \tag{6.76}$$

where $h(x',\, y')$ is the pixel value function of the secret image and serves as the intensity distribution. To construct a transform relation, the two phases, φ_A and φ_B, are introduced into the equation

$$A(u,\, v)\exp[\mathrm{i}\varphi_A(u,\, v)] = \mathcal{G}^{\alpha}\{B(x,\, y)\exp[\mathrm{i}\varphi_B(x,\, y)]\}. \tag{6.77}$$

These two phase functions can be determined by applying a phase retrieval method. Here, two compensation phases are calculated as

$$\Delta\varphi_A(u,\, v) = \phi_A(u,\, v) - \varphi_A(u,\, v),$$
$$\Delta\varphi_B(x,\, y) = \phi_B(x,\, y) - \varphi_B(x,\, y). \tag{6.78}$$

The phases $\Delta\varphi_A$ and $\Delta\varphi_B$ will be uploaded by a mask or spatial light modulator into the system for balancing the phase distribution at the left and right side of the planes of input and P_1. Here, $\Delta\varphi_A$ and $\Delta\varphi_B$ are the encrypted image and key, respectively.

An image encryption scheme based on chaotic mapping and 1D FrFT [22] is given in figure 6.12. Here the thin cylinder lens is used for implementing 1D FrFT by

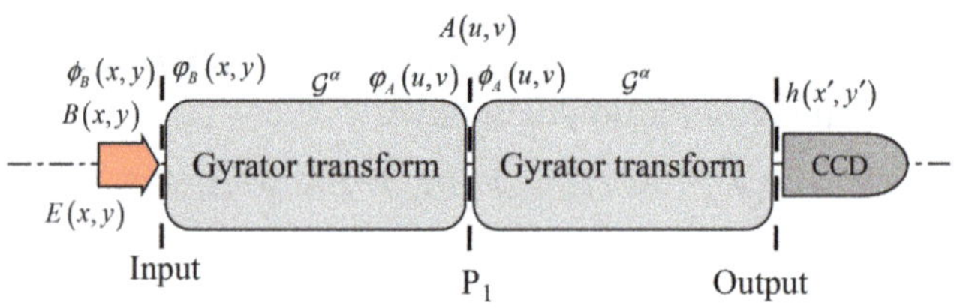

Figure 6.11. Algorithm illustration using a non-uniform beam.

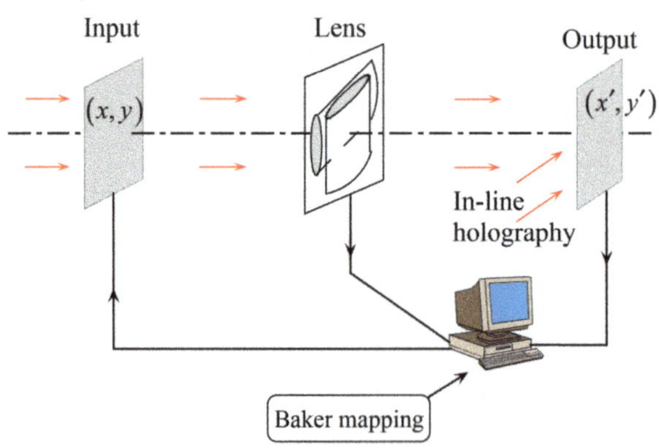

Figure 6.12. An encryption system using baker mapping and 1D FrFT.

a mechanical part rotating around the optical axis. The 1D transform can be randomly performed in the x- or y-direction. The transform is expressed as

$$\begin{aligned} G_{x,0}(x', y) &= \mathcal{F}^{a,0}[g(x, y)](x', y) \\ &= \int g(x, y) \exp\left[-\mathrm{i}2\pi \frac{(x^2 + x'^2)\cos\phi_a - 2xx'}{\sin\phi_a} \right] \mathrm{d}x \end{aligned} \tag{6.79}$$

$$\begin{aligned} G_{0,y}(x, y') &= \mathcal{F}^{a,0}[g(x, y)](x, y') \\ &= \int g(x, y) \exp\left[-\mathrm{i}2\pi \frac{(y^2 + y'^2)\cos\phi_a - 2yy'}{\sin\phi_a} \right] \mathrm{d}y \end{aligned} \tag{6.80}$$

where $\phi_a = a\pi/2$ and a is the order of transform. Here $\tan\phi_a = d/f$, d and f are the object distance and focal length, respectively. The chaotic mapping, such as baker mapping or the Arnold transform, is adopted for changing the position of the image pixel. To enhance security, chaotic mapping and 1D transform are utilized during the data scrambling process.

In random phase encoding, a pair of masks being conjugate to each other is needed for the encryption process and decryption. Sometimes, it is not convenient in data storage and transmission. The quality of the recovered image is reduced if the conjugate relation is not satisfied exactly. Here, we consider introducing an odd function, being regarded as the phase distribution of the mask, as

$$R(x, y) = \begin{cases} R(x, y), & \text{if } x > 0, \\ 0, & \text{if } x = 0, \\ -R(x, y), & \text{if } x < 0, \end{cases} \tag{6.81}$$

where the piecewise definition has a symmetrical axis $x = 0$. The relation $R(-x, y) = -R(x, y)$ will be used for obtaining a conjugate phase mask in the image retrieval. It will be a flipping operation about the axis. An advantage is that the same phase mask can be employed in both the encryption and decryption. A drawback is that the function $R(x, y)$ has half of the random data due to the symmetry.

By using the random phase mask (RPM), a pair of encryption and decryption schemes are illustrated in figure 6.13. The original image I_0 is imported by a spatial light modulator (SLM). Here the spherical mirror (SM) is utilized for implementing FrFT and flipping the optical path. The focal length of SM is equal to $r/2$. In this optical system, the FrFT is achieved three times. In figure 6.13(b), a convex SM and concave lens are selected for generating a negative focal length.

6.5 Double image encryption

As a special case of multi-image encryption, double image encryption can hide two images with one key matrix data simultaneously. In this section, several double image encryption approaches will be introduced.

In figure 6.14, two images with amplitude distribution are related to each other with FrFT [24]. The phase functions ϕ_1 and ϕ_2 are utilized to balance the transform. The value of the two phases can be obtained by using iterative phase retrieval. The orders α and β are used to generate the output of this encryption algorithm. The amplitude A_0 and phase ϕ_0 are the encrypted image and main key. Here, the transform angles α and β are a pair of additional keys for recovering the two secret images. In this method, random data is not employed for the encrypted pattern. The patterns of A_0 and ϕ_0 are random, which is satisfactory for information security.

Several double encryption schemes [25–27] are designed with the help of data encoding in the complex domains of the light field. The original secret images are considered as the real part and imaginary part (or amplitude and phase) in order to be hidden through an optical random system, as shown in figure 6.15. The two images will be coupled in the data of the encrypted pattern. Here the cross-talk

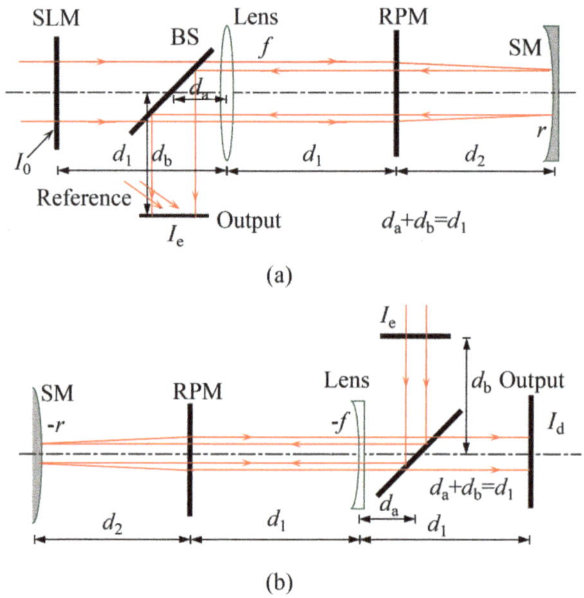

Figure 6.13. An optical system of single random phase encoding.

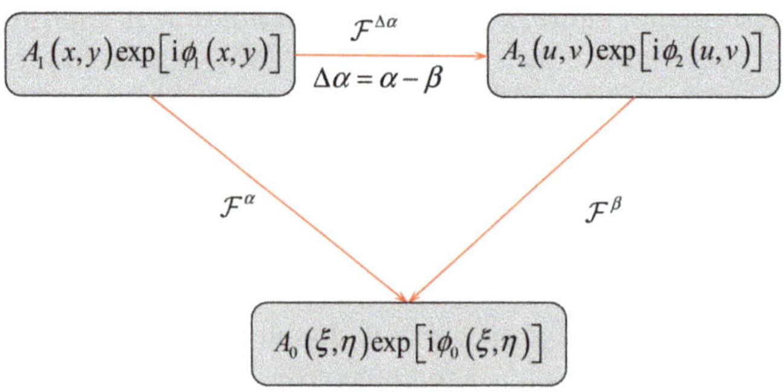

Figure 6.14. A double image encryption method using FrFT.

effect, which is the reason why the retrieved images are degraded, is included in this type of double encryption technique.

6.6 Color image encryption

A color image can be decomposed into three concolorous images (red, green and blue). Some encryption methods are constructed by using three channels to deal with the component images. It is a natural idea to explore a new security algorithm. Moreover, color images are more popular in the various applications of digital pictures from information collection systems, such as cameras. Therefore, protecting the color pattern is a practical information security issue.

Differing from parallel processing in three channels, a color blend operation [28] is defined in figure 6.16. Here, the discrete cosine transform (DCT) is used to replace the fractional transform for generating an output in the real number domain. Other

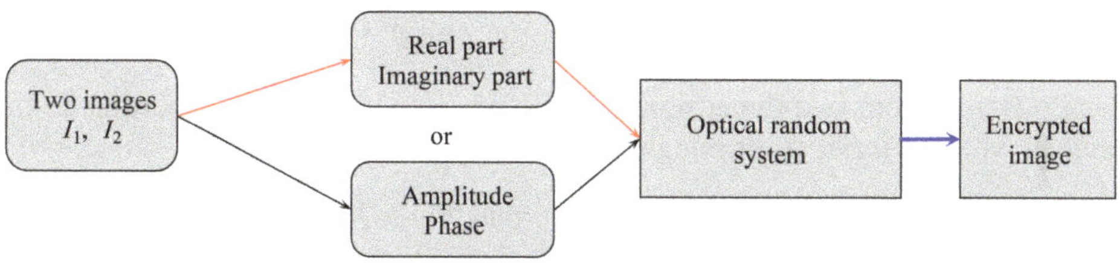

Figure 6.15. The route of double image encryption.

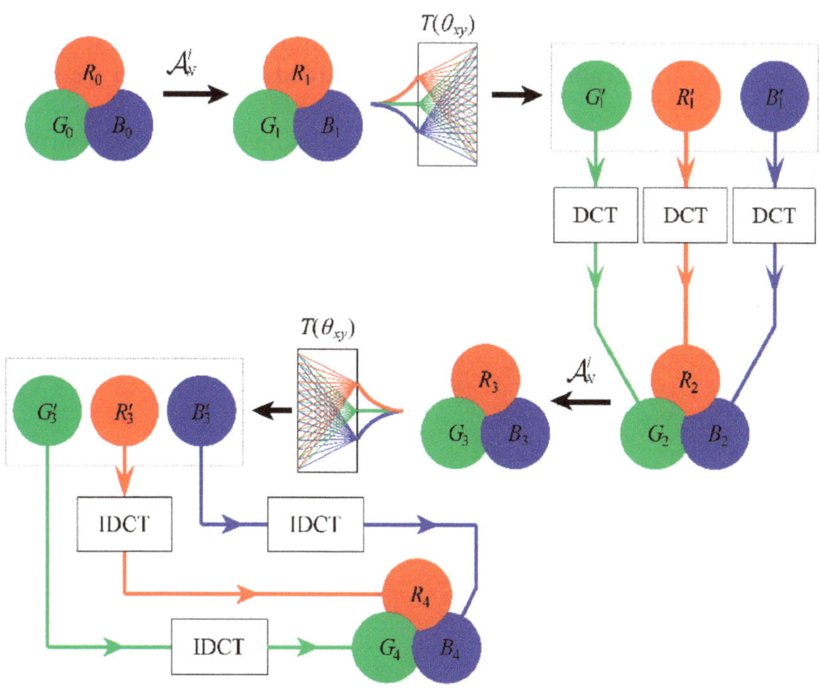

Figure 6.16. A color image encryption flowchart based on color blend operations.

transforms in the real number domain, such as the Hartley transform, can also be employed in this scheme. The Arnold transform is used for scrambling the image pixel distribution. Before encrypting the color data, the color blend is introduced to couple with each other between the concolorous images. A matrix having 3 × 3 elements is expressed as

$$\begin{aligned} T(\theta) &= \begin{bmatrix} t_{11} & t_{12} & t_{13} \\ t_{21} & t_{22} & t_{23} \\ t_{31} & t_{32} & t_{33} \end{bmatrix} \\ &= \begin{bmatrix} \cos\theta & \sin\theta & 0 \\ -\sin\theta & \cos\theta & 0 \\ 0 & 0 & 1 \end{bmatrix} \begin{bmatrix} 1 & 0 & 0 \\ 0 & \sin\theta & -\cos\theta \\ 0 & \cos\theta & \sin\theta \end{bmatrix} \\ &= \begin{bmatrix} \cos\theta & \sin^2\theta & -\sin\theta\cos\theta \\ -\sin\theta & \sin\theta\cos\theta & -\cos^2\theta \\ 0 & \cos\theta & \sin\theta \end{bmatrix} \end{aligned} \tag{6.82}$$

where θ is a random two-dimensional phase function, in which the coordinates are matched with the input color image. The inverse matrix of $T(\theta)$ is

$$T^{-1}(\theta) = \begin{bmatrix} \cos\theta & -\sin\theta & 0 \\ \sin^2\theta & \sin\theta\cos\theta & \cos\theta \\ -\sin\theta\cos\theta & -\cos^2\theta & \sin\theta \end{bmatrix}. \tag{6.83}$$

The inverse matrix will be used for decrypting the color secret image. The final encrypted pattern will be received by two units of transforms. Here the phase θ and the parameters of the Arnold transform are regarded as the key to the encryption algorithm.

A color encryption method [29] can be designed using the spherical coordinates in figure 6.17. Here the red, green and blue components serve as the three elements of a vector, like u and v. The zenith angle $\Delta\varphi$ and azimuth angle $\Delta\theta$ are used to rotate from the vector v to the vector u. The color image can be encoded with the random data map of the angles, $\Delta\varphi$ and $\Delta\theta$. Here the optical transform selected is the Hartley transform due to its real number domain. The color vector rotation is

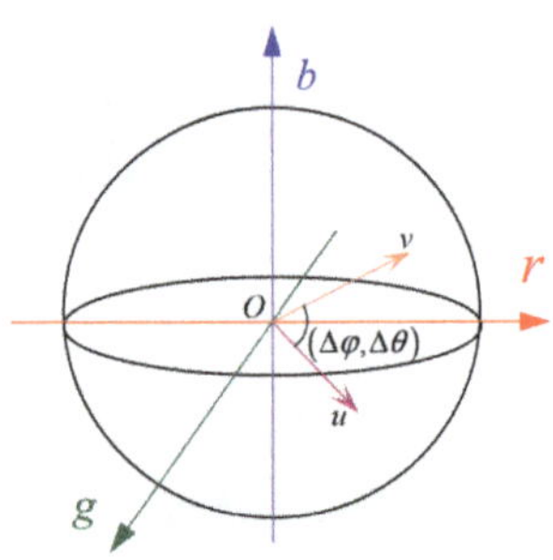

Figure 6.17. Rotation operation model on the color vector in the spherical coordinates which are composed of three color values (red, green and blue).

similar to the random phase encoding technique in the aspect of pattern scrambling. This rotation operation is performed digitally on a computer.

In figure 6.18, a color image encryption system [30] is given with three channels. The phase mask M1 is placed at the common channel to modulate the wavefront of the three beams. The secret image will be shown after diffraction along the distance d. Here, the phase masks q_r, q_g and q_b are used for phase modulation of the uniform illumination beam at the corresponding channel. The image composed of the phases (q_r, q_g and q_b) is regarded as the encrypted image. The phase M1, which can be shifted and rotated, is the main key of this encryption method. The data of M1 is directly obtained by using random numbers from a computer. The optical paths (L_r, L_g and L_b) and distance d serve as the additional keys to enhance the security of the encryption scheme. The phase distribution of the encrypted image can be recovered by using the iterative algorithm in figure 6.19. The phase mask M2 is not considered for figure 6.18. The symbol L → R denotes the diffraction from the left side to the right side in figure 6.19(a). The symbol R → L is the reverse process. After many iterations, the evaluated values of q_r, q_g and q_b can be received.

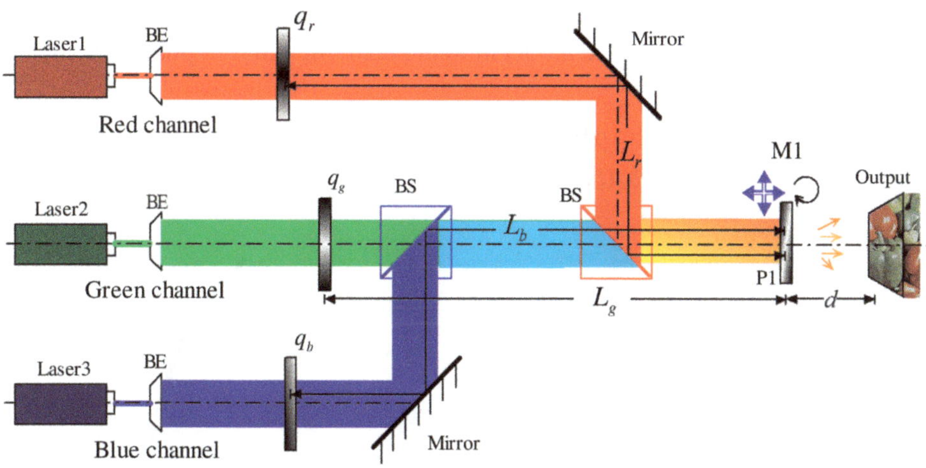

Figure 6.18. A color encryption system using cascaded phase encoding.

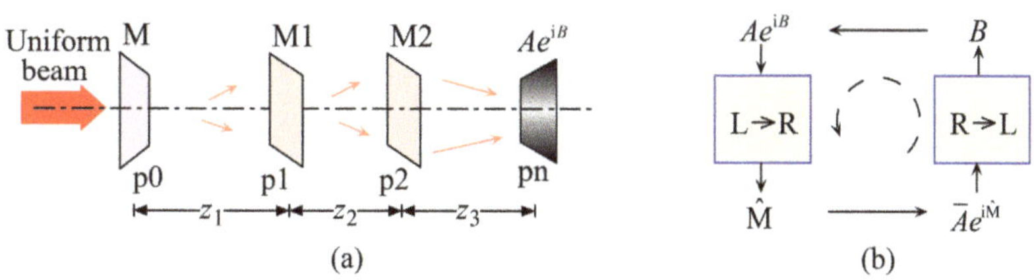

Figure 6.19. The reconstruction of the phase distribution: (a) optical system model, (b) iterative algorithm flowchart.

6.7 Multi-image encryption

An optical system of multiple image encryptions based on interference [31] is designed and displayed in figure 6.20. Here, the expanded beam from the laser is separated into two parts in order to light two random phase masks $R1$ and $R2$. After the plane P0, some receiving planes Pk ($k = 1, 2, \ldots, 5$) are placed and limited on a part of a circle (or spherical surface), in which the projection of these planes are just the chord line. The angle θ and distance l is also used to determine the position of every plane Pk.

In figure 6.20, the function of the light field G_k at the observing plane Pk is expressed as follows

$$G_k(u, v) = \sqrt{I_k(u, v)} \exp [\mathrm{i} \cdot \mathrm{rand}_k(u, v)], \tag{6.84}$$

where I_k is the original secret image, which is the intensity information detected directly. The function 'rand' is used to generate the phase values in the interval $[0, 2\pi]$. The phase data is not requested in the image decryption process. The complex function distribution g_0 at the plane P0 can be given by a superposition format as

$$g_0(x, y) = \sum_k g_k(x, y), \tag{6.85}$$

where the function g_k is the reverse diffraction result from the secret image located at the plane Pk. In the interference model in figure 6.20, the random phases $R1$ and $R2$ [31, 32] can be calculated as

$$\begin{aligned} g_0(x, y) = c_0\{&\exp [\mathrm{i} \cdot R1(\xi, \eta)] + \exp [\mathrm{i} \cdot R2(\xi, \eta)]\} \\ &\otimes h_d(\xi, \eta) \end{aligned}. \tag{6.86}$$

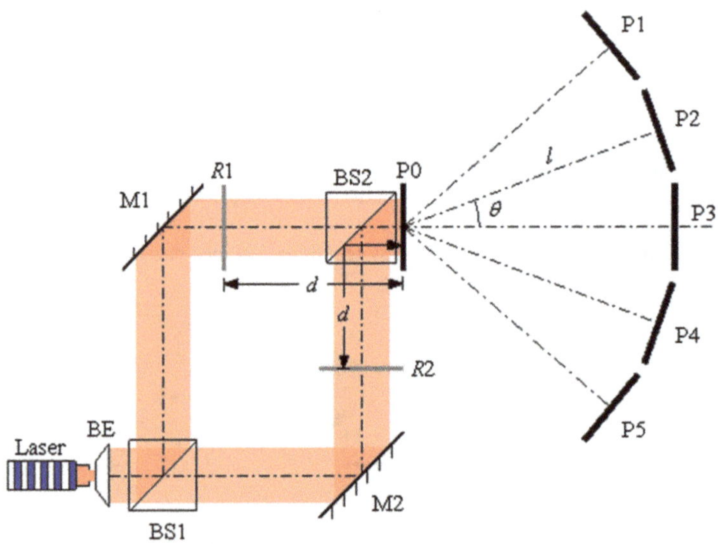

Figure 6.20. A multiple image encryption system using orientation multiplexing.

The parameter c_0 is used to balance the total energy of the input and output in the interference system. The function h_d is the point spread function and is similar to h_l in equation (6.92). The distance d represents the optical path from the mask $R1$ (or $R2$) to the plane P0.

The two planes P0 and P2 are utilized for the explanation of light propagation and are shown in figure 6.20. The distance between the two centers of the planes is l, which is also the radii for the cylinder (or the sphere) (see figure 6.20). The parameter θ denotes the included angle between the normal lines of the two planes for the x-axis and u-axis. Here, the y-axis and v-axis are parallel.

The relation of the input $g(x, y)$ and output $G(u, v)$ is expressed by a Fresnel integral as

$$G(u, v) = \iint g(x, y)\frac{\exp\left[\mathrm{i}2\pi l'/\lambda\right]}{\mathrm{i}l'\lambda} \times \exp\left[\mathrm{i}\pi\frac{(x\cos\theta - u)^2 + (y - v)^2}{l'\lambda}\right]\mathrm{d}x\mathrm{d}y, \tag{6.87}$$

where λ is the wavelength of the illumination light. Here $l' = l + x\sin\theta$. For the sake of simplicity, equation (6.87) can be re-written into a convolution format under paraxial approximation as follows

$$G(u, v) = q_1\{[g \cdot q_2] \otimes [h_l \cdot q_3]\}, \tag{6.88}$$

where the symbol '$\otimes$' denotes the convolution operator. The functions q_1, q_2, q_3 and h_l are listed as follows

$$q_1(u, v) = \exp\left(-\frac{\mathrm{i}4\pi}{l\lambda}u^2\sin^2\frac{\theta}{2}\right), \tag{6.89}$$

$$q_2(x, y) = \exp\left[\frac{\mathrm{i}4\pi x}{l\lambda}\left(l\sin\theta - x\sin^4\frac{\theta}{2}\right)\right], \tag{6.90}$$

$$q_3(x, y) = \exp\left[\frac{4\mathrm{i}\pi x^2}{l\lambda}\sin^2\frac{\theta}{2}\right] \tag{6.91}$$

and

$$h_l(x, y) = \frac{1}{\mathrm{i}l\lambda}\exp\left\{\frac{\mathrm{i}\pi}{l\lambda}[l^2 + x^2 + y^2]\right\}, \tag{6.92}$$

where h_l is the point spread function (PSF) of the Fresnel diffraction between two parallel planes. The functions q_1 and q_2 play an important role of the modulation term with the output and input of the optical system in figure 6.21, respectively. If the term $\exp(\mathrm{i}4\pi\sin\theta/\lambda)$ is neglected, equation (6.87) is a classical optical ABCD system. The function q_3 is a modulation factor for the function h_l of two parallel planes with the distance l. Thereby the fast Fourier transform algorithm (FFT) can

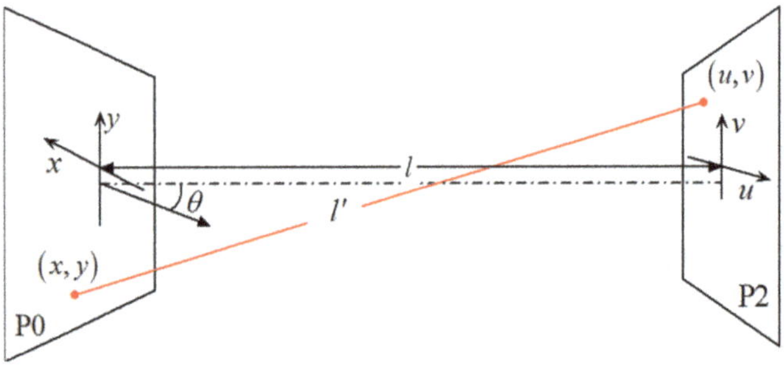

Figure 6.21. A diffraction model between two planes with an angle θ.

be easily employed for calculating and analyzing equation (6.87) in the discrete case. When $\theta = 0$, equation (6.87) is a standard Fresnel diffraction process. If the y-axis and v-axis are not parallel and have an angle ϕ, a set of expressions of the phase function $q_k(k = 1, 2, 3)$ can be deduced with the variables y and v similarly according to equations (6.89)–(6.91) by replacing the angle θ. With the help of the functions q_k depending θ and ϕ, the case in which the observed planes are located on a sphere can be analyzed. In particular, a Fourier transform relation of the product $h_l q_3$ is supplied as

$$\mathcal{F}\left[h_l \cdot q_3\right] = \frac{\exp\left(\frac{\mathrm{i}\pi l}{\lambda}\right)}{\sqrt{1 + 4\sin^2\frac{\theta}{2}}} \exp\left[-\mathrm{i}\pi l\lambda\left(\frac{u^2}{1 + 4\sin^2\frac{\theta}{2}} + v^2\right)\right], \tag{6.93}$$

where the symbol '$\mathcal{F}$' is the Fourier transform. Equation (6.93) will be regarded as an analytic result for the calculation of convolution in the discrete case.

6.8 Optical image hiding scheme and asymmetric encryption scheme

In this section a color image hiding scheme and an asymmetric image encryption scheme based on FrFT are introduced [33, 34]. A color image hiding algorithm by using the Gerchberg–Saxton algorithm in the fractional Fourier domain is described first. In this hiding encryption scheme, the secret image can be embedded either in a single component or the multi-components of the host color image. Moreover, the sequence and combination of embedding the secret information is variable, which can increase the security of the scheme. Firstly, the three components of the secret image are converted into random images by employing the 3D Arnold transform synchronously. Subsequently, the scrambled secret image is hidden into the amplitude of the fractional Fourier spectrum. The Gerchberg–Saxton phase retrieval algorithm in the fractional Fourier domain is utilized to calculate the phase information at the input plane of the FrFT system. To embed the secret image, a computer is required due to the enormous and repetitive calculation in this algorithm. One can extract the secret information by implementing an electro-optical hybrid setup. The different parameters in the 3D Arnold transform and the retrieved phase can be regarded as the extra keys and main key of this hiding scheme, respectively. The flowchart of the color image hiding encryption scheme is illustrated in figure 6.22.

The fractional Fourier transform can be achieved in an optical system. Three kinds of optical beams (red, green and blue) will be utilized for encoding the components of the color host and secret image. The host color image and the main key $\varphi(x, y)$ serve as the input data of the electro-optical hybrid setup as displayed in figure 6.23. At the output plane, an in-line holography technique is introduced to record the phase information. The other calculations are performed on a computer. Moreover, the SLM and CCD will accomplish the data communication between the computer and optical system.

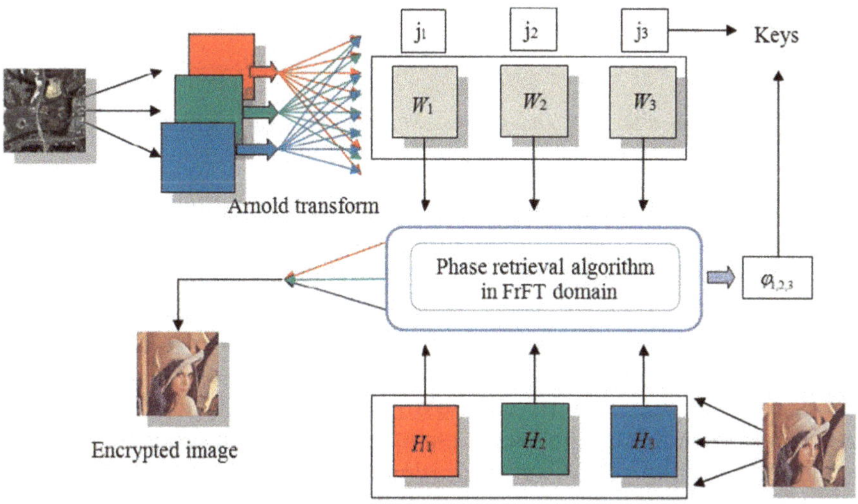

Figure 6.22. A flowchart of the color image hiding algorithm.

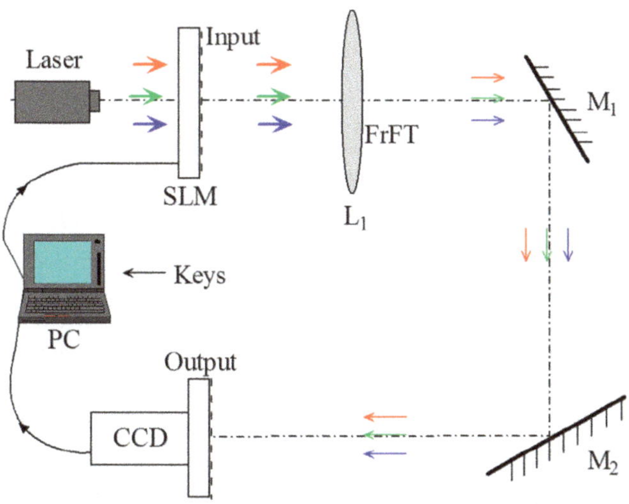

Figure 6.23. The electro-optical setup of the proposed encryption scheme.

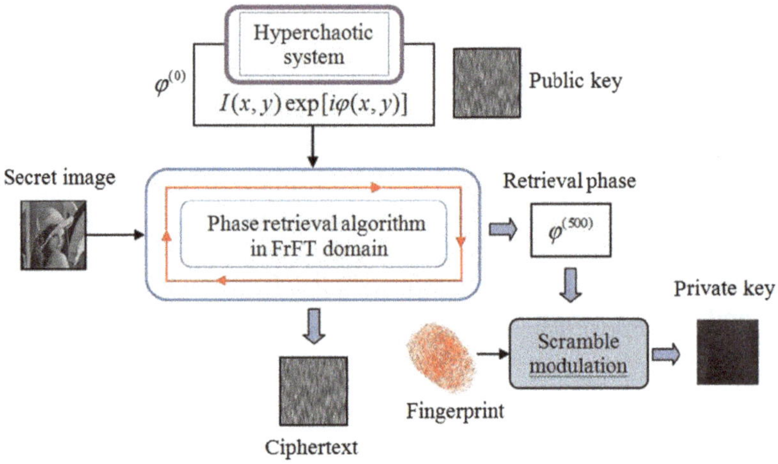

Figure 6.24. A flowchart of the asymmetric encryption algorithm.

Moreover, a gray image asymmetric cryptosystem using fingerprints based on an iterative fraction Fourier transform is introduced as follows. A hyperspectral phase generated by a 4D Lorenz system (shown as equation (6.94)) is considered as the public key in the proposed encryption system, and the private key is submerged by the retrieved phase and fingerprint. In the encryption process, the secret information is hidden in the hyperchaotic phase. Subsequently, the private key can be obtained by a reversible operation. To decrypt the original image, the ciphertext and private key are imported into the input plane of fractional Fourier system. This system is also applicable for information authentication because the fingerprint is used both in the encryption and decryption approach. A flowchart of the encryption scheme is illustrated in figure 6.24.

The 4D Lorenz system in this encryption scheme can be modeled as follows

$$\begin{cases} \dot{x} = a(y - x), \\ \dot{y} = cx - y - xz + u, \\ \dot{z} = xy - bz, \\ \dot{u} = -kx. \end{cases} \tag{6.94}$$

where a, b, c and k are constant parameters. When $a = 35$, $b = 8/3$, $c = 28$ and $k = 5$, the Lyapunov exponents of the hyperchaotic system are $\lambda_1 = 0.3997$, $\lambda_2 = 0.3113$, $\lambda_3 = 0$ and $\lambda_4 = -14.3776$.

The cryptosystem can be implemented with an electro-optical setup, as shown in figure 6.25. The fractional Fourier transform and its inverse transform can be achieved in an optical system. The corresponding function of amplitude and phase are imported into the spatial light modulator (SLM). The hyperchaotic system and the scramble operation can be performed on a computer and the SLM and CCD will accomplish the data communication between the computer and optical system.

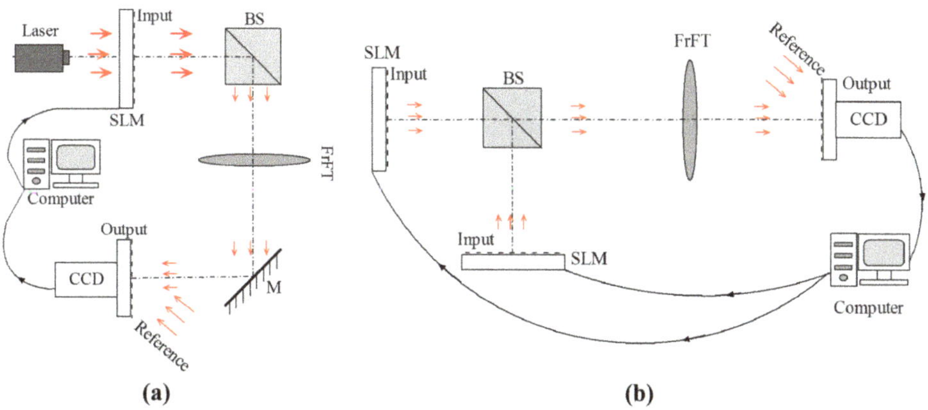

Figure 6.25. The electro-optical setup of the proposed cryptosystem: (a) the encryption setup and (b) the decryption setup.

6.9 Hyperspectral image encryption

A hyperspectral image is a special type of image format that contains rich information both in spatial and spectral domains. Researchers are able to extract plentiful spectral information from a hyperspectral image, which makes it possible to identify and discriminate targets compared with other remote data, such as the multispectral image and ordinary color image. In this section, a hyperspectral image encryption approach will be introduced as follows.

An optical spectrum encryption algorithm for a hyperspectral image is introduced, in which the spatial and spectrum information can be encrypted simultaneously [35]. Firstly, an original hyperspectral image is divided into a single band and baker mapping is utilized to changing the pixel sequence of each band. Subsequently, the scrambled images generated by the baker mapping operation are separated into real and imaginary parts of the complex function expressing light field. These complex functions are then put into the input plane of the optical transform system, which is composed of FrFT and gyrator transform. A random binary vector is designed and introduced as the parameters in the optical transformation to strengthen the security of the proposed encryption system. A flowchart of the spectrum encryption is illustrated in figure 6.26.

In this hyperspectral image encryption cryptosystem, the amplitudes and phases of the complex function mentioned above will be encoded into the optical transform by the modulation of the spatial light modulator (SLM). An intact encryption algorithm is achieved by an electro-optical hybrid setup, as illustrated in figure 6.27. The complicated calculation in baker mapping and new logical map are accomplished by a high-performance computer in the hybrid setup, while the FrFT and gyrator transform are achieved by the optical system. Hundreds of beams encode every single band of the hyperspectral image. The encrypted field light is recorded by an in-line holography technique at the output plane of the hybrid system.

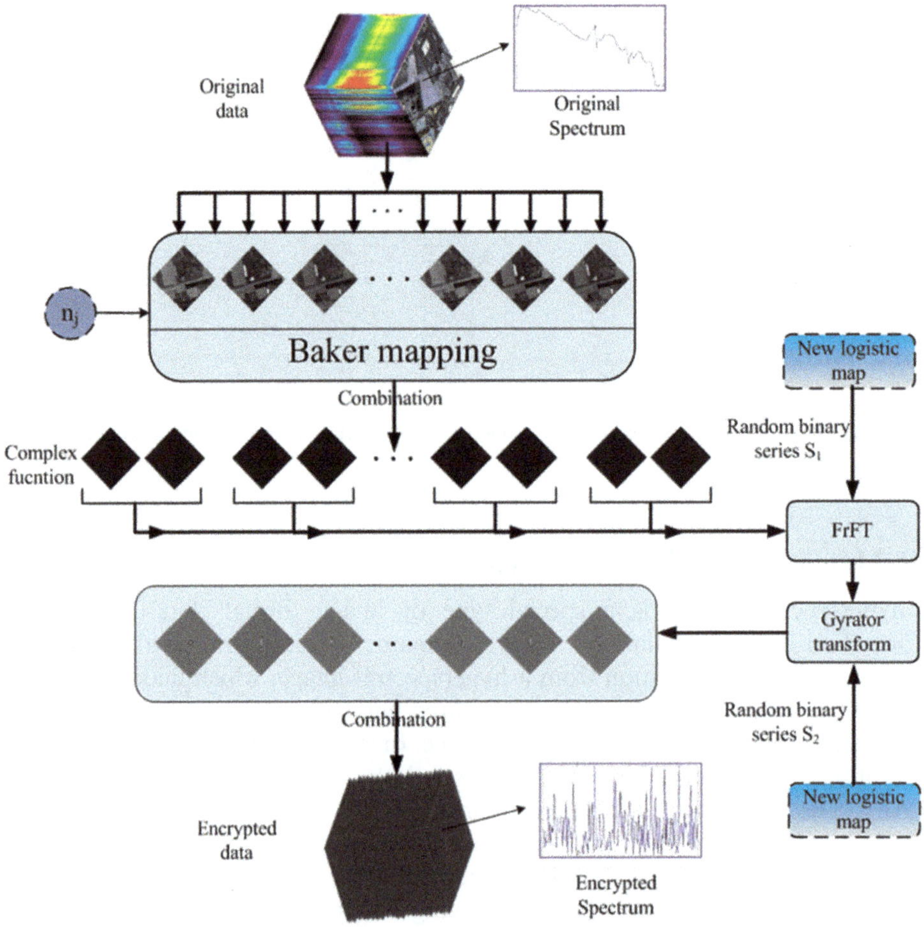

Figure 6.26. A flowchart of the hyperspectral cube encryption algorithm.

6.10 Other information techniques

In this section, the technique of watermarking and image sharing will be introduced. Watermarking is used for protecting the copyright of some digital productions. Image sharing (as an encryption method) has many different encryption outputs, which are held by potential users.

An optical watermarking scheme [36] is given in figure 6.28. Here, the functions W and I are the watermarked image and host image, respectively. Generally, the two images need to be adjusted for satisfying energy conservation as

$$\iint |I(x, y)|^2 \,\mathrm{d}x\mathrm{d}y = \iint |W'(u, v)|^2 \,\mathrm{d}u\mathrm{d}v \tag{6.95}$$

where W' is obtained by using a coefficient. Subsequently the Arnold transform is employed for changing the pixel value distribution. The scrambled watermark W_a is the output of the gyrator transform as follows

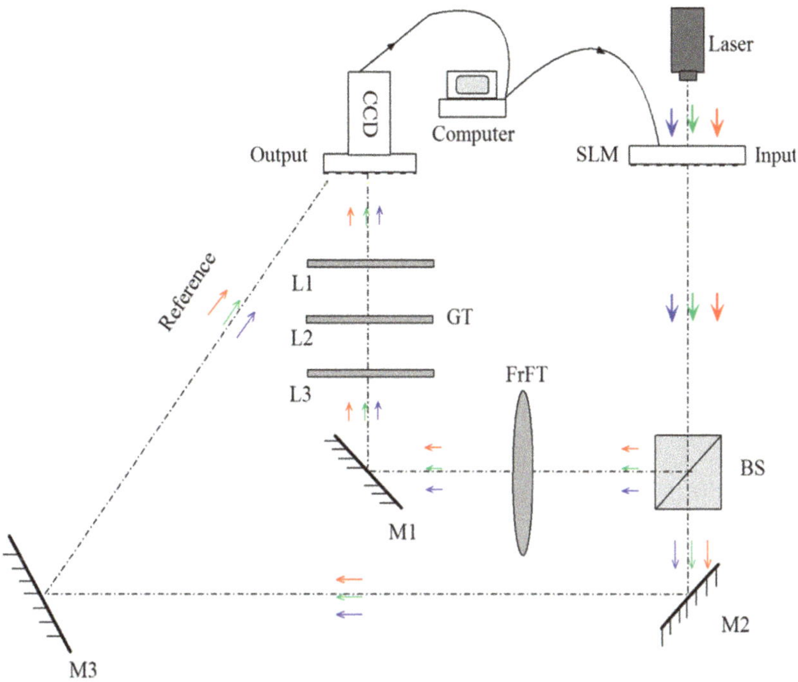

Figure 6.27. The electro-optical setup of the hyperspectral image encryption scheme.

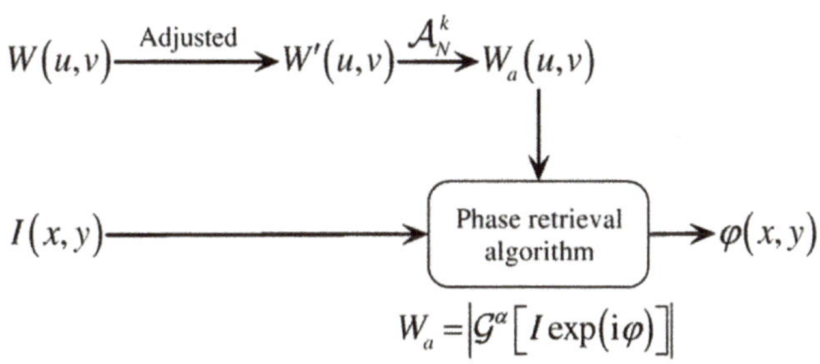

Figure 6.28. The principle of optical watermarking.

$$W_a = |\mathcal{G}^{\alpha}\,[I\ \exp\,(\mathrm{i}\varphi)]|, \tag{6.96}$$

where the evaluated value of phase φ will be calculated by the iterative phase retrieval algorithm.

According to figure 6.28, an optical system of this watermarking scheme is designed and shown in figure 6.29. The host image is displayed in SLM. After the gyrator transform, the random watermark W_a is received by a CCD. The original watermark W is recovered by inverse Arnold transform on a computer. This method is a kind of optical lossless watermarking.

In image sharing, the encoded data is hidden in n random patterns. In figure 6.30, image sharing is explained for encryption and decryption. When the number of shadow images is more than t, the original image can be retrieved, which is called the

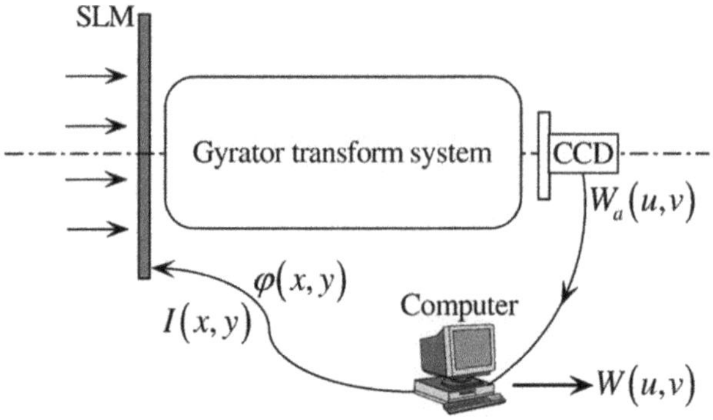

Figure 6.29. An optical watermarking system.

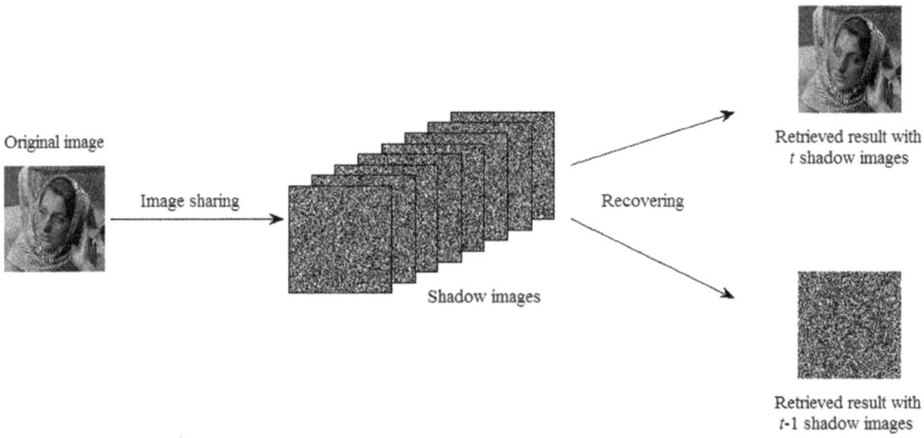

Figure 6.30. An illustration of image sharing.

(t, n) threshold. This encryption technique is of benefit for many users controlling the data of a hidden image. Here, a sharing method [37] is introduced by using discrete fractional random transform (DFRT). The original image X is converted into the DFRT domain as $Y = \mathcal{R}\{X\}$ for enhancing security. And then the shadow images are obtained by taking the following matrix equations as

$$\begin{aligned} S_1 &= c_{1,1}K_1 + c_{1,2}K_2 + \cdots + c_{1,t-1}K_{t-1} + c_{1,t}Y, \\ S_2 &= c_{2,1}K_1 + c_{2,2}K_2 + \cdots + c_{2,t-1}K_{t-1} + c_{2,t}Y, \\ &\vdots \\ S_n &= c_{n,1}K_1 + c_{n,2}K_2 + \cdots + c_{n,t-1}K_{t-1} + c_{n,t}Y, \end{aligned} \tag{6.97}$$

where $K_i(i = 1, 2, ..., t - 1)$ is a random pattern, which has the same pixel numbers with image X. Here the coefficient $c_{j,i}(i = 1, 2, ..., t - 1; j = 1, 2, ..., n)$ is for embedding Y into the random shadow images. Thereby the (t, n) threshold scheme can be satisfied according to Cramer's rule. The image decoding process is performed by solving equation (6.97).

Acknowledgments

The work in this chapter was supported by the National Natural Science Foundation of China (Nos. 11104049, 11047153, 10674038, 10974039, 61377016, 61575055 and 61575053), the Program for New Century Excellent Talents in University (No. NCET-12-0148), the China Postdoctoral Science Foundation (Nos. 2013M540278 and 2015T80340), the Fundamental Research Funds for the Central Universities (No. HIT.BRETIII.201406), the Scientific Research Foundation for the Returned Overseas Chinese Scholars, State Education Ministry, China.

References

[1] Healy J J, Kutay M A, Ozaktas H M and Sheridan J T 2016 *Linear Canonical Transforms: Theory and Applications* (New York: Springer)

[2] McBride A C and Kerr F H 1987 On Namias's fractional Fourier transform *IMAJ Appl. Math.* **39** 159–75

[3] Rodrigo J A, Alieva T and Calvo M L 2007 Experimental implementation of the gyrator transform *J. Opt. Soc. Am.* A **24** 3135–9

[4] Lohmann A W 1993 Image rotation, wigner rotation, and the fractional Fourier transform *J. Opt. Soc. Am.* A **10** 2181–6

[5] Özaktaş H M, Zalevsky Z and Kutay M A 2000 *The Fractional Fourier Transform: With Applications in Optics and Signal Processing* (Wiley) 117–83

[6] Guo C, Tan J and Liu Z 2015 Precision influence of a phase retrieval algorithm in fractional Fourier domains from position measurement error *Appl. Opt.* **54** 6940–7

[7] Liu Z and Liu S 2007 Randomization of the Fourier transform *Opt. Lett.* **32** 478–80

[8] Pei S C and Yeh M H 1997 Improved discrete fractional Fourier transform *Opt. Lett.* **22** 1047–9

[9] Yang X, Tan Q, Wei X, Xiang Y, Yan Y and Jin G 2004 Improved fast fractional Fourier-transform algorithm *J. Opt. Soc. Am.* A **21** 1677–81

[10] Liu Z, Chen D, Ma J, Wei S, Zhang Y, Dai J and Liu S 2011 Fast algorithm of discrete gyrator transform based on convolution operation *Optik* **122** 864–7

[11] Liu Z, Zhao H and Liu S 2005 A discrete fractional random transform *Opt. Commun.* **255** 357–65

[12] Liu Z, Ahmad M A and Liu S 2008 A discrete fractional angular transform *Opt. Commun.* **281** 1424–9

[13] Pei S C and Yeh M H 2001 The discrete fractional cosine and sine transforms *IEEE Trans. Signal Process.* **49** 1198–207

[14] Liu Z, Guo Q and Liu S 2006 The discrete fractional random cosine and sine transforms *Opt. Commun.* **265** 100–5

[15] Refregier P and Javidi B 2000 Optical image encryption based on input plane and Fourier plane random encoding *Opt. Lett.* **20** 767–9

[16] Unnikrishnan G, Joseph J and Singh K 2000 Optical encryption by double-random phase encoding in the fractional Fourier domain *Opt. Lett.* **25** 887–9

[17] Situ G and Zhang J 2004 Double random-phase encoding in the Fresnel domain *Opt. Lett.* **29** 1584–6

[18] Zhang Y, Zheng C and Tanno N 2002 Optical encryption based on iterative fractional Fourier transform *Opt. Commun.* **202** 277–85
[19] Liu Z, Xu L, Lin C, Dai J and Liu S 2011 Image encryption scheme by using iterative random phase encoding in gyrator transform domains *Opt. Lasers Eng.* **49** 542–6
[20] Liu Z, Ahmad M A and Liu S 2007 Image encryption scheme based on the commutation and anti-commutation rules *Opt. Commun.* **279** 285–90
[21] Liu Z, Xu L, Lin C and Liu S 2010 Image encryption by encoding with a nonuniform optical beam in gyrator transform domains *Appl. Opt.* **49** 5632–7
[22] Liu Z, Li S, Liu W and Liu S 2013 Opto-digital image encryption by using baker mapping and 1-D fractional Fourier transform *Opt. Lasers Eng.* **51** 224–9
[23] Liu Z, Dai J, Sun X and Liu S 2010 Single phase encoding method based on the fractional Fourier transform *Optik* **121** 1748–51
[24] Liu Z and Liu S 2007 Double image encryption based on iterative fractional Fourier transform *Opt. Commun.* **275** 324–9
[25] Tao R, Xin Y and Wang Y 2007 Double image encryption based on random phase encoding in the fractional Fourier domain *Opt. Express* **15** 16067–79
[26] Liu Z, Chen H, Liu T, Li P, Dai J, Sun X and Liu S 2010 Double-image encryption based on the affine transform and the gyrator transform *J. Opt.* **12** 035407
[27] Liu Z, Guo Q, Xu L, Ahmad M A and Liu S 2010 Double image encryption by using iterative random binary encoding in gyrator domains *Opt. Express* **18** 12033–43
[28] Liu Z, Xu L, Liu T, Chen H, Li P, Lin C and Liu S 2011 Color image encryption by using arnold transform and color-blend operation in discrete cosine transform domains *Opt. Commun.* **284** 123–8
[29] Liu Z, Dai J, Sun X and Liu S 2010 Color image encryption by using the rotation of color vector in Hartley transform domains *Opt. Lasers Eng.* **48** 800–5
[30] Liu Z, Guo C, Tan J, Liu W, Wu J, Wu Q, Pan L and Liu S 2015 Securing color image by using phase-only encoding in Fresnel domains *Opt. Lasers Eng.* **68** 87–92
[31] Liu Z, Tan J, Liu W, Wu J, Wu Q and Liu S 2014 A diffraction model of direction multiplexing method for hiding multiple images *J. Mod. Opt.* **61** 1127–32
[32] Zhang Y and Wang B 2008 Optical image encryption based on interference *Opt. Lett.* **33** 2443–5
[33] Chen H, Du X, Liu Z and Yang C 2015 Optical color image hiding scheme by using Gerchberg–Saxton algorithm in fractional Fourier domain *Opt. Lasers Eng.* **66** 144–51
[34] Chen H, Tanougast C, Liu Z, Sieler L and Ramenah H 2017 Optical image asymmetric cryptosystem using fingerprint based on iterative fraction Fourier transform *Opt. Quant. Electron.* **49** 157
[35] Chen H, Du X and Liu Z 2016 Optical spectrum encryption in associated fractional Fourier transform and gyrator transform domain *Opt. Quant. Electron.* **48** 12
[36] Liu Z, Xu L, Guo Q, Lin C and Liu S 2010 Image watermarking by using phase retrieval algorithm in gyrator transform domain *Opt. Commun.* **283** 4923–7
[37] Liu Z, Liu S and Ahmad M A 2010 Image sharing scheme based on discrete fractional random transform *Optik* **121** 495–9

IOP Publishing

Advanced Secure Optical Image Processing for Communications

Ayman Al Falou

Chapter 7

Privacy protection by multimodal biometric encryption

Debesh Choudhury

Privacy protection of the biometric data of citizens is of deep concern, because leakage of information from a biometric database can cause serious damage. We present a method for the privacy protection of biometric data using encryption techniques. We have developed a Fourier domain cryptosystem for privacy protection using multimodal biometric encryption. One modality of biometrics (such as a fingerprint) is used as the plaintext, which is encrypted by another modality of biometrics (such as a face). A private key is synthesized from the encrypted biometric signature by complex spatial Fourier processing. The encrypted biometric signature can further be encrypted by other biometric modalities, and the corresponding private keys can be synthesized. The resulting biometric signature is privacy protected since the encryption keys are provided by a human being, and hence those are private keys. Moreover, the decryption keys are synthesized using those private encryption keys. The encrypted signatures are decrypted using synthesized private keys and inverse complex spatial Fourier processing. Computer simulations demonstrate the feasibility of the proposed technique.

7.1 Why is encryption of information important?

Secure communication and sharing of secret information often require special expensive channels. Alternatively, the information may be encrypted and may be exchanged between users through public channels. In the past, secret keys used to be distributed through secure channels, generally by registered post or courier, which are slow and costly. This is known as symmetric key cryptography. The invention of public key cryptography by Diffie and Hellman [1], and independently by Merkle [2] have radically changed the approaches of key distribution in cryptography. A public key cryptosystem [1] has two keys, one for encryption and one for decryption. The two keys affect inverse operations and hence are related, but the decryption key cannot be easily derived from the encryption key [1]. The public key cryptography uses different keys for encryption and decryption, and hence is known as asymmetric key cryptography. The field has been continually enriched by the research works of Rivest *et al* [3], Merkle and Hellman [4] and McEliece [5]. The Rivest–Shamir–Adleman public key scheme (widely known as the RSA scheme) relies on the fact that it is easy to generate two large primes and multiply them together, but it is very difficult to factor the result [3]. Merkel and Hellman have used the trapdoor knapsack problems, which are also very difficult to solve [4]. McEliece's cryptosystem is based on algebraic coding theory where a random Goppa code is chosen to synthesize the encryption key by permutation, the secret decryption key being the permutation and choice of Goppa code. In all these seminal cryptosystems, the secret decryption keys are required to be created from the public encryption keys following some secret mathematical rules or puzzles [6].

7.2 Motivation for biometrics and privacy protection

Biometrics refer to physiological human traits such as the face, fingerprint, iris, palm prints etc, and behavioral human traits such as gait, voice, walking style, gesture, typing rhythm etc. Biometric signatures are unique to individuals and hence are reliable for identifying human beings. Biometric authentication is gaining more acceptance day by day in a variety of applications in governmental programs, as well as in personal verification problems [7, 8]. Biometrics can provide a way to synthesize keys for asymmetric encryption. Biometric data is also more difficult to be stolen or attacked making the encryption process inherently more secure and robust than storing text keys [9]. Moreover, as biometric keys are more secure, they also provide a way for personal identity verification because only the individual with the matching biological traits would be able to produce the correct biometric key [10].

A biometric template is a representation of several unique characteristics of an individual, and hence it contains sensitive personal and private information. Biometric identification requires a verifier to search for matches in the entire database of biometric signatures of the whole population. This can naturally open a possible security threat: that the verifier can steal biometric templates from the database. A malicious verifier can acquire biometric templates and attempt severe impersonation attacks. Thus, privacy protection is essential in the biometric authentication process because the biometric template data may be misused by dishonest verifiers [11, 12]. Security and privacy are two essential requirements of biometric systems, but they mutually hinder each other. Biometric encryption has been an easy route for privacy protection and surmounting security threats [13], although there are several other methods available for securing the privacy of information [12].

We utilized the principles of asymmetric key cryptography to encrypt and decrypt biometric signatures using Fourier domain cryptography, popularly known as double random phase encoding [14] and phase truncated Fourier transform [17]. Encrypted biometric signatures are inherently privacy protected by the respective asymmetric keys derived from the multimodal biometric traits of humans [19].

7.3 The proposed cryptosystem

The double random phase encoding cryptographic protocol relies on encryption of the plaintext using two random phase functions by two Fourier transformations [14–16]. The plaintext is multiplied by a random phase function (first private key) and Fourier transformed, which is again multiplied by a second random phase function (second private key) and inverse Fourier transformed. Since this algorithm utilizes two random phase functions for encrypion, it is called double random phase encoding (DRPE). If we have information specified by a spatial function $f(x, y)$, and have two independent white sequences $r_1(x, y)$ and $r_2(x, y)$, the encrypted function $g(x, y)$ after DRPE will be given by

$$g(x, y) = [f(x, y)\exp\{j2\pi r_1(x, y)\}] \circledast \exp\{j2\pi r_2(x, y)\} \tag{7.1}$$

where x, y are the space coordinates, $j = \sqrt{-1}$ and $\circledast$ is a convolution operation. The flow diagrams of encryption and decryption algorithms based on DRPE are given in figure 7.1 where FT denotes a Fourier transform operator.

We propose an asymmetric cryptosystem that utilizes one modality of biometric signature as the plain text image, and another modality of the biometric signature used as a private encryption key. The second biometric modality image is also encrypted using the DRPE technique. The plain text image is multiplied by the complex exponential function of the DRPE encrypted second biometric signature, and the product is Fourier transformed. The phase part of the Fourier transform is extracted and is preserved as a private decryption key [17]. The amplitude part of the Fourier transform is utilized for further encryption by a third biometric signture. The product of the amplitude part of the Fourier transform and the exponential function of the DRPE encrypted third biometric key is inverse Fourier transformed. Then, again the phase part of it is extracted and preserved as a second private decryption key. The amplitude part of the Fourier transform is the encrypted privacy protected biometric signature.

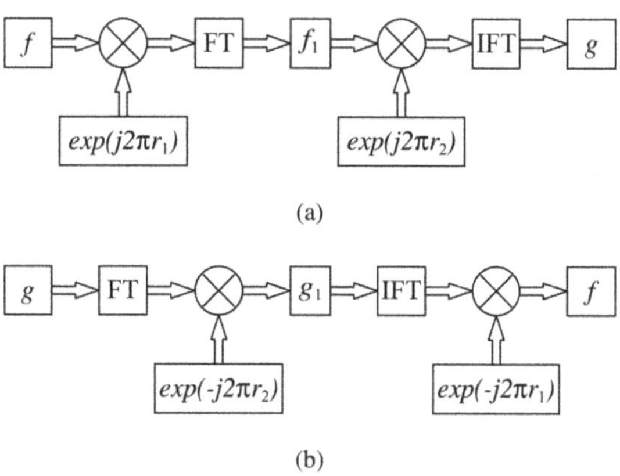

Figure 7.1. Double random phase encoding: (a) encryption algorithm, and (b) decryption algorithm. Reproduced from [19].

Let b_1 be the first biometric trait image of a human (such as a face), and b_{E2}, b_{E3} be his/her other two biometric trait images (such as, fingerprint and iris) but are encrypted by DRPE. If PT is a phase truncation operator and PR a phase reservation operation, we can express our cryptosystem using the flow diagram of figure 7.2(a). The two private phase keys PK_1 and PK_2 are extracted by the PR operation. The final encrypted image PPBS represents the privacy protected biometric signature.

The encrypted biometric signature PPBS can be decrypted using the synthesized private phase keys PK_1 and PK_2 as shown in the flow diagram of figure 7.2(b). These encrypted privacy protected biometric signatures can be preserved in the database. These data can be used for authentication by a cross-correlation operation [18].

7.4 Simulation results

The feasibility of the proposed cryptosystem is tested by implementing the algorithm on image data of the human face and fingerprints. To check our algorithm, we have used a few face images from the Yale Face Database B and have associated a few fingerprint images from the Bologna fingerprint database. The face image is used as a plaintext image, and the two fingerprint images are encrypted by the DRPE method and the encrypted versions of the fingerprint images are utilized as encryption keys to encrypt the face image. The fingerprint images of a human subject's right and left thumbs, and a human subject's face are shown in figures 7.3(a), (e) and (i), respectively, which are the plain images in our simulation experiment. The corresponding cipher images of the fingerprint images by DRPE are shown in figures 7.3(b) and (f). These encrypted images are utilized to produce the final privacy protected biometric signature PPBS by our proposed encryption algorithm and this is shown in figure 7.3(j). The histograms of the plain images are shown in figures 7.3(c), (g) and (k), respectively. The histograms of the cipher images

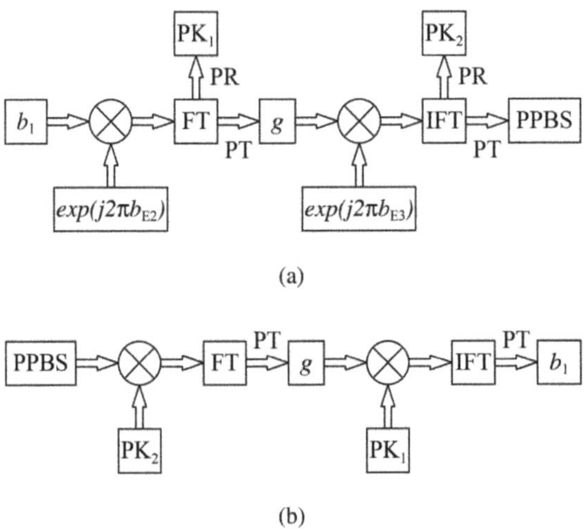

Figure 7.2. The proposed asymmetric cryptosystem: (a) encryption algorithm, and (b) decryption algorithm. Reproduced from [19].

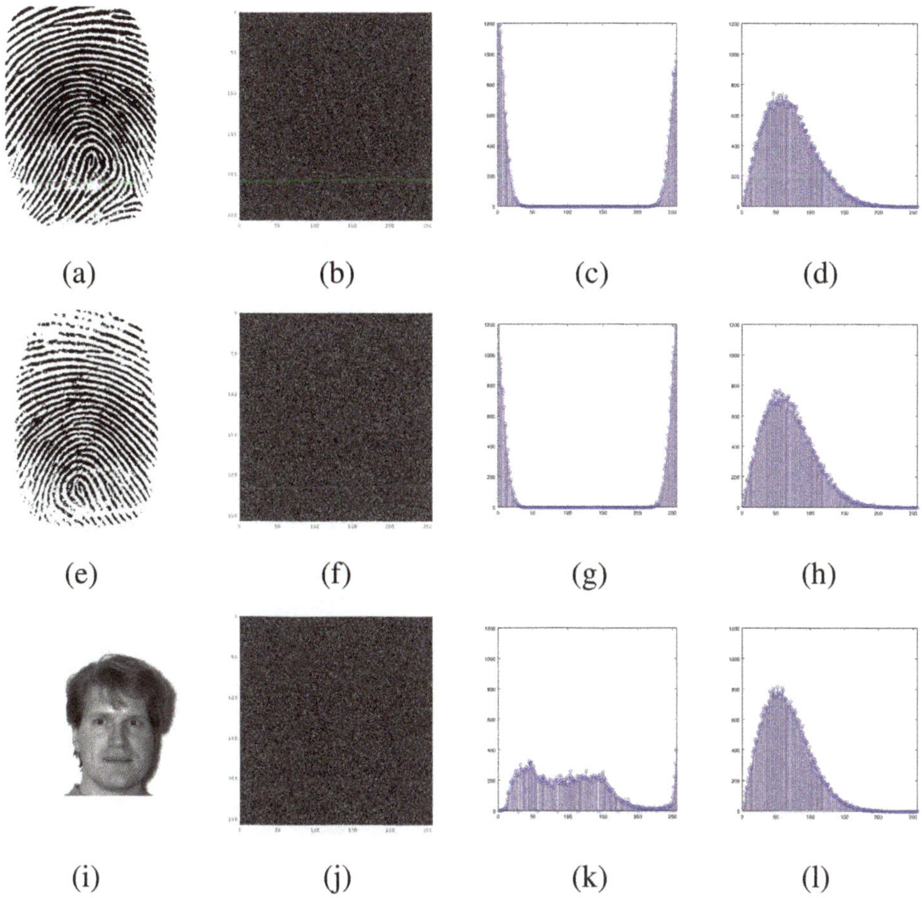

Figure 7.3. Results: (a) fingerprint of right thumb, (e) fingerprint of left thumb, (i) face of the human subject; (b), (f) cipher images of the fingerprints by DRPE, (j) final privacy protected cipher image of the human face; (c), (g), (k) histograms of the plain images; (d), (h), (l) histograms of the cipher images. Reproduced from [19].

are shown in figures 7.3(d), (h) and (l), respectively. It is evident from the results that there remain significant differences between the histograms of the plain images and those of the cipher images. According to the theory of classical cryptology it is good because this can be considered as diffusion, which diffuses the statistical characteristics of the original image to the whole space. In addition, we also find that the cipher image of different original images have similar histograms after being encrypted with independent encryption keys. This can be considered as 'confusion' in classical cryptology. It is difficult for attackers to obtain useful information from the statistical characteristics, thus it can have considerable capacity for resisting statistical cryptanalysis.

The final privacy protected biometric signature can be decrypted by the phase truncated Fourier transform algorithm of figure 7.2 using the appropriate private phase keys PK_1 and PK_2. This result is shown in figure 7.4. With the correct phase keys, the original plain image of the human subject is extracted as shown in figure 7.4(a). When one of the phase keys, here the second phase key PK_2, is used incorrectly, the algorithm produces an incorrect extraction, which is a random image instead of the original image of a face, as shown in figure 7.4(b).

The final encrypted PPBS can be used for authentication using cross-correlation [18]. A high correlation peak authenticates the original cipher image and confirms no 'man-in-the-middle' attacks. The height of the correlation peak determines whether it is a correct recognition or otherwise. Figure 7.5 shows the three-dimensional plots of the cross-correlation functions. A successful biometric recognition is confirmed by the high correlation peak of figure 7.5(a) which is 0.99. When the PPBS of figure 7.3(j) is cross-correlated with a false signature, such as that of

Figure 7.4. Decryption result: (a) decrypted image with correct phase keys, (b) no decryption with a false key. Reproduced from [19].

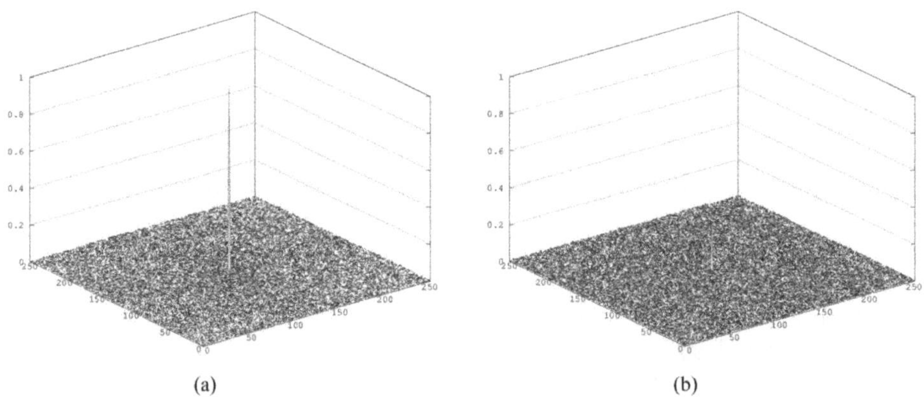

Figure 7.5. Plots of cross-correlation: (a) authentication of correct cipher biometric signature (correlation peak height = 0.99), (b) biometric rejection, no authentication (correlation peak height = 0.2). Reproduced from [19].

figure 7.4(b), a very low correlation peak is produced, as shown in figure 7.5(b), (which is equal to 0.2), and it signifies biometric rejection. The implementation of all the algorithms has been carried out using GNU Octave [20].

7.5 Conclusion

An asymmetric cryptosystem is proposed for privacy protection in biometric pattern recognition. Facial biometrics are encrypted by fingerprint biometrics and double random phase encoding. Two private phase keys are generated by phase truncated Fourier transform. The encrypted final biometric image is privacy protected. The encrypted biometric signature is decrypted by using the synthesized two private phase keys. The encryption keys and the decryption keys are different. Authentication of the encrypted biometric signature is also carried out by a cross-correlation operation. The simulation results verify the feasibility of the proposed cryptosystem.

References

[1] Diffie W and Hellman M E 1976 New directions in cryptography *IEEE Trans. Inf. Theory* **IT-22** 644–54

[2] Merkle R C 1978 Secure communication over an insecure channel *Commun. ACM* **21** 294–9

[3] Rivest R L, Shamir A and Adleman L 1978 On digital signatures and public key cryptosystems *Commun. ACM* **21** 120–6

[4] Merkle R C and Hellman M E 1978 Hiding information and signatures in trap-door knapsacks *IEEE Trans. Inf. Theory* **IT-24** 525–30

[5] McEliece R J 1978 *A public key system based on algebraic coding theory JPL DSN Progress Report*

[6] Hellman M E 2002 An overview of public key cryptography *IEEE Commun. Mag.* 42–9

[7] Ashbourn J 2004 *Practical Biometrics: From Aspiration to Implementation* (Springer)

[8] Boulgouris N V, Plataniotis K N and Micheli-Tzanakou E (ed) 2009 *Biometrics : Theory, Methods, and Applications* (Wiley-IEEE Press)

[9] Uludag U, Pankanti S, Prabhakar S and Jain A K 2004 Biometric cryptosystems: issues and challenges *Proc. IEEE* **92** 948–60

[10] Jain A K, Bolle R and Pankanti S 1999 *Biometrics: Personal Identification in Networked Scoiety* (Norwell, MA: Kluwer)

[11] Ratha N, Connel J and Bolle R 2001 Enhancing security and privacy in biometrics-based authentication systems *IBM Syst. J.* **40** 614–34

[12] Campisi P (ed) 2013 *Security and Privacy in Biometrics* (London: Springer)

[13] Roberge C S D, Stoianov A, Giuroy R and Kumar B V K V 1999 *Biometric encryption. ICSA Guide to Cryptography* chapter 2

[14] Javidi B 1997 Securing information with optical technologies *Phys. Today* **50** 27–32

[15] Refregier P and Javidi B 1995 Optical image encryption based on input plane and Fourier plane random encoding *Opt. Lett.* **20** 767–9

[16] Goodman J W 2003 *Introduction to Fourier Optics* 3rd edn (Roberts & Company)

[17] Qin W and Peng X 2010 Asymmetric cryptosystem based on phase-truncated Fourier transforms *Opt. Lett.* **35** 118–20

[18] VanderLugt A 1964 Signal detection by complex spatial filtering *IEEE Trans. Inf. Theory* **IT-10** 139–45

[19] Choudhury D 2016 Fourier domain asymmetric cryptosystem for privacy protected multi-modal biometric security *Proc. SPIE* **9845** 98450G

[20] GNU Octave, https://www.gnu.org/software/octave/ accessed Nov 15, 2017.

IOP Publishing

Advanced Secure Optical Image Processing for Communications

Ayman Al Falou

Chapter 8

Nonlinear techniques for secure optical encryption and multifactor authentication

María Sagrario Millán, Elisabet Pérez-Cabré and Juan Manuel Vilardy

List of abbreviations

CCA:	Chosen-ciphertext attacks
COA:	Ciphertext-only attacks
CPA:	Chosen-plaintext attacks
DRPE:	Double-random phase encoding
FD:	Fourier domain
FrFD:	Fractional Fourier domain
FrFT:	Fractional Fourier transform
FrD:	Fresnel domain
FrT:	Fresnel transform
FT:	Fourier transform
GD:	Gyrator domain
GS:	Gerchberg–Saxton
GT:	Gyrator transform
ID:	Identity
JFPD:	Joint Fresnel power distribution
JGPD:	Joint gyrator power distribution
JPS:	Joint power spectrum
JTC:	Joint-transform correlator
KPA:	Known-plaintext attacks
MOEA:	Multifactor optical encryption-authentication
NIR:	Near infrared
PhC:	Photon-counting
RCM:	Random complex mask
RMSE:	Root mean square error
RPM:	Random phase mask

doi:10.1088/978-0-7503-1457-2ch8

8.1 Introduction: optical implementations of double-random phase encoding (DRPE)

Double-random phase encoding (DRPE) is an important and highly successful technique for optical image encryption proposed by Réfrégier and Javidi [1]. The DRPE uses two random phase masks (RPMs); the first and second RPMs are placed at the input plane and the Fourier plane, respectively, in order to encode an original image (the image to encrypt) into a stationary white noise pattern (encrypted image) [1–3]. The second RPM placed in the Fourier plane represents the key of the DRPE when the original image is encoded in amplitude at the input plane of the encryption system [1], whereas the two RPMs become keys of the security system if the original image is encoded in phase [4]. Later, the DRPE was further extended from the Fourier domain (FD), to the Fresnel domain (FrD) [5–8], the fractional Fourier domain (FrFD) [9–11], and the Gyrator domain (GD) [12, 13], with the purpose of increasing the security strength of the optical encryption technique.

The DRPE and their extensions use integral transformations that can be implemented using optoelectronic setups [1–3, 10, 14]. The initial optical setup of the DRPE was implemented through a holographic system based on the optical processor $4f$ [1]. Soon after, however, a joint-transform correlator (JTC) architecture presented some advantages, such as no requirement for interferometric recording of the phase distribution or complex conjugation of the key phase mask [15].

In cryptanalysis, it is always assumed that attackers are trying to determine the security key of an encryption process and they have access to some resources (for instance, a pair of signals such as an original image and its encrypted image, or access to either the encryption or the decryption system). In chosen-plaintext attacks (CPA), the attacker introduces an adequate input image (skillfully designed chosen plaintext) in either the encryption [16] or decryption [17] process, and they get the corresponding output image in order to deduce the security key. The known-plaintext attacks (KPA) and the ciphertext-only attacks (COA) are iterative processes, which are based on a heuristic hybrid algorithm [18, 19] and/or the Gerchberg–Saxton (GS) algorithm [20, 21]. Concerning the security of the DRPE proposed by Réfrégier and Javidi, it has been proved that the DRPE is vulnerable to CPA [16, 17], KPA [16], and COA [22]. This weakness is due to the linear property of the DRPE algorithm [16]. The DRPE implemented with a JTC architecture, taken as a computational algorithm, would be also vulnerable to CPA [23], KPA [24], and COA [25].

The security strength of the optical implementation of encryption/decryption systems stems from the ability of optics to process the information in a hyperspace of states, where physical variables such as amplitude, phase, polarization, wavelength, distance of propagation, spatial position, and fractional spatial frequency domain can all be used to serve as additional degrees of freedom for data encryption in order to protect the secret of the original input image and to increase the security over the encrypted image. Moreover, optical processing has the valuable property of inherent parallelism, which allows for fast encryption and decryption of large volumes of complex data. In addition, high-resolution optical materials can store encoded data in small volumes.

Encryption and decryption processes can be further complemented with authentication. Thus, while an encryption/decryption system aims to decrypt a signal (primary image or plaintext) embedded in a formerly encrypted image (ciphertext), so that the recovered information replicates the original signal (information recovery), an encryption/authentication system would additionally compare the decrypted signal—which would not need to be visually disclosed—with a reference signal and recognise their similarity. Pattern recognition algorithms based on matched filtering and optical correlation [14] have been successfully applied to these optical encryption-authentication systems [26, 27]. Additionally, encryption and authentication of multiple signals is another important achievement in optical security applications that increases system reliability because its response does not rely on the verification of a single factor but a set of factors. They all must obtain positive authentication to provide a final validation. Multifactor optical encryption-authentication (MOEA) systems have been proposed for encryption and authentication of a set of images as well as their joint validation [26, 28, 29]. Since the quality of the retrieved images is not the main goal, some compression can be achieved by different procedures that, in addition, may contribute to further securing the encryption system.

In this chapter, we present several JTC architectures for nonlinear information encryption, decryption and authentication systems. These JTC architectures are developed in different processing domains, such as Fourier, fractional Fourier, Fresnel and Gyrator domains. Some of these processing domains can add new security keys in order to improve the security of the image encryption–decryption process. The probability density functions of the random phase masks also play a role to increase the security level of the overall system. Additional nonlinear modifications can significantly enhance the quality of the retrieved image as well as security. We also review recent nonlinear modifications of encryption-authentication systems that permit one to simultaneously encrypt and verify multiple pieces of information for highly secure optical systems.

8.2 Nonlinear techniques for image retrieval (decryption) based on the JTC

8.2.1 JTC modified optical implementation

A practical modification of the optical JTC implementation of the DRPE has been proposed [30, 31] in a way that the whole information (primary image and two random phase masks) can be easily introduced in the same input plane of the JTC system.

Using one-dimensional notation for the sake of simplicity, let $f(x)$ be the real input image encoded in amplitude with values in the interval [0,1], $r(x)$ and $h(x)$ be two RPMs given by

$$r(x) = \exp\{i2\pi m(x)\}, \qquad h(x) = \exp\{i2\pi n(x)\}, \tag{8.1}$$

where $m(x)$ and $n(x)$ are normalized positive functions randomly generated, statistically independent and uniformly distributed in the interval [0,1]. Figure 8.1 shows the optical encrypting scheme based on a linear JTC architecture (*via 1*), and the

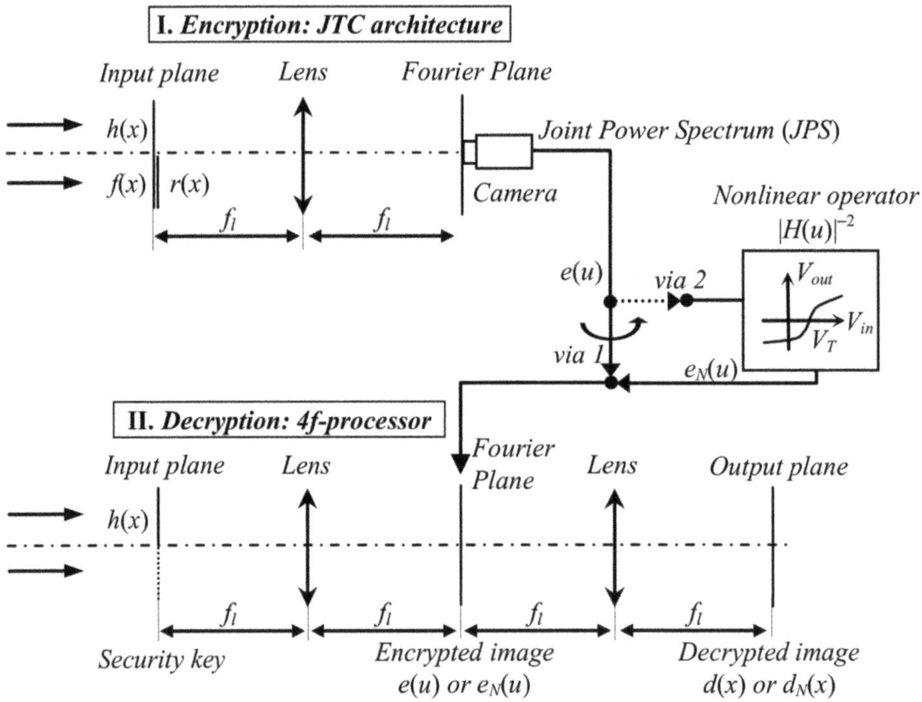

Figure 8.1. Schematic representation of the optical setup. The encryption system is based on a linear (*via 1*) or a nonlinear (*via 2*) JTC architecture. A 4*f*-processor is used for decryption. Reproduced with permission from [32].

optical decrypting scheme based on a 4*f*-processor. In the encryption process, the RPM-I $r(x)$ is placed on the real image $f(x)$ and then, the product $f(x)r(x)$ and the RPM-II $h(x)$ are placed side by side in the input plane of JTC at coordinates $x = a$ and $x = -a$, respectively [15, 30]. The joint power spectrum (JPS), also called the encrypted power spectrum $e(u)$, is specified by

$$\begin{aligned} e(u) = \text{JPS}(u) &= |\text{FT}[\{f(x)r(x)\}*\delta(x-a) + h(x)*\delta(x+a)]|^2 \\ &= |F(u)*R(u)|^2 + |H(u)|^2 \\ &\quad + [F(u)*R(u)]^* H(u)e^{i2\pi(2a)u} + [F(u)*R(u)]H^*(u)e^{-i2\pi(2a)u}, \end{aligned} \tag{8.2}$$

where the functions represented by capital letters correspond to the Fourier transforms (FTs) of functions represented in lowercase letters, the variable u denotes the spatial frequency coordinate, the symbol (*) indicates the convolution operation and the superscript * represents the complex conjugation operation. Equation (8.2) demonstrates that the encrypted image is real, and thus, it can be captured and stored by a conventional intensity sensor, such as a CCD camera. The security key is represented by the RPM-II $h(x)$ and the RPM-I $r(x)$ is used to spread the information content of the original image $f(x)$ onto the encrypted image $e(u)$.

By following the described procedure, a simplified optical setup permits one to encrypt and decrypt information based on the JTC architecture [30, 31]. However, poor image quality is commonly obtained in the decryption stage, as will be analysed in section 8.2.3, and some vulnerability to certain intruder attacks still affects the encryption system [32].

8.2.1.1 Cryptanalysis

In this section, the vulnerability of the system is presented taking into consideration a simplified implementation in the FD [30–32]. The CPA can be applied to the encryption system presented above, in order to obtain the FT of the security key $H(u)$ and hence, by inverse Fourier transforming, RPM-II $h(x)$ [23]. This attack introduces a couple of chosen plaintexts in the encryption system to obtain the FT of the RPM-II $h(x)$. The first chosen plaintext and its corresponding encrypted image are

$$f_1(x) = 0, \qquad e_1(u) = |H(u)|^2. \tag{8.3}$$

Therefore, when a null image is introduced in the encryption system, the square modulus of the FT of the security key is obtained. The second chosen plaintext is a shifted Dirac delta function, and its corresponding encrypted image is

$$\begin{aligned} f_2(x) &= \delta(x - x_p), \\ e_2(u) &= 1 + |H(u)|^2 + 2\,|H(u)| \cos\left\{2\pi\left[\phi(u) - m(x_p) + \left(2a + x_p\right)u\right]\right\}, \end{aligned} \tag{8.4}$$

where $\phi(u)$ is the phase of $H(u)$ and $m(x_p)$ represents a constant value. The argument of the cosine function in equation (8.4) is mainly related to $\phi(u)$ with some additional phase terms. Provided $|H(u)|$ is obtained from equation (8.3), the phase of $H(u)$ can be retrieved from equation (8.4) by following a phase-shifting procedure similar to the one indicated in [23]. Therefore, if the two chosen plaintexts represented by equations (8.3)–(8.4) are introduced into the encryption system, the attacker can have access to the complete complex information of the FT of the security key.

The KPA and COA were also designed to be applied to the described JTC-based encryption system in the FD [24, 25]. The KPA and COA are iterative processes based on a heuristic hybrid algorithm and/or the GS algorithm. The heuristic hybrid algorithm produces a first key based on an initial guess of the RPM key and, a plaintext (original input image) with its corresponding ciphertext (encrypted image). The first key along with the plaintext and its corresponding ciphertext are introduced in the GS algorithm adapted to the JTC architecture, in order to obtain either the final RPM key in the case of the KPA or the original input image in the case of the COA.

8.2.2 Increasing security by changing the processing domains

For some optical encryption systems, the classical implementation of the DRPE in the spatial and Fourier domains using the processor 4*f* and the JTC architecture along with several keys given by the physical variables of these optical processes does not sufficiently protect the secret of the original input image or the RPMs keys [8, 16, 17, 22–25, 33]. Therefore, it is necessary to introduce nonlinear modifications in the encrypted function [32] and to use different processing domains in the optical implementations of the DRPE with the purpose of improving the security of these encryption systems: from the FD to the FrD [8, 34], FrFD [11, 35] and GD [13].

8.2.2.1 Fourier domain (FD)

As we have already seen in section 8.2.2.1, the encryption system based on JTC (given by equation (8.2)) is vulnerable to some plaintext attacks [23–25]. We propose a nonlinear modification of the encryption system based on JTC in order to overcome these flaws in security. We introduce a nonlinearity that consists of dividing the JPS by $|H(u)|^2$. Thus, the nonlinearly modified encrypted information becomes

$$\begin{aligned} e_N(u) &= \frac{\mathrm{JPS}(u)}{|H(u)|^2} \\ &= \frac{|F(u)*R(u)|^2}{|H(u)|^2} + 1 + [F(u)*R(u)]^* \frac{H(u)}{|H(u)|^2} e^{i2\pi(2a)u} \\ &+ [F(u)*R(u)] \frac{H^*(u)}{|H(u)|^2} e^{-i2\pi(2a)u}. \end{aligned} \tag{8.5}$$

If $|H(u)|^2$ is equal to zero for a particular value of u, this intensity value will be substituted by a small constant to avoid singularities when computing $e_N(u)$. Equation (8.5) represents the new encrypted image when the JPS is nonlinearly modified. Figure 8.1, which corresponds to the encrypting and decrypting schemes based on a JTC architecture, also shows the nonlinear modification of this security system (*via 2*).

Now, let us test again the resistance of the described nonlinear encryption system against CPA [23]. According to equation (8.5) and the CPA, the first chosen plaintext and its corresponding nonlinear encrypted image become

$$f_1(x) = 0, \qquad e_{N1}(u) = 1. \tag{8.6}$$

Unlike the first JTC architecture (*via 1* of figure 8.1), if the nonlinear process (*via 2*) is chosen, it will not be possible to obtain any information about the RPM-II $h(x)$ from the encrypted image given by equation (8.6). The second chosen plaintext is a shifted Dirac delta function, and its corresponding nonlinear encrypted image is

$$\begin{aligned} f_2(x) &= \delta(x - x_p), \\ e_{N2}(u) &= \frac{1}{|H(u)|^2} + 1 + \frac{2}{|H(u)|} \\ &\cos\left\{2\pi\left[\phi(u) - m(x_p) + (2a + x_p)u\right]\right\}, \end{aligned} \tag{8.7}$$

where both the modulus and phase functions of $H(u)$ will be unknown. Therefore, these CPA will not allow the attacker to easily obtain any information about either $h(x)$ or $H(u)$. In conclusion, using the nonlinear term introduced into the JPS, it permits an improvement in the security of the encrypted image. Due to this, the encrypted image becomes more robust against CPA.

For the decryption process (figure 8.1), the RPM-II $h(x)$ is placed at coordinate $x = -a$ in the input plane of a 4f-processor [30–32], and, consequently, in the Fourier plane, the encrypted power spectrum $e_N(u)$ is illuminated by $H(u)\exp[i2\pi au]$. The resulting product is given by

$$\begin{aligned} g_N(u) = e_N(u)H(u)e^{i2\pi au} &= |F(u)*R(u)|^2 \frac{H(u)}{|H(u)|^2}e^{i2\pi au} + H(u)e^{i2\pi au} \\ &+ [F(u)*R(u)]^* \frac{H^2(u)}{|H(u)|^2}e^{i2\pi(3a)u} + [F(u)*R(u)]\frac{H^*(u)H(u)}{|H(u)|^2}e^{-i2\pi au}. \end{aligned} \tag{8.8}$$

When the inverse FT is applied to the simplified fourth term of equation (8.8), the obtained decrypted distribution is

$$d_N(x) = \mathrm{FT}^{-1}\{[F(u)*R(u)]e^{-i2\pi au}\} = f(x)r(x)*\delta(x-a). \tag{8.9}$$

The intensity of equation (8.9) produces the original image intensity $|f(x)|^2$ at coordinate $x = a$.

The DRPE implemented with a JTC architecture has been extended from the FD to the FrD [8, 34, 36–39], the FrFD [11, 35, 40–42] and the GD [13, 43, 44]. In the nonlinear approaches [8, 11, 13, 34, 35], the nonlinear operations introduced in the JTC architecture and the extension to other domains have improved the security of the overall encryption–decryption system by introducing new parameters that must be adjusted prior to obtaining the hidden information.

8.2.2.2 Fresnel domain (FrD)

The encrypted image in the FrD for the described nonlinear encryption system based on a JTC is [8]

$$\begin{aligned} e_z(u) &= \frac{\mathrm{JFPD}_z(u)}{|h_z(u)|^2} \\ &= \frac{|t_z(u)|^2}{|h_z(u)|^2} + 1 + t_z^*(u)\frac{h_z(u)}{|h_z(u)|^2}\exp\left\{\frac{i\pi}{\lambda z}(4x_0)u\right\} \\ &+ t_z(u)\frac{h_z^*(u)}{|h_z(u)|^2}\exp\left\{\frac{-i\pi}{\lambda z}(4x_0)u\right\}, \end{aligned} \tag{8.10}$$

where the functions $h_z(u)$ and $t_z(u)$ are the Fresnel transforms (FrTs) of the functions $h(x)$ and the product $f(x)r(x)$, respectively. The joint Fresnel power distribution (JFPD) at parameters λ (wavelength of the wave plane used as illumination), z (distance of propagation), and $v_0 = -x_0 = -a$ is given by

$$\mathrm{JFPD}_z(u) = |\mathrm{FrT}_{\lambda,z}\{g(x-x_0) + c(x+x_0)\}|^2, \tag{8.11}$$

where the operator $\mathrm{FrT}_{\lambda,z}$ denotes the FrT at parameters λ and z, the two non-overlapping data distributions, $g(x)$ and $c(x)$, are placed side-by-side in the input plane of the JTC at coordinates $x = x_0 = a$ and $x = -x_0 = -a$, respectively. These distributions are

$$g(x) = \exp\left\{\frac{i2\pi\nu_0(x - \nu_0)}{\lambda z}\right\} f(x)r(x), \quad c(x) = \exp\left\{\frac{-i2\pi\nu_0(x + \nu_0)}{\lambda z}\right\} h(x). \quad (8.12)$$

The linear phase terms presented in equation (8.12) are symmetrically introduced to ensure the complete overlapping of both Fresnel spectra at a distance of propagation z where the JFPD is recorded.

The nonlinear JTC-based encryption system in the FrD can be applied by means of a lensless optical system [8, 34], which minimizes the optical hardware requirements and is easier to implement. Figure 8.2 (part I) shows the optical encryption scheme based on a nonlinear JTC architecture in the FrD. The encrypted image consists of a real-valued distribution that can be computed from the intensity distributions of the JFPD and $|h_z(u)|^2$ previously acquired by a camera. The nonlinear expression of equation (8.10) is the natural extension of equation (8.5) from operating in the FD to the FrD. The security keys needed for decryption are the RPM $h(x)$, the wavelength λ and the distance of propagation z. Therefore, the extension from the FD to the FrD allows the introduction of two new parameters: the physical parameters λ and z, thus improving the security of the encryption system [8].

To further improve the security of this system in the FrD, we can replace the RPM $h(x)$ by a random complex mask (RCM) described by

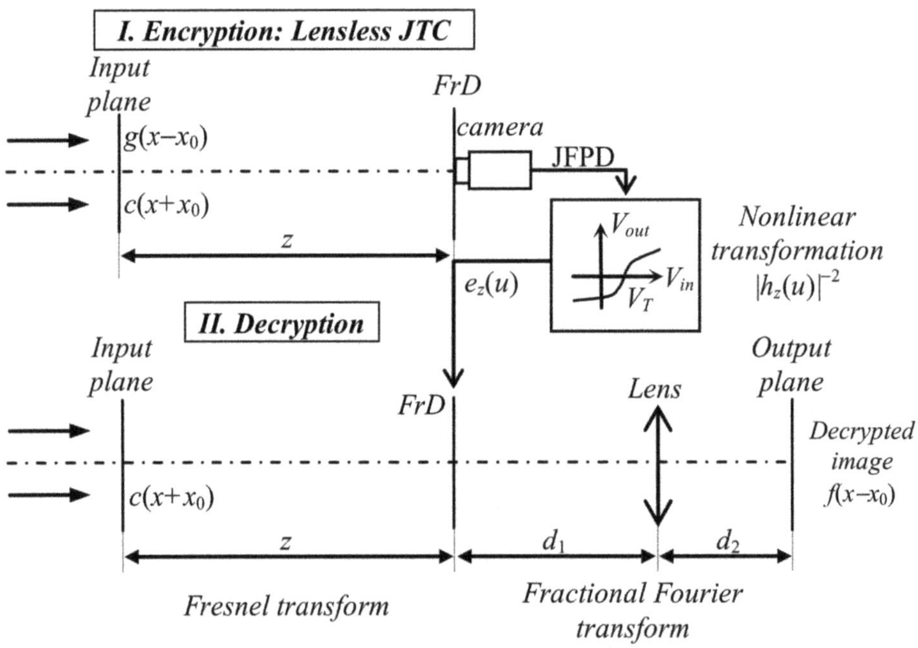

Figure 8.2. Schematic representation of the optical setup for the encryption and decryption system using the Fresnel domain. The encryption system is composed of a nonlinear JTC architecture in the FrD and the decryption system is based on an optical FrT combined with an optical fractional Fourier transform. Reproduced with permission from [8]. Copyright 2014 Optical Society of America.

$$h(x) = s(x)\exp\{i2\pi n(x)\}. \tag{8.13}$$

The random codes $s(x)$ and $n(x)$ can be generated using different probability random distribution (uniform and/or non-uniform) [8]. The random code $s(x)$ represents a new key that increases the security for the described encryption system. Finally, the nonlinear term introduced in equation (8.10) and the use of uniform and non-uniform random distributions in the generation of the random codes of the RPM $r(x)$ and the RCM $h(x)$ make the nonlinear JTC-based encryption system in the FrD more resistant to the CPA and KPA [8].

In the decryption method (figure 8.2, part II), the data distribution $c(x)$ is placed at coordinate $x = -x_0 = -a$ and, consequently, the encrypted image $e_z(u)$ located at a distance of propagation z in the FrD is illuminated by $\mathrm{FrT}_{\lambda,z}\{c(x + x_0)\}$. This initial step of the decryption process can be expressed by

$$\begin{aligned} d_z(u) &= e_z(u)\mathrm{FrT}_{\lambda,z}\{c(x + x_0)\} \\ &= \exp\left\{\frac{i\pi}{\lambda z}(2x_0u - x_0^2)\right\}\frac{h_z(u)}{|h_z(u)|^2}|t_z(u)|^2 \\ &+ \exp\left\{\frac{i\pi}{\lambda z}(2x_0u - x_0^2)\right\}h_z(u) \\ &+ \exp\left\{\frac{i\pi}{\lambda z}(6x_0u - x_0^2)\right\}\frac{h_z^2(u)}{|h_z(u)|^2}t_z^*(u) \\ &+ \exp\left\{\frac{-i\pi}{\lambda z}(2x_0u + x_0^2)\right\}\frac{h_z^*(u)h_z(u)}{|h_z(u)|^2}t_z(u). \end{aligned} \tag{8.14}$$

The fourth term on the right side of previous equation is the most interesting term since it retains the information to be decrypted. Therefore, if we calculate the FrT of that term, with parameters λ and $-z$, and then, take the absolute value, we obtain the decrypted image at coordinate $x = x_0 = a$

$$\tilde{f}(x - x_0) = \left|\mathrm{FrT}_{\lambda,-z}\left\{\exp\left[\frac{-i\pi}{\lambda z}(2x_0u + x_0^2)\right]t_z(u)\right\}\right|. \tag{8.15}$$

Note that, if we use $c(x)$ shifted to the position of coordinate $x = -x_1$ with $v_0 = -x_1$ for the decryption process, the decrypted image can be recovered still, but at coordinate $x = 2x_0 - x_1$, that is,

$$\tilde{f}(x - 2x_0 + x_1) = \left|\mathrm{FrT}_{\lambda,-z}\left\{\exp\left[\frac{-i\pi}{\lambda z}(2(2x_0 - x_1)u + x_1^2)\right]t_z(u)\right\}\right|. \tag{8.16}$$

The previous equation demonstrates that the encryption–decryption system based on a nonlinear JTC in the FrD preserves the shift-invariance property of the complex key mask $c(x)$ for decryption, in the same way as the FD-JTC encryption system does [2]; this shift-invariant property of the described encryption–decryption system is a consequence of the definition of the JFPD given in equation (8.11).

Let us remark that the original image cannot be retrieved in the decryption stage unless the nonlinear operation has been introduced as described in equation (8.10). This result makes the difference between the nonlinearity proposed to apply in the FD, through equation (8.5), and the nonlinearity proposed to apply in the FrD, through equation (8.10). The nonlinearity applied in the FD is not essential for image decryption, although it potentially improves its quality, as we will see in section 8.2.3. However, the nonlinearity becomes essential for image decryption when the FrD is involved and, additionally, the key mask of the system is a fully complex-valued random distribution in the input plane of the JTC.

The FrT at parameters λ and $-z$ contained in equation (8.16) can be implemented optically by taking advantage of the relationship between the FrT and the fractional Fourier transform (FrFT) at this purpose [8]. Figure 8.2 (part II) also depicts the optical decryption scheme based on an optical FrT combined with an optical FrFT. In the latter, the distances d_1, d_2 and the focal length of the lens define the value of the fractional order that simulates the required optical inverse FrT [8, 10].

8.2.2.3 Fractional Fourier domain (FrFD)

A generalized formulation of nonlinear JTC-based encryption systems using the FrFT and its combination with a nonzero-order JTC was introduced with the purpose of increasing the security of the processor [11, 35]. This generalized formulation uses the following fractional Fourier operators: the FrFT, fractional traslation, fractional convolution and fractional correlation [11]. The mathematical expression of the encrypted image in the FrFD for the encryption system based on a JTC is a generalization of equation (8.5) and equation (8.10) because FT and FrT can be expressed in terms of FrFT by taking particular values of the fractional order α [10, 45].

The JTC-based encryption system in the FrFD preserves the shift-invariance property of the RPM or RCM $h(x)$ for the decryption system and the retrieval of the original input image. The system performance has been described in detail using the definitions of fractional correlation and fractional convolution [11].

The input and output planes of the JTC-based encryption system are related by an FrFT. The decryption process uses two successive optical FrFTs. The fractional order α of the FrFT provides an additional degree of freedom that can improve the security of the system and allow further control of the encryption and decryption stages. The optical FrFT can be implemented using a single lens positioned between the input and output planes, where the fractional order α can be defined in terms of the distances d_1 (distance between the input plane and the lens position), d_2 (distance between the lens position and the output plane) and the focal length (f_l) of the lens [10, 45]

$$\alpha = \arccos\left(\frac{\sqrt{(f_l - d_1)(f_l - d_2)}}{f_l}\right). \tag{8.17}$$

8.2.2.4 Gyrator domain (GD)

A fully-phase nonzero-order JTC architecture in the GD can be used in order to improve resistance to brute force attacks, CPA, KPA and COA of the security system [13]. In this case the combination of two nonlinearities along with the rotation angle of the Gyrator transform (GT) improves the security of the encryption scheme. Because rotations are involved in GT, we use 2D-notation in the mathematical expressions of this section. The nonlinear encrypted image for the fully-phase nonzero-order JTC architecture in the GD is

$$\begin{aligned} e_\alpha(u, v) &= \frac{\mathrm{JGPD}_\alpha(u, v) - |t_\alpha(u, v)|^2 - |h_\alpha(u, v)|^2}{|h_\alpha(u, v)|^2} \\ &= t_\alpha^*(u, v)\frac{h_\alpha(u, v)}{|h_\alpha(u, v)|^2} \exp\{\mathrm{i}2\pi(2x_0)v \csc \alpha\} \\ &\quad + t_\alpha(u, v)\frac{h_\alpha^*(u, v)}{|h_\alpha(u, v)|^2} \exp\{-\mathrm{i}2\pi(2x_0)v \csc \alpha\}, \end{aligned} \tag{8.18}$$

where α denotes the rotation angle of the GT and variables u and v are the coordinates in the GD. Functions $h_\alpha(u, v)$ and $t_\alpha(u, v)$ correspond to the GTs at parameter α of the functions $h(x, y)$ and the product $f_{ph}(x, y)r(x, y)$, respectively. Function $f_{ph}(x, y)$ is the original input image encoded in phase: $f_{ph}(x, y) = \exp[\mathrm{i}2\pi f(x, y)]$. In (8.18), the joint Gyrator power distribution (JGPD) at parameter α is defined as [13]

$$\mathrm{JGPD}_\alpha(u, v) = |\mathrm{GT}^\alpha\{g(x - x_0, y) + c(x + x_0, y)\}|^2, \tag{8.19}$$

where the operator GT^α represents the GT at parameter α of two non-overlapping data distributions, $g(x, y)$ and $c(x, y)$, placed side-by-side in the input plane of the JTC at coordinates $(x, y) = (x_0, 0) = (a, 0)$ and $(x, y) = (-x_0, 0) = (-a, 0)$, respectively. These data distribution are given by

$$\begin{aligned} g(x, y) &= \exp\{-\mathrm{i}2\pi x_0 y \cot \alpha\} f_{Ph}(x, y)r(x, y), \\ c(x, y) &= \exp\{\mathrm{i}2\pi x_0 y \cot \alpha\} h(x, y). \end{aligned} \tag{8.20}$$

The pure linear phase terms presented in $g(x, y)$ and $c(x, y)$ (equation (8.20)) are symmetrically introduced to ensure the complete overlapping of the GTs of the two data distributions displayed on the input plane of the JTC, at the camera plane where the JGPD is captured. The two RPMs $r(x, y)$ and $h(x, y)$ are given by the following equations

$$r(x, y) = \exp\{\mathrm{i}2\pi m(x, y)\}, \qquad h(x, y) = \exp\{\mathrm{i}2\pi n(x, y)\}. \tag{8.21}$$

Therefore, the input plane of the JTC-based encryption scheme is defined using the two phase-only images $g(x, y)$ and $c(x, y)$. Function $f(x, y)$, and the random codes $m(x, y)$ and $n(x, y)$ are images with $M \times N$ pixels. Figure 8.3 shows a representation of the input plane of the JTC-based encryption scheme (figure 8.4). With $x_0 = M/2$ pixels, the spatial region of the input plane is fully covered by the input functions with no blank space. This modification allows a more efficient use of the zone available for data display at the input plane of the encryption system.

Figure 8.4 depicts the optical encryption scheme (part I) based on a fully phase nonzero-order JTC architecture and a GT, and the optical decryption scheme (part II) based on two successive GTs. The security keys needed for decryption are the two RPMs $r(x, y)$ and $h(x, y)$, and the rotation angle α of the GT. In this case, more specifically, the two RPMs $r(x, y)$ and $h(x, y)$ are keys because the original input image was encoded in phase.

In the decryption process (figure 8.4, part II), we perform the product between the encrypted data distribution $e_\alpha(u, v)$ and the GT at parameter α of $c(x + x_0, y)$, this product is given by

$$\begin{aligned} d_\alpha(u, v) &= e_\alpha(u, v)\mathrm{GT}^\alpha\{c(x + x_0, y)\} \\ &= t_\alpha^*(u, v)\frac{h_\alpha^2(u, v)}{|h_\alpha(u, v)|^2}\exp\{\mathrm{i}2\pi(3x_0)v \csc \alpha\} \\ &+ t_\alpha(u, v)\frac{h_\alpha^*(u, v)h_\alpha(u, v)}{|h_\alpha(u, v)|^2}\exp\{-\mathrm{i}2\pi x_0 v \csc \alpha\}. \end{aligned} \tag{8.22}$$

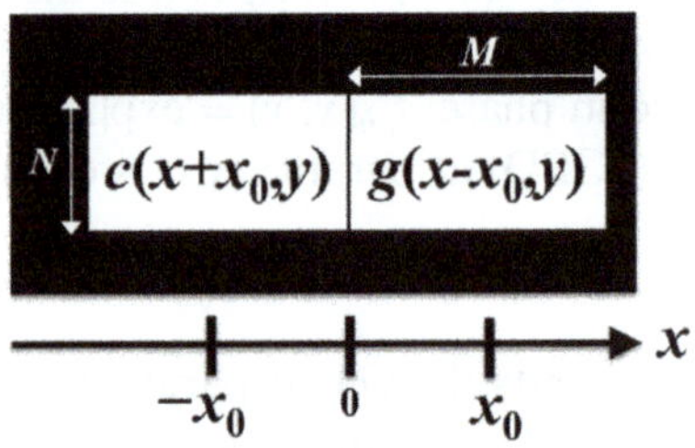

Figure 8.3. Placement of the data distributions in the input plane of the JTC architecture for the encryption system. Reprinted from [13], with permission from Elsevier, copyright 2017 Elsevier Ltd.

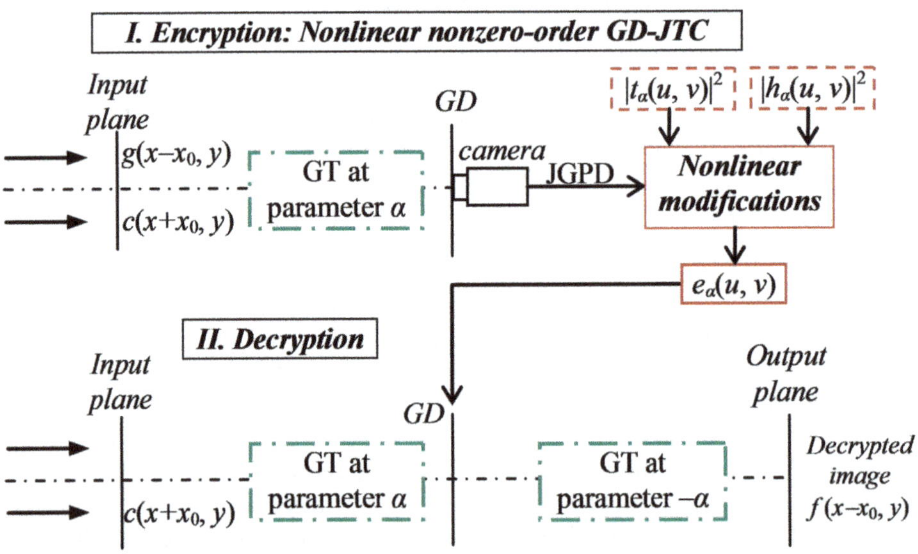

Figure 8.4. Scheme of the optical setup composed by an encryption scheme based on a fully phase nonzero-order JTC in the GD and a decryption scheme based on two successive GTs. Reprinted from [13], with permission from Elsevier, copyright 2017 Elsevier Ltd.

In the optical implementation, the two terms of the previous equation would separate angularly from each other. By Gyrator transforming at parameter $-\alpha$ the second term of equation (8.22) and then, subtracting the phase shift ($-i2\pi x_0 y \cot\alpha$), multiplying by the complex conjugate of $r(x—x_0, y)$ and finally, extracting the argument of the resulting phase (function arg{•}), we obtain a version of the decrypted image at coordinate $(x, y) = (x_0, 0) = (a, 0)$ given by

$$2\pi\tilde{f}(x - x_0, y) = \arg\{GT^{-\alpha}\{\exp\{-i2\pi x_0 v \csc\alpha\} t_\alpha(u, v)\} \exp\{i2\pi x_0 y \cot\alpha\} r^*(x - x_0, y)\}. \quad (8.23)$$

The JTC-based encryption system in the GD also preserves the shift-invariance property with respect to the location of the RPM $h(x, y)$ for the decryption system and the retrieval of the original input image [13].

The simulation results for the encryption and decryption schemes in the GD are shown in figure 8.5. The original image to encrypt $f(x, y)$ and the random distribution code $m(x, y)$ of the RPM $r(x, y)$ are displayed in figure 8.5(a) and (b), respectively. The random distribution code $n(x, y)$ of RPM $h(x, y)$ has different values but a similar appearance to the image presented in figure 8.5(b). The images $f(x, y)$, and the random codes $m(x, y)$ and $n(x, y)$ are 512 × 512 pixel size (M = N = 512). The encrypted image $e_\alpha(u, v)$ for the rotation angle $\alpha = 0.885\pi$ in the GT is depicted in figure 8.5(c); the digital GT was implemented using the fast algorithm of discrete GT based on the convolution operation [46]. The final distributions in phase for the output plane of the decryption scheme with the correct value keys, the rotation angle α of GT, and the RPMs $r(x, y)$ and $h(x, y)$, are shown in figure 8.5(d). The decrypted image presented in figure 8.5(e) is the magnified region of interest, centered at position $(x, y) = (x_0, 0) = (a, 0)$, of figure 8.5(d). Figure 8.5(f) shows an unsuccessful decryption produced when the term in the denominator of equation (8.18) is not introduced in the encrypted function, even though the correct values keys are used in the decryption scheme. The resulting image is noisy and does not disclose the original image. The noisy distribution of figure 8.5(g) corresponds to the result obtained when the key of the RPM $r(x, y)$ is wrong although the rest of the decryption process is correct. When an incorrect RPM $h(x, y)$ or value of the rotation angle α are used in the decryption scheme, the results after the decryption process are noisy distributions similar to figure 8.5(g). Therefore, the results provided prove that the all correct parameters and keys (the rotation angle α, and the RPMs $r(x, y)$ and $h(x, y)$) are required in the decryption scheme for the correct retrieval of the original image.

The GT has a period equal to 2π with respect to the rotation angle α. The rotation angle can be expressed as $\alpha = p\pi/2$, where p has a period equal to 4. The sensitivity on the rotation angle p of the GT for the decrypted images has been examined in [13] by introducing a small error in this, and then by calculating the root mean square error (RMSE) metric between the original image $f(x, y)$ and the decrypted image $\tilde{f}(x, y)$ to measure the level of protection of the encrypted image $e_\alpha(u, v)$. The RMSE is defined as

Figure 8.5. (a) Original image $f(x, y)$ to encrypt. (b) Random distribution code $m(x, y)$ of the RPM $r(x, y)$. (c) Encrypted image $e_\alpha(u, v)$ for the rotation angle $\alpha = 0.885\pi$. (d) Phase values of the output plane for the decryption scheme with the correct keys, rotation angle α of GT and keys RPMs $r(x, y)$ and $h(x, y)$. (e) Magnified region of interest of the output plane (d) corresponding to the correctly decrypted image. Incorrectly decrypted images when: (f) the nonlinear denominator $|h_\alpha(u, v)|^2$ of equation (8.18) is not introduced in the encrypted function, (g) an incorrect RPM $r(x, y)$ key is used. Reprinted from [13], with permission from Elsevier, copyright 2017 Elsevier Ltd.

$$\mathrm{RMSE} = \left(\frac{\sum_{x=1}^{M} \sum_{y=1}^{N} [f(x, y) - \tilde{f}(x, y)]^2}{\sum_{x=1}^{M} \sum_{y=1}^{N} [f(x, y)]^2} \right)^{\frac{1}{2}}. \tag{8.24}$$

Figure 8.6 shows the RMSE versus the relative error of p for the image retrieval and it was found [13] that the system was sensitive to a variation of 10^{-7} in p. This sensitivity was higher than the sensitivity to the fractional order of the FrFT used in similar JTC-based encryption systems, being sensitive to variations of 10^{-4} in such parameter as reported in reference [11]. This feature provides evidence of the improvement achieved in securing the encryption–decryption systems studied so far.

Concerning cryptanalysis, the described nonlinear JTC-based encryption system in the GD is resistant to the COA. Let us recall that the COA uses an iterative phase retrieval algorithm for retrieving an arbitrary plaintext by using only a ciphertext, whenever the plaintext was encoded in amplitude at the input plane of the JTC-based encryption scheme [25]. The COA fails when it is applied to the encryption scheme described above, because the plaintext is encoded in phase at the input plane of the JTC architecture, the encryption system is nonlinear and this system is defined in the GD [13]. It is significantly harder to gain any success by attacking the encryption system due to the difficulty of retrieving the phase information (the plaintext encoded in phase) from a single intensity (ciphertext) using the phase retrieval algorithm [22, 25]. The COA is not able to retrieve the correct RPMs $r(x, y)$ and $h(x, y)$ either [22].

8.2.3 Increasing the quality of the decrypted image

8.2.3.1 Nonlinear encrypted distribution

Encryption based on a JTC where the whole information (primary image and two random phase masks) was easily introduced in the same input plane of the JTC system [30, 31] presents some practical advantages as mentioned in section 8.2.1. However, it also presents some drawbacks, such as some vulnerability to certain intruder attacks [32], as shown in section 8.2.2, and the poor quality of the image retrieved after decryption [32]. Let us now review the origin of this image degradation and study some modifications of the encryption algorithms based on JTC architecture that permit one to significantly increase the quality of the retrieved image with no loss of security.

In a modified JTC-encryption system, the encrypted distribution or JPS $e(u)$ is given by (8.2). In the decryption process (figure 8.1, *via 1*), the RPM $h(x)$ is placed at

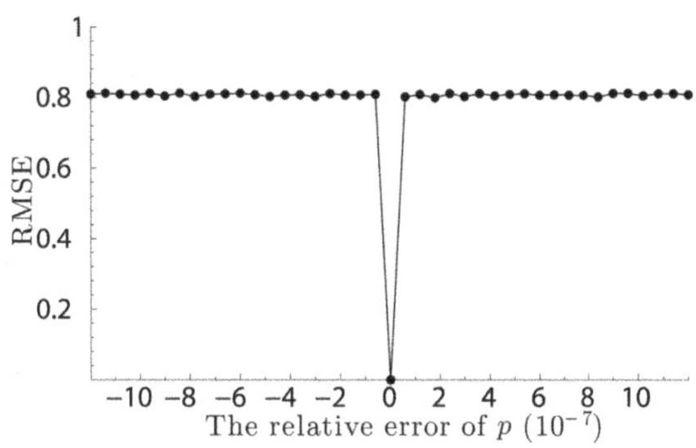

Figure 8.6. Results obtained for the RMSE metric when the decryption process is performed using a small relative error on p. Reprinted from [13], with permission from Elsevier, copyright 2017 Elsevier Ltd.

coordinate $x = -a$ and, consequently, in the Fourier plane, the encrypted power spectrum $e(u)$ is illuminated by $H(u)\exp[i2\pi au]$, as shown in figure 8.1. This product can be mathematically developed into four terms, of which the one of interest is $[F(u)*R(u)]H^*(u)H(u)e^{-i2\pi au}$ that, after Fourier transforming, leads to

$$\begin{aligned} d(x) &= \mathrm{FT}^{-1}\{[F(u)*R(u)]H^*(u)H(u)e^{-i2\pi au}\} \\ &= f(x)r(x)*\{h(x) \otimes h(x)\}*\delta(x-a), \end{aligned} \tag{8.25}$$

where symbol ($\otimes$) indicates correlation. Although RPM-I $r(x)$ is a phase-only function, the intensity distribution of $d(x)$ at coordinate $x = a$ would no longer be the intensity of the original function $f(x)$ as was obtained in [1, 15]. Note that in (8.25) the product $f(x)r(x)$ appears convolved by the complex-valued autocorrelation of RPM-II $h(x)$. The more this autocorrelation approaches a Dirac delta function (i.e. $h(x) \otimes h(x) \cong \delta(x)$) [30], the more the intensity distribution of $d(x)$ will resemble the original image intensity $|f(x)|^2$. However, as a result of such modification in the input plane, the decrypted image is frequently affected by poor quality. This drawback can be overcome by introducing the nonlinear operation in the encrypted function [32] described in section 8.2.2.1. This nonlinearity consists of dividing the JPS by the squared magnitude of the Fourier transform of the RPM-II (figure 8.1, *via 2*) (equation (8.5)). In the decryption process, the resulting distribution at the output plane is then described by (8.9).

The intensity of equation (8.9) produces the original image intensity $|f(x)|^2$ at coordinate $x = a$. Unlike (8.25), the autocorrelation of RPM-II is no longer contained in (8.9). It is worth mentioning that the introduced nonlinearity allows the whole system to approach the output result of DRPE more closely as was originally formulated in [1].

The simulation results of the encryption and decryption process are shown in figure 8.7: the image to be encrypted (original image) is presented in (a), the random distribution code $n(x)$ embedded in the RPM-II $h(x)$ in (b), and a truncated linear plot of the autocorrelation of the RPM-II $h(x)$ in (c). The decrypted image in (d) shows the result when the linear process (8.2) is followed. To express the quality of decrypted image, the RMSE metric (equation (8.24)) is commonly used.

The RMSE between the original image and the decrypted image (figure 8.7(d)) using the linear process (*via 1*) is 0.687. The quality of the decrypted image compared to the original image is poor, due to the fact that the autocorrelation of the RPM $h(x)$ is not purely a Dirac delta function as depicted in figure 8.7(c). The autocorrelation of the RPM $h(x)$ has a noisy background that has an influence on the quality of the decrypted image. The decrypted image in (e) depicts the result obtained when the nonlinearity of equation (8.5) is applied (*via 2*). The RMSE between the original image and the decrypted image (figure 8.7(e)) using the non-linear process (*via 2*) is 0.061. This metric reveals the great improvement achieved in the quality of the decrypted image (equation (8.9)) due to the removal of the autocorrelation term.

The nonlinear modification of the JPS can be implemented using the optoelectronic setup of figure 8.1 (top part I, JTC) by following the procedure proposed in [47, 48]. The encrypted image given by (8.5) can be optically implemented by a two-

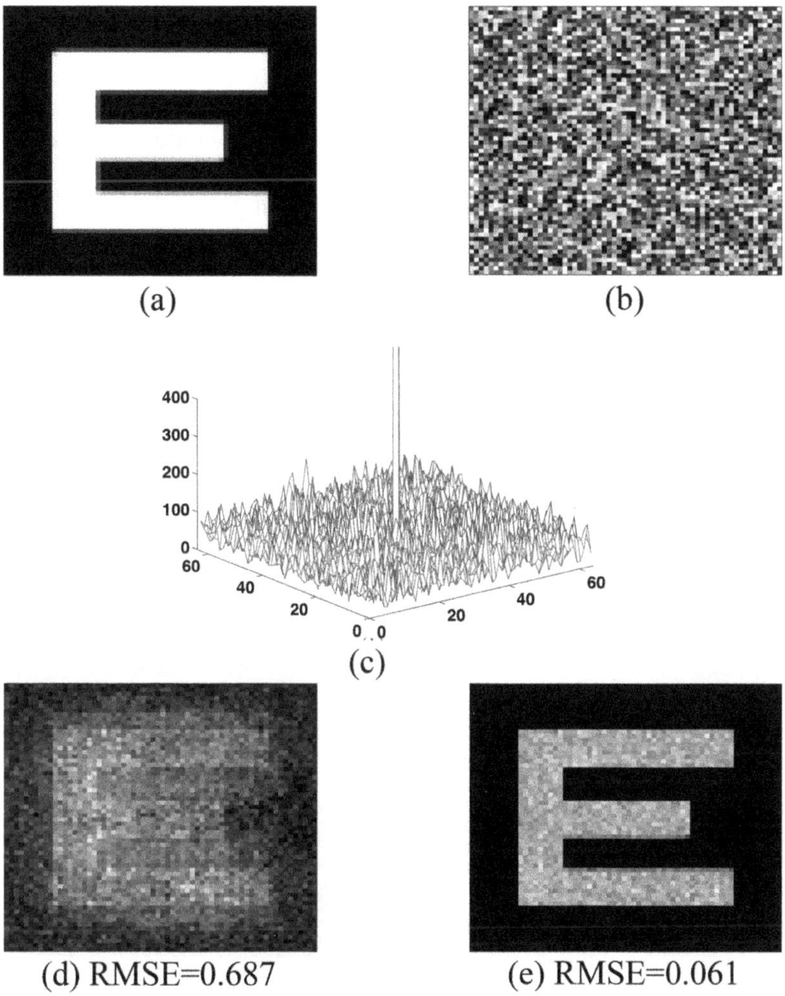

Figure 8.7. (a) Original image $f(x)$. (b) Random function $n(x)$ from the RPM-II $h(x)$, (c) autocorrelation of $h(x)$. Decrypted image using: (d) linear JTC (*via 1* in figure 8.1), (e) nonlinear JTC (*via 2* in figure 8.1). Reproduced with permission from [32].

step JTC. In the first step, the power spectrum of the security key $|H(u)|^2$ is captured. Then, the JPS of (8.2) is captured in the second step. Finally, the JPS is digitally divided by $|H(u)|^2$, and, thus, the encrypted image is computed. This encrypted distribution, along with the key, is the only information to be transmitted; therefore, this method does not increment the amount of data to be sent prior to the decryption stage.

Further improvement in the quality of the decrypted image can be obtained by using a nonzero-order joint transformation combined with the nonlinearity introduced in (8.5). Still in the Fourier domain, the removal of the zero order from JPS(u) is implied, so that the nonzero-nonlinear-encrypted image becomes

$$e_{NN}(u) = \frac{\mathrm{JPS}(u) - |F(u) * R(u)|^2 - |H(u)|^2}{|H(u)|^2}. \tag{8.26}$$

With this modification, useless terms would be suppressed in the output plane and an increase in the energy efficiency of the decrypted image would be expected.

Figure 8.8(a) shows the optoelectronic implementation of the encryption system with a phase-only spatial light modulator in the input plane. The main elements of the system are labelled in the setup. Figure 8.8(b) shows, in grayscale, the phase distribution displayed on the spatial light modulator, with the RPMs and the original image. The camera captures a sequence of intensity signals: $|H(u)|^2$, $|F(u) * R(u)|^2$, and JPS(u) (figure 8.9). The experimental encrypted function $e_{\mathrm{exp}}(u)$ of figure 8.9 is digitally computed from the captured intensity distributions using (8.26). The decryption process is digitally implemented to produce the result depicted at the bottom of figure 8.9.

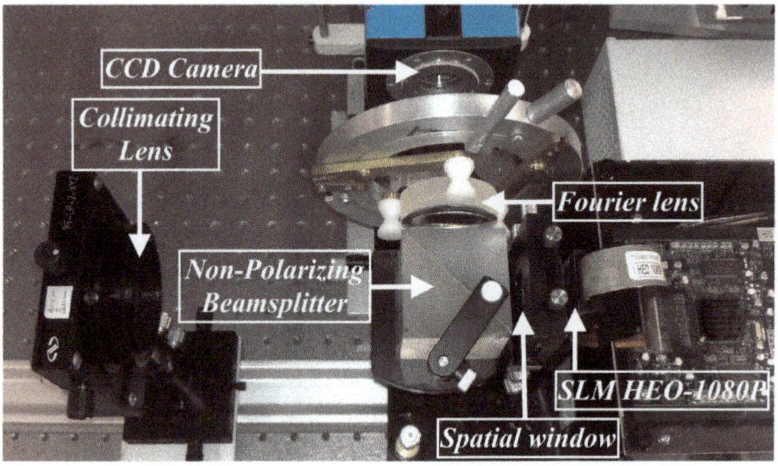

(a)

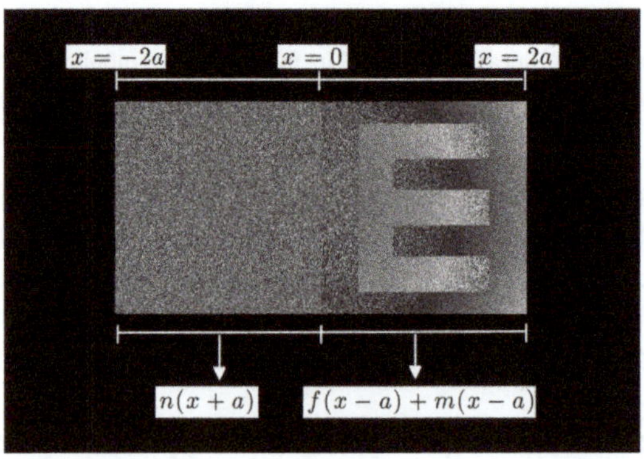

(b)

Figure 8.8. (a) Experimental setup of nonlinear encryption system based on JTC. (b) Grayscale representation of the input plane: phase distribution addressed to the SLM.

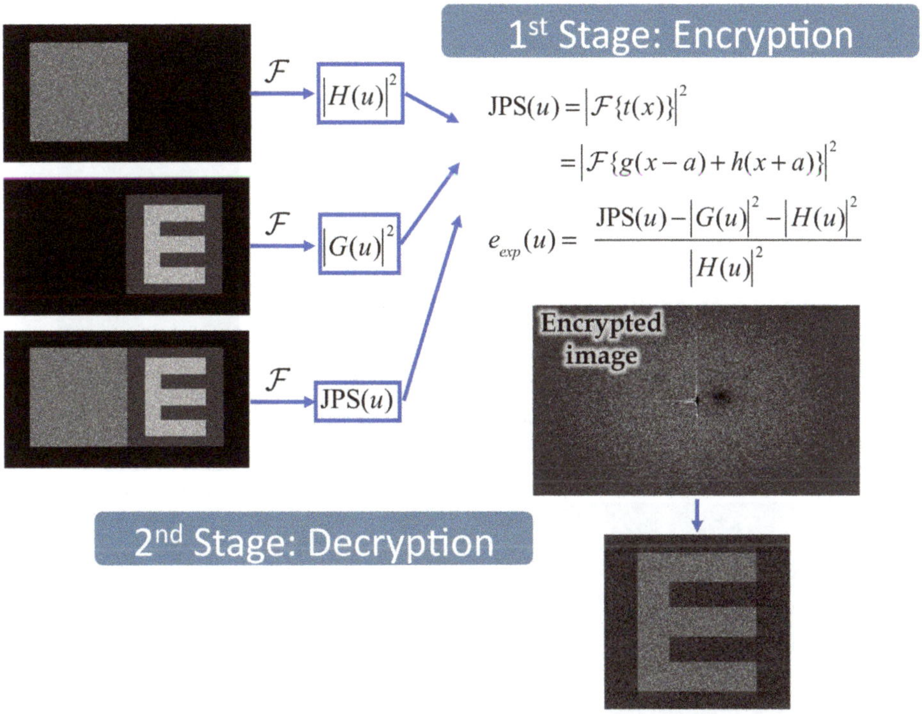

Figure 8.9. Encryption (top) and decryption (bottom) stages of the example implemented experimentally with the optoelectronic system of figure 8.8(a). In the encryption stage, a sequence of signals (top left) are displayed on a phase only spatial light modulator and their respective intensity spectrum are acquired by a camera to compute JPS(u) and the experimental encrypted image $e_{\mathrm{exp}}(u)$ according to equation (8.26). The decrypted image (bottom) is obtained by digital calculation.

Improving the quality of experimental decrypted images is also a matter of growing interest. Recently, for instance, a new procedure for noise reduction in the decryption process of experimental DRPE setups has allowed the use of grayscale images with a higher dynamic range in optical encryption [49].

The modifications of (8.5) and (8.26) have been also adapted to the FrD and FrFD using a generalized formulation [11] as described in section 8.2.2. Figure 8.10 illustrates in sequential steps the benefit of applying a nonzero-order nonlinear encrypted function in the FrFD for the fractional order $p = 1.5$ ($\alpha = p\pi/2 = 3\pi/4$). The original image (a) is shown along with a sequence of output planes obtained after following either the linear decryption process (b), or a non-linear decryption process, in the FrFD (c), or a nonzero-order nonlinear decryption process in the FrFD (d). Figure 8.10(e)–(g) shows the zoomed decrypted images with the respective RMSE values of ≈0.7, 0.187, and 0.012. This example demonstrates the improvement in energy and quality in this sequence of decrypted images with the introduced nonzero-order nonlinearities.

In addition to better image quality retrieval, this technique shows a certain resistance to the presence of some types of degradation of the encrypted function [11]. Following

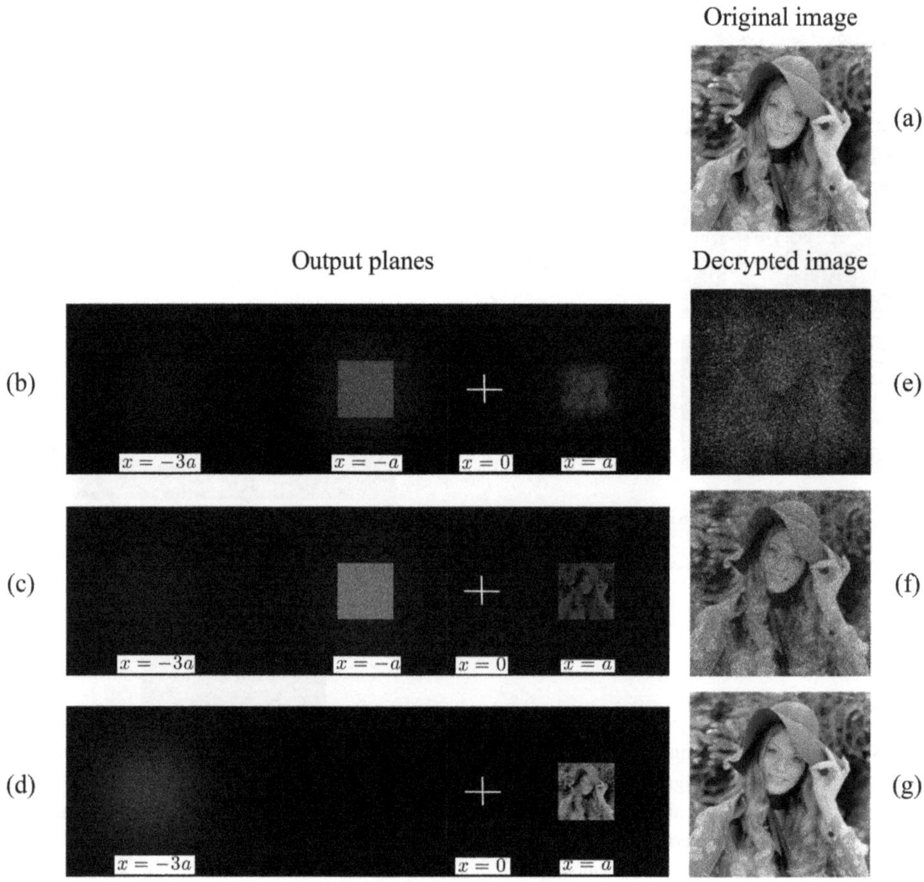

Figure 8.10. Decryption results obtained from an encrypted function in the fractional Fourier domain for the fractional order $p = 1.5$ ($\alpha = 3\pi/4$). (a) Original image. Output plane obtained after (b) a linear decryption, (c) a nonlinear decryption, and (d) a nonzero-order nonlinear decryption. (e)–(g) zoomed decrypted images of (b)–(d), respectively. Reproduced with permission from [11].

the original image of figure 8.10(a), for instance, figure 8.11 depicts the decrypted images obtained after corrupting the nonzero-order nonlinear encrypted image in the FrFD with $p = 1.5$ ($\alpha = 3\pi/4$), with Gaussian white noise with zero mean and variance of $\sigma^2 = 0.2$ in the form of additive noise (figure 8.11(a)), multiplicative noise (figure 8.11(b)), and 25% occlusion of the top left corner (figure 8.11(c)). With the respective RMSE values of 0.251 (figures 8.11(a)), 0.238 (figures 8.11(b)), and 0.406 (figure 8.11(c)), the original image could be still recognized in all of them.

The nonlinear JTC encryption system described in section 8.2.2 and in this section simultaneously increases the decrypted image quality and improves its overall security. Cryptanalysis of these encryption systems demonstrates their resistance against a variety of system attacks. The nonlinearity introduced in the encrypted function through (8.5) (or analogous equations in a generalized formulation [11])

(a) (b) (c)

Figure 8.11. Decrypted images after corrupting the nonzero-order nonlinear encrypted distribution with: (a) Gaussian white noise with zero mean and variance of $\sigma^2 = 0.2$ in the form of additive noise; (b) multiplicative noise; and (c) 25% occlusion of the top left corner. Reproduced with permission from [11].

makes the system more resistant to chosen and known plaintext attacks in the various FD, FrD, FrFD and GD [8, 11, 13, 32].

8.2.3.2 Influence of the probability density function of random phase masks

We have additionally analyzed the influence of the probability density function used to generate the two RPMs in the system's resistance against attacks. Let us illustrate this with two cases of study concerning the nonlinear JTC-based encryption system in the FD [32]. In both cases, the original image to be encrypted is represented by figure 8.7(a), its corresponding nonlinear encrypted image is calculated by using (8.5), and the modulus and phase of the FT of the security key $H(u)$ are denoted by $|H(u)|$ and $\phi(u)$, respectively. Let us consider different types of random distributions to generate the RPMs.

The first case studied considers uniform random distributions so that all grey levels have the same probability. For example, the images shown in figure 8.12(a) and (b) depict $m(x)$ and $n(x)$ with the named *Uniform* and *Beta* random distributions, respectively, and their histograms, which reveal their relative uniformity. These random codes were used to obtain the two RPMs and therefore, the encrypted image. Let us assume that the system is attacked and, as a result, partial access to the security key $H(u)$ is achieved. In a first experiment, we consider that only the modulus $|H(u)|$ is accessed and perform the decryption process using only this information and setting $\phi(u) = 0$ (or any constant value). In the decrypted image (figure 8.12(c)) the original image cannot be discerned from the noisy background. In a second experiment, we consider that the phase information $\phi(u)$ of the security key has been accessed and perform again the decryption process assuming $|H(u)| = 1$ (or any constant value). Now, in the decrypted image (figure 8.12(d)) the original image can be clearly distinguished from the noisy background even though it is poor quality. The results for the decrypted images indicate that the phase of the security key is more relevant in the decryption process than the modulus information, whenever $m(x)$ and $n(x)$ use uniform random distributions.

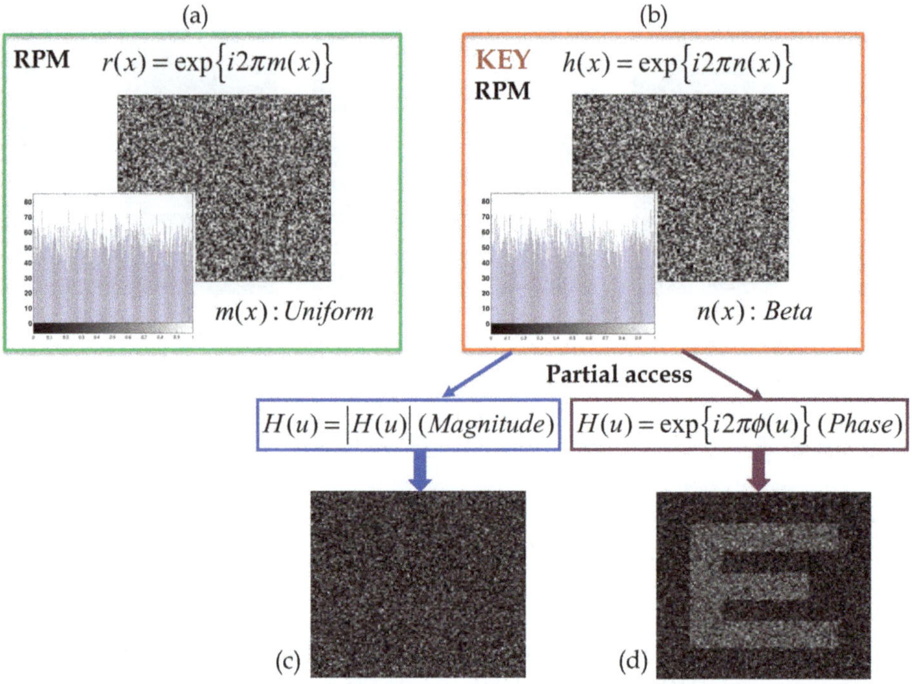

Figure 8.12. Scheme of the first case of study (see the text for details).

The second case of study considers non-uniform random distributions to generate the RPMs. In particular, *Weibull* and *Chi-Square* random distributions are considered to generate $m(x)$ and $n(x)$, respectively. These images of random distribution codes are shown in figure 8.13(a) and (b), along with their histograms. The nonlinear encrypted function (equation (8.5)) is computed again for the same original image with the new RPMs. For the decryption process, let us consider the full access on $\phi(u)$, which has been revealed as the most relevant piece of information if only partial data on the FT of the security key are available. The corresponding decrypted image, assuming $|H(u)| = 1$, is shown in figure 8.13(c). This result demonstrates that the retrieval of the original image is much harder when $m(x)$ and $n(x)$ are generated by non-uniform random distributions. We have repeated the latter experiment for different combinations of random distributions (within Uniform, Weibull and Chi-Square) to generate $m(x)$ and $n(x)$, and the results also differ [32]. For instance, figure 8.13(d) shows the decrypted image when the random distributions of figure 8.13(a) and (b) swap in their roles, that is, when $m(x)$ and $n(x)$ are given by *Chi-Square* and *Weibull* random distributions, respectively. In figure 8.13(d), the original image can be identified, although with some difficulty. The result is clearly different from figure 8.13(c). Summing up, the probability density function does play a role in the security of the encryption system. The encryption system is asymmetric with respect to the random distributions used for the two RPMs, and the secret of the encrypted image is better protected when a non-uniform random distribution is used in the generation of RPM-II.

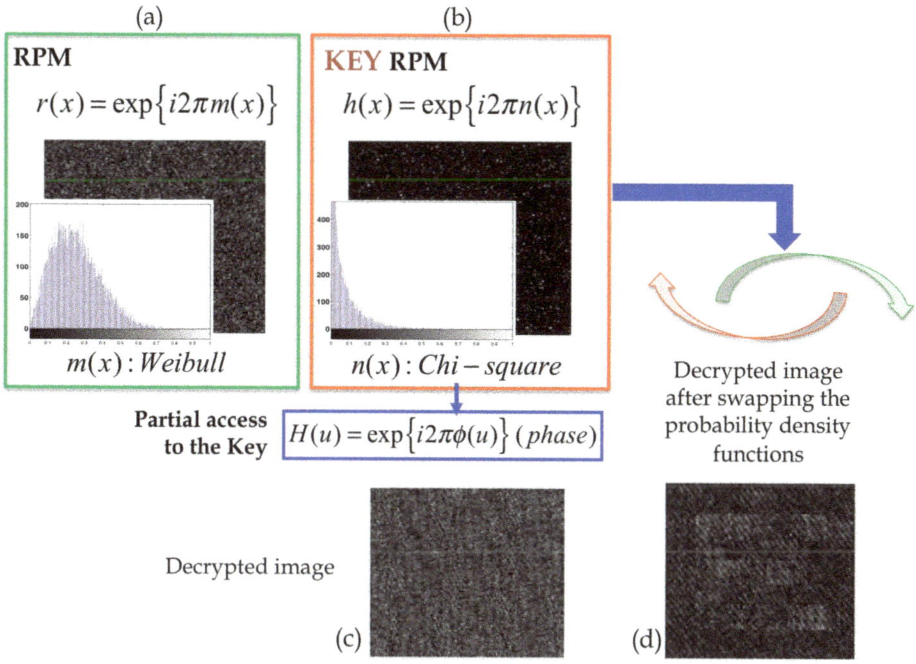

Figure 8.13. Scheme of the second case of study (see the text for details).

8.3 Nonlinear techniques for information authenticaton and compression

Unlike previous sections, which drew our attention to optical security systems designed to decrypt a signal (primary image or plaintext) embeded in a encrypted distribution (ciphertext), so that the recovered information replicated the original image (information recovery), we are now focusing on systems intended to perform a more complex operation of authentication. These security systems must compare the decrypted signal, which sometimes does not visually reveal the original image, with a reference signal and recognize their similarity. Pattern recognition algorithms based on matched filtering and optical correlation techniques have been successfully applied to these systems for encryption and validation purposes [26, 27, 29].

Since the quality of the decrypted image is no longer the main goal, some compression can be achieved by different procedures. For instance, binarization of the amplitude or phase distribution of the encrypted signal has been explored [50]. A successful technique that achieved compression of the encrypted distribution without negatively affecting the final signal validation was proposed in 2011 [27]. Photon-counting imaging techniques were integrated with optical encryption algorithms. The generated encrypted function consists of a sparse distribution, which permits bandwidth reduction and increases robustness against intruder attacks [51]. Other combinations of photon-counting (PhC) imaging techniques with different encryption techniques have been recently proposed [52, 53].

A different approach that also achieves information compression corresponds to the multifactor optical encryption-authentication (MOEA) [26, 54]. The main goal of the MOEA technique is the digital encoding of several signals at the same time in the same ciphering plane. By doing this, the MOEA technique already compresses the information of up to four signals into a single encrypted distribution with the same spatial resolution as the original primary images. The MOEA allows a significant reduction of the amount of information to be transmitted, as well as increased security since the whole set of primary signals has to be simultaneously validated to obtain a positive authentication output. The MOEA procedure permits not only validation of a person's identity but also, for instance, identification of a person's vehicle or the place the person is permitted to access.

Compression of multifactor optical encryption is even higher when this technique is applied together with the aforementioned photon-counting imaging techniques [29]. Some of these proposals are reviewed in the following subsections.

8.3.1 Photon-counting information verification

In photon-counting (PhC) imaging systems, images are captured under photon starving conditions by controlling the expected number of incident photons, N_p, in the entire scene [27, 55]. Thus, in general, a photon-limited image has less information than the original counterpart. The probability of counting l_j photons at pixel x_j can be shown to be Poisson distributed [56]

$$P_d(l_j; \alpha_j) = \frac{[\alpha_j]^{l_j} e^{-\alpha_j}}{l_j!}, \quad l_j = 0, 1, 2\ldots, \tag{8.27}$$

where l_j is the number of photons detected at pixel x_j and the Poisson parameter α_j is given by $\alpha_j = N_p g(x_j)$ with $g(x_j)$ being the normalized irradiance at pixel x_j such that $\sum_{j=1}^{M} g(x_j) = 1$ and M the total number of pixels in the scene.

8.3.1.1 Single plaintext verification

A photon-limited version of an encrypted distribution consists of a sparse representation of the encrypted information (figures 8.14(a)–(c)). In general, this sparse representation is used for decryption and, as a result, a noisy decrypted signal, which is not recognizable by visual inspection, is retrieved. By following this procedure, intruders cannot easily recognize the decrypted image (figure 8.14(d)) retrieved from the sparse encrypted distribution since it is not intended for visualization of the primary image, but for verification of the information by means of optical correlation (figure 8.14(e)). Thus, the integration of photon-counting techniques along with DRPE [51] introduces a nonlinear operation to the encryption procedure that permits an additional layer of information protection by increasing the system security and makes the verification process more robust against unauthorized attacks.

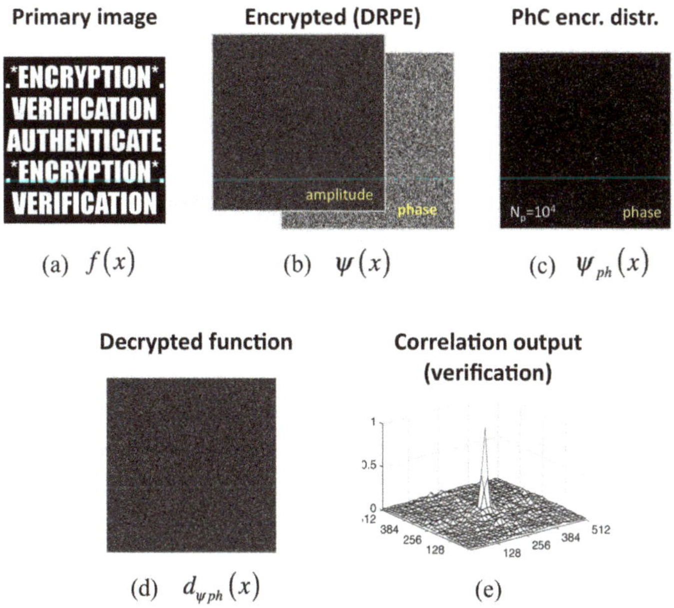

Figure 8.14. (a) Primary image; (b) encrypted distribution using the DRPE technique; (c) photon-limited encrypted distribution with $N_p = 10^4$; (d) decrypted function; (e) correlation output when the decrypted function (d) is compared to the primary image (a) through nonlinear correlation [51].

Taking into account that the encrypted distribution is of a complex nature, commonly, both amplitude and phase must be kept for decryption. However, after applying the photon-counting imaging technique to the encrypted distribution, the amplitude information turns to be a binary distribution, whereas the phase information corresponding to the non-zero amplitude pixels is kept between $[0, 2\pi]$ with its maximum resolution of 8 bits (figure 8.14(c)). Further compression can be achieved if one limits the number of bits used for representing the phase information [51, 57, 58]. In [57, 58], the quality needs in the reconstruction of 3D objects from digital holograms required at least 4–5 bits to obtain an intense correlation peak between the compressed and the uncompressed reconstructed images. The required number of bits increased to about 6–7 bits in the visualization of a good 3D object reconstruction [57, 58]. In contrast, the verification of the encrypted information and its discrimination from very similar but non-authorized 2D signals, permits a stronger reduction in the number of bits (to just 2 bits) [51].

It is worth mentioning that the PhC-DRPE algorithm also introduces the possibility of image retrieval from the correlation output result (figure 8.14(e)). According to the correlation output, taking into account the peak intensity and location, the verification system can also synthesize the retrieved image with the elements of the scene correctly placed [51].

8.3.2 Multifactor optical encryption-authentication (MOEA)

Some approaches permit the storing of multiple primary images, either in an optical memory [5] or in a single encrypted distribution [59–61], with the purpose of sequential and independent one-by-one decryption. Alternatively, security can be reinforced by combining several authenticators in a single encrypted distribution by using the MOEA technique [26, 28, 54, 62], which aims for the simultaneous verification of more than one primary image. Unlike sequential encryption methods, the MOEA technique achieves digital encoding of all the signals at the same time in the same ciphering plane. The original MOEA technique already compresses the information of up to four signals into a single encrypted distribution with the same spatial resolution as the primary images. Encryption and authentication of multiple signals is an important achievement in optical security applications since it increases the system reliability because its response does not rely on the verification of a single factor but on a set of factors [26, 63]. They all must obtain positive authentication to provide a final validation.

The MOEA algorithm was firstly introduced in 2006 [26] to provide simultaneous encryption of up to four factors (named $r(x)$, $s(x)$, $b(x)$ and $n(x)$) into a single complex-valued distribution ($\psi(x)$) given by

$$\psi(x) = t_{r+2b}(x) * t_s(x) * FT^{-1}[t_{2n}(x)]. \tag{8.28}$$

In (8.28) all the factors are phase-encoded, that is, for a general function $a(x)$, $t_a(x) = \exp[j\pi a(x)]$ (in the case of two signals placed together $t_{r+2b}(x) = t_r(x)t_{2b}(x)$), FT^{-1} denotes an inverse Fourier transform and $*$ is the convolution operation. The signals for encryption can be of a different nature: biometrics, logos, traces, random codes, text or others. They are scrambled all together into a dim, noisy-like encrypted function that does not reveal any piece of information of the factors being protected. For instance, let us consider the case depicted in figure 8.15(a) for which a security control requires the joint verification of the driver's identity through

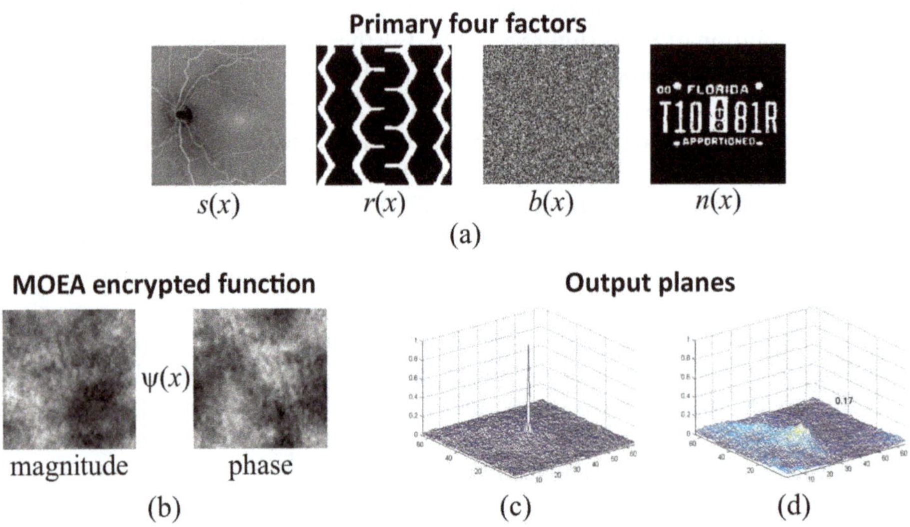

Figure 8.15. Example of MOEA techniques: (a) four original factors to be encrypted; (b) magnitude and phase of the encrypted function; output planes for (c) the correct simultaneous validation of the four factors; (d) a negative validation.

their retina scan $s(x)$ along with the tyre pattern of the vehicle $r(x)$, the place intended to be accessed through the code $b(x)$, and the vehicle plate $n(x)$ [64]. Figure 8.15(b) shows the amplitude and phase distributions of the encrypted information obtained through equation (8.28).

The MOEA procedure permits the simultaneous optical authentication of a whole set of factors hidden in function $\psi(x)$ by means of their comparison with *in situ* captured images and information obtained from a database. An optical processor composed of a joint transform and a 4*f* correlator linked through a nonlinear operation (figure 8.16 using MOEA in the input plane as depicted with a red solid line) provides a sharp and intense output peak only when all the factors are verified positively. Otherwise, when one or more checked images differ from the factors previously encrypted the output does not reveal the presence of any signal. In the previous example of figure 8.15, part (c) depicts the output plane obtained for a positive validation of the whole set of signals and (d) the corresponding output for a negative authorization when one or more of the signals do not resemble the correct ones.

The encrypted function $\psi(x)$ can be further manipulated to obtain a distortion-invariant identity (ID) tag [62, 65] for remote surveillance or tracking of vehicles or moving objects. Additionally, the ID tag can be designed to require near infrared (NIR) illumination to retrieve its content. This makes it invisible to the naked eye, to

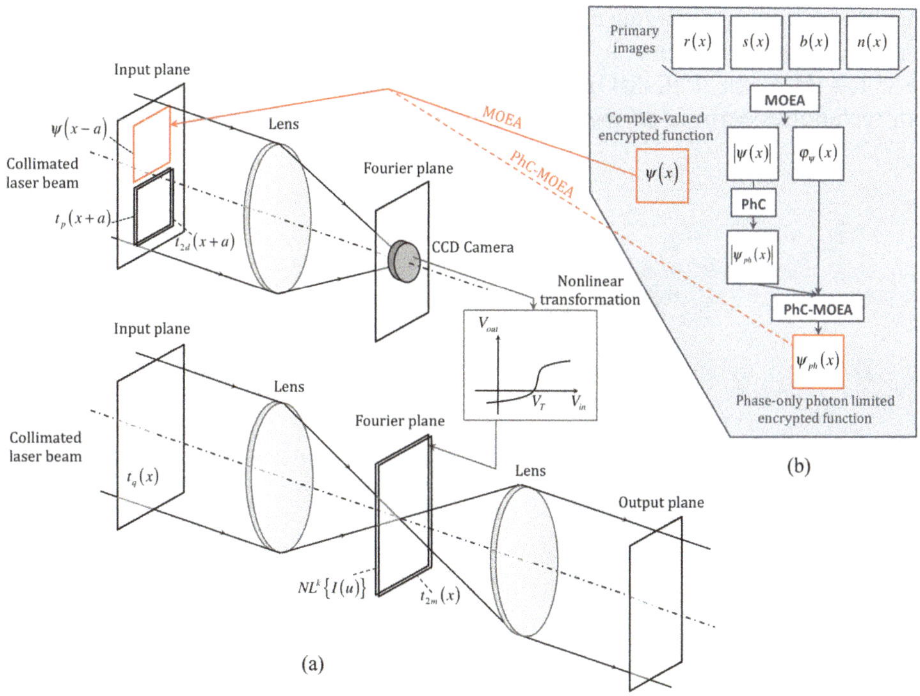

Figure 8.16. (a) Optical processor for the MOEA decryption-authentication system. (b) The encrypted function $\psi(x - a)$ at the input plane can either correspond to the multifactor encryption (MOEA) or to the multifactor encryption followed by the photon-counting procedure (PhC-MOEA).

most common inspection cameras and, therefore, more secure. Furthermore, information redundancy on the designed ID tag has been proved to allow verification robustness against scratches or data loss due to handling or wear damage [28, 54].

8.3.2.1 Photon-counting multifactor information compression

Similar to other ciphering methods, the resultant encrypted function $\psi(x)$ is a complex-valued distribution that involves certain difficulties when trying to reproduce it by commonly available optical means (cameras, spatial light modulators). The separate preservation of the magnitude and phase information of the encrypted distribution is usually required to obtain satisfactory image retrieval or information verification. In practice, however, further reduction or simplification of the encrypted content to transmit can be necessary. Photon-counting imaging techniques have been successfully implemented for the encryption of single primary images [51] achieving certain information compression as aforementioned in section 8.3.1. A recent application of the photon-counting imaging technique to the MOEA (PhC-MOEA depicted with a dashed red line in figure 8.16(b)) shows the preservation of the good qualities of the multifactor encryption, and sheds light on a more powerful and secure system in comparison to the original version [29].

To illustrate these properties with an example, let us consider the retinal scans of two different eyes and two random phase codes as input factors (figure 8.17(a)). They have been merged together in an MOEA distribution by applying (8.28) (figure 8.17(b)). The photon-counting imaging technique (equation (8.27)) is applied to the multifactor encrypted function so that a sparse phase-only distribution is generated for the encrypted data, the PhC-MOEA distribution (figure 8.17(c)). The integration of both techniques permits an increased capacity for signal hiding with simultaneous

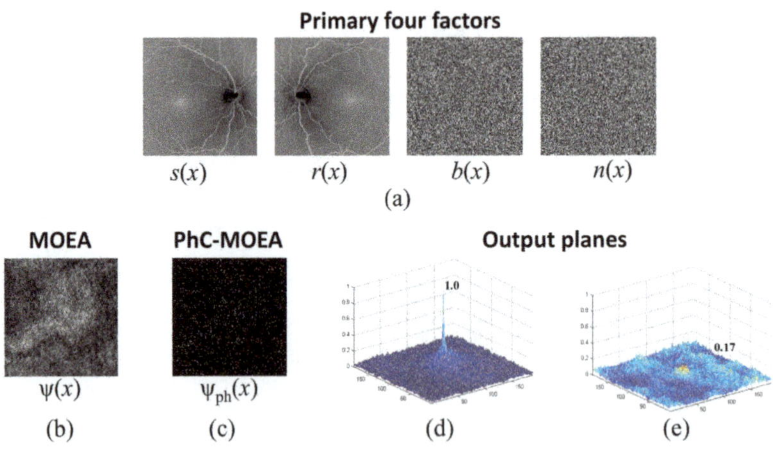

Figure 8.17. Example of the PhC-MOEA technique: (a) four input factors; (b) complex-valued multifactor encrypted distribution (only its magnitude is depicted); (c) photon-counting-MOEA sparse phase-only encrypted function; output planes when (d) a positive authentication of the four factors is achieved and (e) a negative validation occurs for any mismatch.

data reduction to better fulfill the general requirements of protection, storage and transmission.

Only when the four input signals in the validation stage correspond to the four factors previously encrypted, does the output plane of the processor consist of a sharp and intense autocorrelation peak that accounts for the positive simultaneous validation of the whole set of factors (figure 8.17(d)). Any mismatch among the four factors involved in the encrypted distribution and the four checked signals leads to a drastic decrease of the output correlation peak indicating a negative validation (figure 8.17(e)) [29].

These results show the satisfactory performance of the PhC-MOEA integration in terms of multifactor authentication. In this section, it will be pointed out that the proposed integration also benefits the system security.

Cryptanalysis of the original MOEA method has been carried out in terms of CPA and chosen-ciphertext attacks (CCA) [29]. Let us consider a CPA where an intruder uses two common plaintexts such as dot functions and that the two random phase codes are known by the processor in the encryption stage. In this case, the encrypted function keeps its noisy appearance. When the resulting ciphertext is introduced in the decryption-verification stage, with the retinal scans captured *in situ*, and the RPMs obtained from the processor database, a negative validation is achieved (figure 8.18(a)), which indicates a robust behaviour of the MOEA scheme against the described CPA. However, the output plane partially reveals the biometric signal used as one of the factors (figure 8.18(a)). In this case, the attacker

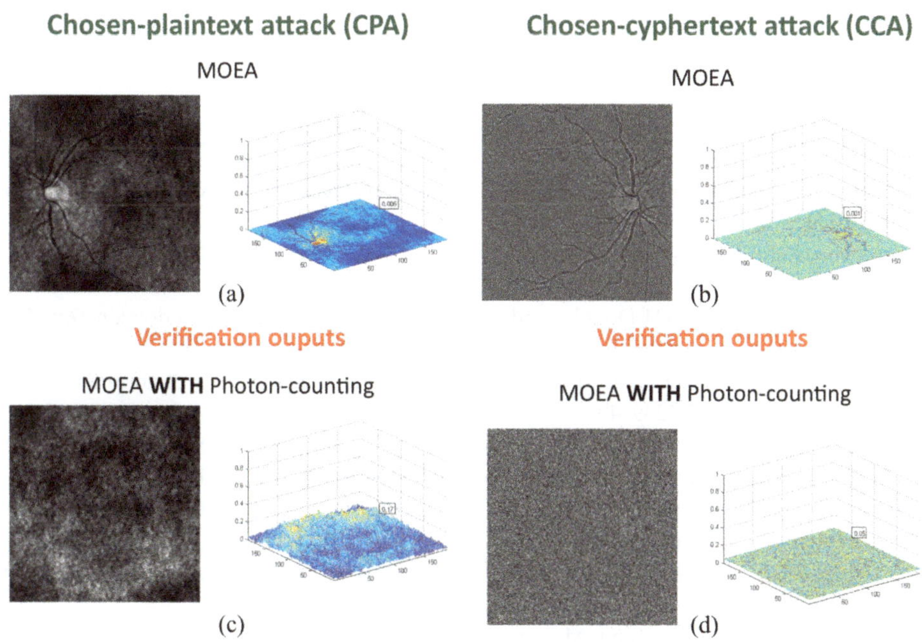

Figure 8.18. Cryptanalysis of the MOEA and PhC-MOEA: verification outputs for the MOEA when (a) a CPA or (b) a CCA occurs; verification outputs for the PhC-MOEA when (c) a CPA or (d) a CCA takes place.

can recognize at least the nature of the information used for encryption and verification. This disclosure can be considered to be a weakness of the system. A similar situation is obtained for CCA. In such a situation, the attacker can obtain a valid encrypted function with the information of four correct factors, and has access to the decryption machine. In this verification stage, if the intruder alters two signals with two dot functions and the processor has the correct RPMs from a database, again the output plane does not allow authentication, but discloses a dim reproduction for the retinal scan (figure 8.18(b)). If a PhC-MOEA scheme is used instead of the original MOEA, this unacceptable information disclosure is completely avoided. This stronger resistance to CPA and CCA attacks is illustrated in figure 8.18(c) and (d). This shows the output planes for the PhC-MOEA processes corresponding to the analogous cases described for the MOEA technique when either a CPA or a CCA occurs, respectively. In both cases a noisy appearance of the output planes is recovered for a different set of factors (invalid authentication) with no disclosure of the content of any factor.

Although the multifactor authentication process is not substantially altered by the attacks, its integration with the photon-counting imaging technique prevents possible partial disclosure of any encrypted factor, thus increasing the security level of the overall process.

References

[1] Réfrégier P and Javidi B 1995 Optical image encryption based on input plane and Fourier plane random encoding *Opt. Lett.* **20** 767–9

[2] Millán M S and Pérez-Cabré E 2011 *Optical data encryption Optical and Digital Image Processing: Fundamentals and Applications* ed G Cristóbal, P Schelkens and H Thienpont (Wiley) pp 739–67

[3] Chen W, Javidi B and Chen X 2014 Advances in optical security systems *Adv. Opt. Photon.* **6** 120–55

[4] Towghi N, Javidi B and Luo Z 1999 Fully phase encrypted image processor *J. Opt. Soc. Am.* A **16** 1915–27

[5] Matoba O and Javidi B 1999 Encrypted optical memory system using three-dimensional keys in the Fresnel domain *Opt. Lett.* **24** 762–4

[6] Situ G and Zhang J 2004 Double random-phase encoding in the Fresnel domain *Opt. Lett.* **29** 1584–6

[7] Hennelly B M and Sheridan J T 2004 Random phase and jigsaw encryption in the Fresnel domain *Opt. Eng.* **43** 2239–49

[8] Vilardy J M, Millán M S and Pérez-Cabré E 2014 Nonlinear optical security system based on a joint transform correlator in the Fresnel domain *Appl. Opt.* **53** 1674–82

[9] Unnikrishnan G, Joseph J and Singh K 2000 Optical encryption by double-random phase encoding in the fractional Fourier domain *Opt. Lett.* **25** 887–9

[10] Ozaktas H M, Zalevsky Z and Kutay M A 2001 *The Fractional Fourier Transform: With Applications in Optics and Signal Processing* (Wiley)

[11] Vilardy J M, Torres Y, Millán M S and Pérez-Cabré E 2014 Generalized formulation of an encryption system based on a joint transform correlator and fractional Fourier transform *J. Opt.* **16** 125405

[12] Rodrigo J A, Alieva T and Calvo M L 2007 Applications of gyrator transform for image processing *Opt. Commun.* **278** 279–84
[13] Vilardy J M, Millán M S and Pérez-Cabré E 2017 Nonlinear image encryption using a fully phase nonzero-order joint transform correlator in the Gyrator domain *Opt. Lasers Eng.* **89** 88–94
[14] Goodman J W 1996 *Introduction to Fourier Optics* (McGraw-Hill)
[15] Nomura T and Javidi B 2000 Optical encryption using a joint transform correlator architecture *Opt. Eng.* **39** 2031–5
[16] Frauel Y, Castro A, Naughton T J and Javidi B 2007 Resistance of the double random phase encryption against various attacks *Opt. Express* **15** 10253–65
[17] Carnicer A, Montes-Usategui M, Arcos S and Juvells I 2005 Vulnerability to chosen-cyphertext attacks of optical encryption schemes based on double random phase keys *Opt. Lett.* **30** 1644–6
[18] Liu W, Yang G and Xie H 2009 A hybrid heuristic algorithm to improve known-plaintext attack on Fourier plane encryption *Opt. Express* **17** 13928–38
[19] Nieto-Vesperinas M, Navarro R and Fuentes F J 1988 Performance of a simulated-annealing algorithm for phase retrieval *J. Opt. Soc. Am.* A **5** 30–8
[20] Gerchberg R W and Saxton O 1972 A practical algorithm for the determination of the phase from image and diffraction plane pictures *Optik* **35** 237–46
[21] Zalevsky Z, Mendlovic D and Dorsch R G 1996 Gerchberg-Saxton algorithm applied in the fractional Fourier or the Fresnel domain *Opt. Lett.* **21** 842–4
[22] Guo C, Liu S and Sheridan J T 2015 Iterative phase retrieval algorithms. Part II: Attacking optical encryption systems *Appl. Opt.* **54** 4709–19
[23] Barrera J F, Vargas C, Tebaldi M and Torroba R 2010 Chosen-plaintext attack on a joint transform correlator encrypting system *Opt. Commun.* **283** 3917–21
[24] Barrera J F, Vargas C, Tebaldi M, Torroba R and Bolognini N 2010 Known-plaintext attack on a joint transform correlator encrypting system *Opt. Lett.* **35** 3553–5
[25] Zhang C, Liao M, He W and Peng X 2013 Ciphertext-only attack on a joint transform correlator encryption system *Opt. Express* **21** 28523–30
[26] Millán M S, Pérez-Cabré E and Javidi B 2006 Multifactor authentication reinforces optical security *Opt. Lett.* **31** 721–3
[27] Pérez-Cabré E, Cho M and Javidi B 2011 Information authentication using photon-counting double-random-phase encrypted images *Opt. Lett.* **36** 22–4
[28] Pérez-Cabré E, Millán M S and Javidi B 2007 Near infrared multifactor identification tags *Opt. Express* **15** 15615–27
[29] Pérez-Cabré E *et al* 2015 Photon-counting multifactor optical encryption and authentication *J. Opt.* **17** 25706
[30] Rueda E, Barrera J F, Henao R and Torroba R 2009 Optical encryption with a reference wave in a joint transform correlator architecture *Opt. Commun.* **282** 3243–9
[31] Barrera J F, Tebaldi M, Ríos C, Rueda E, Bolognini N and Torroba R 2012 Experimental multiplexing of encrypted movies using a JTC architecture *Opt. Express* **20** 3388–93
[32] Vilardy J M, Millán M S and Pérez-Cabré E 2013 Improved decryption quality and security of a joint transform correlator-based encryption system *J. Opt.* **15** 025401
[33] Situ G, Pedrini G and Osten W 2010 Strategy for cryptanalysis of optical encryption in the Fresnel domain *Appl. Opt.* **49** 457–62

[34] Vilardy J M, Millán M S and Pérez-Cabré E 2013 Joint transform correlator-based encryption system using the Fresnel transform and nonlinear filtering *Proc. SPIE* **8785** 87853J

[35] Vilardy J M, Millán M S and Pérez-Cabré E 2014 Images encryption system based on a fractional joint transform correlator and nonlinear filtering *Opt. Pura Apl.* **47** 35–41

[36] Xueju S, Chao L, Xiao Z and Jianjun C 2015 Nonlinear optical cryptosystem based on joint Fresnel transform correlator under vector wave illumination *J. Opt.* **17** 055701

[37] Barrera J F, Osorio A J, Vélez A and Torroba R 2016 Experimental analysis of a joint free space cryptosystem *Opt. Lasers Eng.* **83** 126–30

[38] Tajahuerce E, Matoba O, Verrall S C and Javidi B 2000 Optoelectronic information encryption with phase-shifting interferometry *Appl. Opt.* **39** 2313–20

[39] La Mela C and Iemmi C 2006 Optical encryption using phase-shifting interferometry in a joint transform correlator *Opt. Lett.* **31** 2562–4

[40] Lu D and Jin W 2011 Color image encryption based on joint fractional Fourier transform correlator *Opt. Eng.* **50** 068201

[41] Rajput S K and Nishchal N K 2012 Image encryption and authentication verification using fractional nonconventional joint transform correlator *Opt. Lasers Eng.* **50** 1474–83

[42] Wang Q, Guo Q, Lei L and Zhou J 2013 Optical image encryption based on joint fractional transform correlator architecture and digital holography *Opt. Eng.* **52** 048201

[43] Li H 2009 Image encryption based on gyrator transform and two-step phase-shifting interferometry *Opt. Lasers Eng.* **47** 45–50

[44] Abuturab M R 2013 Noise-free recovery of color information using a joint-extended gyrator transform correlator *Opt. Lasers Eng.* **51** 230–9

[45] Pellat-Finet P 1994 Fresnel diffraction and the fractional-order Fourier transform *Opt. Lett.* **19** 1388–90

[46] Liu Z, Chen D, Ma J, Wei S, Zhang Y, Dai J and Liu S 2011 Fast algorithm of discrete gyrator transform based on convolution operation *Optik* **122** 864–7

[47] Pérez E, Chałasinska-Macukow K, Styczynski K, Kotynski R and Millán M S 1997 Dual nonlinear correlator based on computer controlled joint transform processor: digital analysis and optical results *J. Mod. Opt.* **44** 1535–52

[48] Pérez E, Millán M S and Chałasinska-Macukow K 2002 Optical pattern recognition with adjustable sensitivity to shape and texture *Opt. Commun.* **202** 239–55

[49] Velez A, Barrera J F and Torroba R 2017 Experimental optical encryption of grayscale information *Appl. Opt.* **56** 5883–9

[50] Chen W and Chen X 2016 Digital holography-secured scheme using only binary phase or amplitude as ciphertext *Appl. Opt.* **55** 6740–6

[51] Pérez-Cabré E *et al* 2012 Photon-counting double-random-phase encoding for secure image verification and retrieval *J. Opt.* **14** 94001

[52] Maluenda D *et al* 2015 Optical encryption using photon-counting polarimetric imaging *Opt. Express* **23** 655–66

[53] Carnicer A *et al* 2016 Optical encryption in the longitudinal domain of focused fields *Opt. Express* **24** 6793

[54] Javidi B *et al* 2016 Roadmap on optical security *J. Opt.* **18** 1–39

[55] Yeom S, Javidi B and Watson E 2005 Photon counting passive 3D image sensing for automatic target recognition *Opt. Express* **13** 9310–30

[56] Goodman J W 2015 *Statistical Optics* 2nd edn (Wiley)

[57] Naughton T J *et al* 2002 Compression of digital holograms for three-dimensional object reconstruction and recognition *Appl. Opt.* **41** 4124–32

[58] Naughton T J, McDonald J B and Javidi B 2003 Efficient compression of Fresnel fields for Internet transmission of three-dimensional images *Appl. Opt.* **42** 4758–64

[59] Tao R, Xin Y and Wang Y 2007 Double image encryption based on random phase encoding in the fractional Fourier domain *Opt. Express* **15** 16067–79

[60] Liu Z *et al* 2010 Double-image encryption based on the affine transform and the gyrator transform *J. Opt.* **12** 35407

[61] Rajput S K and Nishchal N K 2014 An optical encryption and authentication scheme using asymmetric keys *J. Opt. Soc. Am.* A **31** 1233–8

[62] Matoba O *et al* 2009 Optical techniques for information security *Proc. IEEE* **97** 1128–48

[63] Javidi B and Horner J L 1994 Optical pattern recognition for validation and security verification *Opt. Eng.* 1752–56

[64] Millán M S, Pérez-Cabré E and Javidi B 2006 High secure authentication by optical multifactor ID tags *Proc. SPIE* **6394** 63940J

[65] Horrillo S, Pérez-Cabré E and Millán M S 2010 Information compression for remote readable ID tags *J. Opt.* **12** 115404

IOP Publishing

Advanced Secure Optical Image Processing for Communications

Ayman Al Falou

Chapter 9

Enhanced single random phase holographic encryption of optical images

P W M Tsang

9.1 Introduction

Image encryption is the art of encoding an optical image into a form (which is referred to as the ciphertext in some literature) that is not recognizable unless the change that has been imposed on it can be reverted. The encryption process can be accomplished by transforming the original image with a certain mathematical formulation, based on a set of parameters that are commonly known as the encryption key. To decode the encrypted data, it is necessary to have knowledge of both the mathematical formulation and the encryption key. Image encryption has been applied in many important areas, such as industrial informatics, commerce, communication, and military defense. It can be envisaged that if the encryption key is exposed in an image encryption method, it could lead to serious monetary loss and breach of security. Therefore, research on high security image encryption methods has been vigorously conducted for many years, with the aim of finding a method to prevent the secret encryption key from being revealed through any form of attack.

However, such a goal is difficult, if not impossible, to attain, as there has also been a lot of research on developing methods to expose the secret key of different image encryption methods through various kinds of attacks. A fundamental approach in obtaining the secret key is to deduce the relation between the source and the encrypted image, from which the changes (and hence the secret keys) that have been applied can be obtained. The emergence of digital holography has instigated a new direction in image encryption. A hologram is a two-dimensional complex-valued image that captures the optical waves of an optical image, comprising of high frequency fringe patterns, that bears little resemblance to the source image it represents. As a result, the method of encrypting the hologram of an image is often more resistant to attack than encryption of the source image directly.

doi:10.1088/978-0-7503-1457-2ch9

One of the most effective holographic encryption techniques developed to date, is based on the double random phase encoding (DRPE) framework [1]. Briefly, a random phase mask $K_1(x, y)$ (first encryption key) is added to the source image $I(x, y)$, resulting in a complex image $I_R(x, y)$. The random phase added image $I_R(x, y)$ is then Fourier transformed to its frequency spectrum $\hat{I}_R(\omega_x, \omega_y)$, and added with a random phase mask $K_2(\omega_x, \omega_y)$, (second encryption key) to give a new spectral image $\hat{I}_{R2}(\omega_x, \omega_y)$. The encrypted ciphertext hologram, $H(u, v)$, is obtained by applying Fourier transform on $\hat{I}_{R2}(\omega_x, \omega_y)$. There are a number of variants on the DRPE method, such as applying the fractional Fourier transform [2], or the Fresnel transform [3], instead of the Fourier transform in obtaining the spectrum of the random phase added image. Recently, techniques that employ the QR-Code [4] for noise-free reconstruction of the encrypted images have also been reported in [5, 6].

However, the DRPE method is inherently susceptible to a plaintext attack. As indicated in some of the literature [7–11], the random phase mask on the transform plane can be deduced from the inverse Fourier transform of the ciphertext hologram corresponding to the source image of a Dirac impulse (i.e. a single dot). Subsequently, attempts have been made to decrease the vulnerability of the DRPE framework, such as space multiplexing [12], Arnold transform [13], and scrambling [14]. More effective measures to prevent attacks on the DRPE method can be accomplished by preserving only the phase information on the encrypted hologram [15]. This approach is more suitable for authentication, as the reconstructed image is either not visible, or of poor quality.

Recently, a highly secure encryption method with low complexity, known as single random phase encryption (SRPE), has been reported [16]. In this method, the source image to be encrypted is geometrically transformed (e.g. scaling and translation) with random parameters, as well as with the addition of random contents that are non-overlapping with the source image. The modified image is then converted into a phase-only hologram, and encrypted with a random phase mask which is taken as the encryption key. In the decoder, the source image will be recovered and observed visually if the encrypted hologram is overlaid with the encryption key phase mask, and illuminated with a coherent beam. The method only involves a single random phase mask in the encryption, and is highly resistant to attacks, as repetitive application of the SRPE on the same image will result in vastly different encrypted holograms. As such, it is not possible to draw a meaningful relation between the encrypted hologram and the source image. Despite the success of this technique, the source image is geometrically distorted after encryption, which is not acceptable in some applications.

In this chapter, a new single random phase method for holographic image encryption, which is referred to as enhanced single random phase encryption (ESRPE), is proposed. Differing from the SRPE method in [16], the proposed method will not lead to any geometrical distortion of the source image.

9.2 The proposed enhanced single random phase encryption (ESRPE) method

In this chapter, a new holographic encryption method that only requires a single random phase mask as the encryption key is proposed. The ciphertext hologram generated by the proposed method is a phase-only hologram that is referred to as the enhanced single random phase encrypted (ESRPE) hologram. The proposed method is comprised of four stages, as shown in figure 9.1, and outlined as follows.

In the first stage, the source image $I(x, y)$ to be encrypted is converted into an off-axis complex hologram based on angles of incidence $\theta_{1;x}$ (along the x direction) and $\theta_{1;y}$ (along the y direction). If $\theta_{1;x} = 0$ and $\theta_{1;y} = 0$, the complex hologram is equivalent to an on-axis or in-line hologram. The source image can be a planar image, or a three-dimensional (3D) image. When the off-axis hologram of the source image is illuminated with a plane wave, the source image will be reconstructed on the focused plane orientated at angles $\theta_{1;x}$ and $\theta_{1;y}$ along the x and y directions, respectively, from the normal of the hologram plane. The hologram that is generated from the source image is known as the source hologram.

In the second stage, at least one secondary image $I_p(x, y)|_{0 \leqslant p < P}$ (where $P \geqslant 1$ is the total number of secondary image(s)) that is randomly generated, or selected from a pool of images that is unknown to the operator, is converted into an off-axis complex hologram. For simplicity of illustration, only one secondary image $I_0(x, y)$ is shown in figure 9.1. Each off-axis hologram is generated based on a reference beam at an angle of incidence $\theta_{2;x;p}$ (along the x direction) and $\theta_{2;y;p}$ (along the y direction). When the off-axis hologram of the secondary image is illuminated with a plane wave, the secondary image will be reconstructed on the focused plane orientated at angles $\theta_{2;x;p}$ and $\theta_{2;y;p}$ along the x and y directions, respectively, from the normal of the hologram plane. The hologram that is generated from a secondary image is known as the secondary hologram. The same applies if there is more than one secondary image, in which case there will be a set of secondary images, and a corresponding set of secondary holograms.

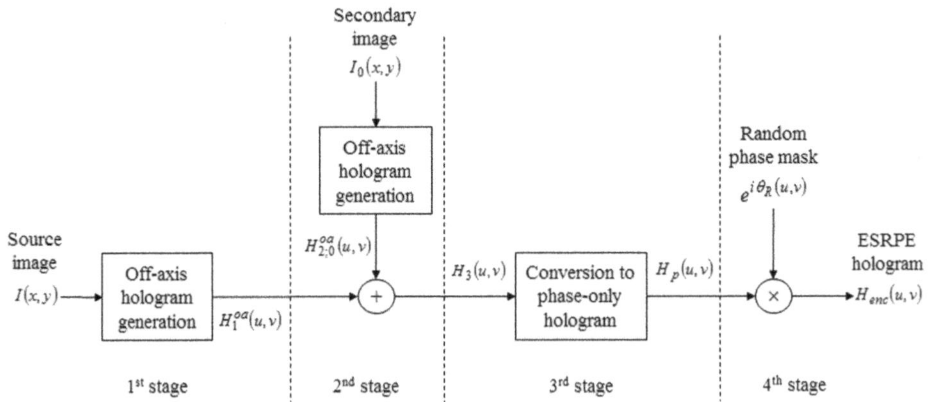

Figure 9.1. Proposed ESRPE method.

In the third stage, the source hologram and the secondary hologram(s) are summed up into a global hologram $H_3(u, v)$, and converted into a global phase-only hologram $H_p(u, v)$ with existing techniques such as (but not limited to) error diffusion [17, 18]. As such, the global phase-only hologram is very different from the source hologram and any of the secondary holograms. In addition, as the secondary image is randomly generated or randomly selected from some unknown source, even for an identical source image, a totally different global phase-only hologram will be generated in each run of the encryption process.

In the fourth stage, a random phase mask $e^{i\theta_R(u,v)}$ is added to the global phase-only hologram, resulting in an enhanced single random phase encrypted (ESRPE) hologram $H_{enc}(u, v)$. The latter is a cipher-text hologram that bears little correlation to the source image or the secondary image(s).

The mixing of the holograms of the source image and one or more secondary images ensures that there is no prior knowledge of what is being encrypted, so that there is no direct relation between the ESRPE hologram and the source image. It also increases the difficulty in deducing the encryption key through a large-scale chosen plaintext attack (e.g. encrypting a large number of source images, each with a small number of incremental changes, to locate the affected regions of the ciphertext hologram).

In the decoder, the ESRPE hologram is overlayed with the conjugate of the random phase mask (encryption key), and illuminated with a coherent beam. The source and the secondary images will be reconstructed at the focused plane along orientations $(\theta_{1;x}, \theta_{1;y})$ and $(\theta_{2;y;p}, \theta_{2;x;p})$ $(0 \leqslant p < P)$, respectively. If the angular separation between $(\theta_{1;x}, \theta_{1;y})$ and the orientation(s) of the secondary holograms $(\theta_{2;y;p}, \theta_{2;x;p})$ is/are large enough, the images will be formed at non-overlapping areas on the focused plane, and the source image can be observed visually at its original position without any geometrical changes. The detail of each stage of the proposed method is described in the following sub-sections.

9.2.1 Stage 1: converting the source image into an off-axis hologram

For simplicity of explanation, the source is assumed to be a two-dimensional planar image $I(x, y)$ that is parallel to the hologram, where x is the horizontal, and y is the vertical discrete co-ordinate axes. The source image is first converted into a complex Fresnel hologram $H_1(u, v)$, as given by

$$H_1(u, v) = \sum_{x=0}^{X-1}\sum_{y=0}^{Y-1} I(x, y) \exp\left[\frac{i2\pi}{\lambda} r_{x;y;u;v}\right], \tag{9.1}$$

where λ is the wavelength of the optical beam. $r_{x;y;u;v}$ is the distance from a point at (x, y) on the global image, to a point (u, v) on the hologram, u is the horizontal, and v is the vertical discrete co-ordinate axes of the hologram plane. Suppose the axial distance between the global image and the hologram is denoted by z_1, and δ is the sampling interval which is assumed to be identical along the horizontal and the vertical directions, we have

$$r_{x;y;u;v} = \sqrt{(x-u)^2\delta^2 + (y-v)^2\delta^2 + z_1^2}. \tag{9.2}$$

Next, the complex hologram is multiplied with an inclined plane wave $R_1(u, v)$ with an angle of incidence $(\theta_{1;x}, \theta_{1;y})$, resulting in an off-axis hologram $H_1^{oa}(u, v)$.

9.2.2 Stage 2: converting a secondary image into an off-axis hologram

For simplicity of explanation, the secondary image is assumed to be a two-dimensional planar image $I_{2;p}(x, y)$ that is parallel to, and at an axial distance $z_{2;p}$ from the hologram. The term $0 \leqslant p < P$ is the index of the secondary image, and P is the number of secondary image(s). Following a similar procedure to stage 1, an off-axis hologram $H_{2;p}^{oa}(u, v)$ is generated for the pth secondary image with the angle of incidence of the plane wave $R_{2;p}(u, v)$ at angles of incidence $(\theta_{2;x;p}, \theta_{2;y;p})$.

9.2.3 Stage 3: generating a global hologram

In this stage, the source and the secondary holograms are summed up into a new hologram $H_3(u, v)$ given by

$$H_3(u, v) = H_1^{oa}(u, v) + \sum_{p=0}^{P-1} H_{2;p}^{oa}(u, v). \tag{9.3}$$

Subsequently, $H_3(u, v)$ is converted into a global phase-only hologram $H_p(u, v)$, for example with the error diffusion method reported in [17, 18]. In $H_p(u, v)$, the magnitude of all the pixels is set to a constant value.

9.2.4 Stage 4: converting the global hologram into a ESRPE hologram

Finally, the global hologram is multiplied with a random phase mask $e^{i\theta_R(u,v)}$, resulting in an ESRPE hologram given by

$$H_{\text{enc}}(u, v) = H_p(u, v)e^{i\theta_R(u,v)}. \tag{9.4}$$

The ESRPE hologram is taken as the encrypted (ciphertext) hologram, with the phase mask $e^{i\theta_R(u,v)}$ as the private encryption key. When the ciphertext hologram is overlaid with the conjugate of the phase mask, and illuminated with a coherent beam, the global image, together with the source image within it, will be reconstructed. The decryption process, which can be expressed mathematically in equation (9.5), can be realized through optical or numerical reconstruction.

$$H_{\text{enc}}(u, v)e^{-i\theta_R(u,v)} = H_p(u, v)e^{i\theta_R(u,v)}e^{-i\theta_R(u,v)} = H_p(u, v). \tag{9.5}$$

A simplified realization is studied in this chapter with $(\theta_{1;x}, \theta_{1;y}) = (0, 0)$, $P = 1$, and the source and secondary images being planar images parallel to the hologram plane.

9.3 Experimental results

The source image 'Lenna' shown in figure 9.2(a) is 768 × 768 pixels in size, and is employed to demonstrate the feasibility of the proposed method. All the holograms involved in the encryption process are assumed to have an identical size of 2048 × 2048 pixels. The pixel size of the source image, the secondary image(s) and the holograms are 8 μm × 8 μm. The hologram plane is located at an axial distance of 0.4 m from the source and the secondary images. A global hologram is generated with the steps described in section 9.2. Without loss of generality, only one secondary image (i.e. $P = 1$) is included in generating the global hologram, with $(\theta_{1;x}, \theta_{1;y}) = (0, 0)$, $(\theta_{2;x;0}, \theta_{2;y;0}) = (1.0°, 1.0°)$, and wavelength λ of 633 nm. Two different secondary images, 'Baboon' and 'Man', as shown in figures 9.2(b) and (c),

(a)

Figure 9.2(a). Source image 'Lenna'. © Playboy Enterprises, Inc.

(b)

Figure 9.2(b). First secondary image 'Baboon'.

(c)

Figure 9.2(c). Second secondary image 'Man'.

which are of the same size as the source image and located at 0.3 m from the hologram plane, are adopted in the experiment. In the generation of the secondary holograms, the intensities of the secondary images are reduced by a factor of 2 to lower their effect on the source hologram.

Steps 2.1–2.3 in section 9.2 are employed to generate the global phase-only holograms based on the uni-directional error diffusion method described in [18], and the dynamic range of the hologram is normalized to [0,255]. The phase-only hologram of the image 'Lenna' alone is shown in figure 9.3(a). The global phase-only holograms corresponding to the mixing of the source hologram with each of the secondary holograms are shown in figures 9.3(b) and (c). The numerical reconstructed images of the three holograms in figures 9.3(a)–(c) are shown in figures 9.4(a)–(c). We observed that the reconstructed 'Lenna' images from the global phase-only holograms are very similar to the one corresponding to the source hologram. In addition, the peak signal-to-noise ratio (PSNR) of the reconstructed images in figures 9.4(b) and (c) are 29.26dB and 29.45dB, which is rather good and consistent with the favorable visual quality. The secondary images, which are formed outside the area of the source image, can be easily discarded with a blocking mask.

Next, in stage 2 the original source image, as well as the global images, are each converted into a complex phase-only hologram with the process described in section 9.2.2. In the final stage, each phase-only hologram is added with a random phase mask, resulting in an enhanced single random phase encrypted (ESRPE) hologram. The numerical reconstructed images of the 2 ESRPE holograms, which are encoded versions of the global phase-only holograms in figures 9.3(b) and (c), are shown in figures 9.5(a) and (b). Evidently, without the private encryption key, the reconstructed image of an ESRPE hologram is comprised of random noise and the source image is not recovered.

Subsequently, the root-mean-square-error (RMSE) between the phase-only hologram of 'Lenna' with the phase-only holograms of the two global images are computed, and the results are 119 and 120, respectively. The results show a

(a)

Figure 9.3(a). Phase-only hologram of source image 'Lenna'. © Playboy Enterprises, Inc.

(b)

Figure 9.3(b). Global phase-only hologram formed by merging the hologram of the source image 'Lenna' and the hologram of the secondary image 'Baboon'.

(c)

Figure 9.3(c). Global phase-only hologram formed by merging the hologram of the source image 'Lenna' and the hologram of the secondary image 'Man'.

(a)

Figure 9.4(a). Reconstructed image from the source hologram in figure 9.3(a). © Playboy Enterprises, Inc.

(b)

Figure 9.4(b). Reconstructed image from the global phase-only hologram in figure 9.3(b). © Playboy Enterprises, Inc.

(c)

Figure 9.4(c). Reconstructed image from the global phase-only hologram in figure 9.3(c). © Playboy Enterprises, Inc.

(a)

Figure 9.5(a). Reconstructed image from the single random phase encrypted hologram in figure 9.3(b).

(b)

Figure 9.5(b). Reconstructed image from the single random phase encrypted hologram in figure 9.3(c).

significant difference between the pair of global phase-only holograms, although both of them contain the source image. As such, it is very unlikely that the secret key could be revealed from the relation between the ESRPE hologram and the source image.

9.4 Conclusion

This chapter reports the ESRPE framework that is employed to encrypt optical images. The method has two major advantages. First, an ESRPE hologram represents the source image as well as unknown, randomly selected content. As such, it is not possible to draw any relation between the encrypted hologram and the source image, hence preventing the secret key from being exposed by attacks. Second, the source image can be recovered from the ESRPE hologram with good fidelity, and not geometrically distorted as in the SRPE scheme. Although this chapter only demonstrates the application of the ESRPE method on a 2D source image, it can be easily extended to the encryption of a 3D image. In addition, the security of the method can also be further strengthened by incorporating 3D, instead of 2D secondary image(s).

References

[1] Refregier P and Javidi B 1995 Optical image encryption based on input plane and Fourier plane random encoding *Opt. Lett.* **20** 767–9

[2] Unnikrishnan G, Joseph J and Singh K 2000 Optical encryption by double-random phase encoding in the fractional Fourier domain *Opt. Lett.* **25** 887–9

[3] Situ G and Zhang J 2004 Double random-phase encoding in the Fresnel domain *Opt. Lett.* **29** 1584–6

[4] ISO/IEC 18004:2006 Information technology. Automatic identification and data capture technique, QR Code 2005 barcode symbology specification; www.iso.org/iso/iso_catalogue/catalogue_tc/catalogue_detail.htm?csnumber=43655

[5] Ren Z, Su P, Ma J and Jin G 2014 Secure and noise-free holographic encryption with a quick-response code *Chin. Opt. Lett.* **12** 010601

[6] Barrera J, Mira A and Torroba R 2013 Optical encryption and QR codes: Secure and noise-free information retrieval *Opt. Express* **21** 5373–8

[7] Zhang Y, Xiao D, Wen W and Liu H 2013 Vulnerability to chosen-plaintext attack of a general optical encryption model with the architecture of scrambling-then-double random phase encoding *Opt. Lett.* **38** 4506–9

[8] Tashima H, Takeda M, Suzuki H, Obi T, Yamaguchi M and Ohyama N 2010 Known plaintext attack on double random phase encoding using fingerprint as key and a method for avoiding the attack *Opt. Express* **18** 13772–81

[9] Frauel Y, Castro A, Naughton T and Javidi B 2007 Resistance of the double random phase encryption against various attacks *Opt. Express* **15** 10253–65

[10] Peng X, Zhang P, Wei H and Yu B 2006 Known-plaintext attack on optical encryption based on double random phase keys *Opt. Lett.* **31** 1044–6

[11] Carnicer A, Montes-Usategui M, Arcos S and Juvells I 2005 Vulnerability to chosen-cyphertext attacks of optical encryption schemes based on double random phase keys *Opt. Lett.* **30** 1644–6

[12] Gong Q, Liu X, Li G and Qin Y 2013 Multiple-image encryption and authentication with sparse representation by space multiplexing *Appl. Opt.* **52** 7486–93

[13] Chen W, Quan C and Tay C J 2009 Optical color image encryption based on Arnold transform and interference method *Opt. Commun.* **282** 3680–5

[14] Zhang Y, Xiao D, Wen W and Liu H 2013 Vulnerability to chosen-plaintext attack of a general optical encryption model with the architecture of scrambling-then-double random phase encoding *Opt. Lett.* **38** 4506–9

[15] Chen W and Chen X 2014 Double random phase encoding using phase reservation and compression *J. Opt.* **16** 205402

[16] Tsang P W M 2016 Single-random-phase holographic encryption of images *Opt. Lasers Eng* **89** 22–8

[17] Tsang P W M, Jiao A and Poon T-C 2014 Fast conversion of digital Fresnel hologram to phase-only hologram based on localized error diffusion and redistribution *Opt. Express* **22** 5060–6

[18] Tsang P W M and Poon T-C 2013 Novel method for converting digital Fresnel hologram to phase-only hologram based on bidirectional error diffusion *Opt. Express* **21** 23680–6

Chapter 10

Single-pixel optical information encoding and authentication

Wen Chen

10.1 Introduction

Due to the popularization of modern techniques, many researchers and scientists have made much effort to develop new infrastructures and algorithms for securing data transmission and storage. It has been well recognized that important products (e.g. private images) should be encoded, e.g. protected using watermarks [1]. If important or confidential information is not protected, it may inevitably induce huge losses to a person, company, city or even a nation [1, 2].

To date, various infrastructures have already been established for information security, such as electronic cryptography and physical methods. However, it is still desirable for advanced information security technologies to be continuously developed to satisfy different requirements in various applications [1, 2]. Since Refregier and Javidi [3] proposed the double random phase encoding approach, many optical structures [4–25] have been designed and some application cases have also been explored. It is illustrated that compared with electronic cryptography, optical information security methods have some remarkable advantages as follows [2–25]: (1) Optical instruments possess parallel-processing and high-speed capabilities. In addition to the software platform, optical hardware provides an effective and promising approach in practice. (2) Higher security can be achieved in optical encoding and decoding systems for some applications. Information can be processed (such as encrypted or hidden) in multiple dimensions, such as phase, light polarization and intensity. (3) Sophisticated optoelectronic devices and systems need to be analyzed before information decoding. An attacker needs to possess a multidisciplinary background in order to conduct a process of decoding the information [2]. (4) Noise in light can be flexibly controlled in most cases, and optical noise is random by nature.

doi:10.1088/978-0-7503-1457-2ch10

Since optical technologies possess these significant advantages and unique characteristics, many research groups have conducted extensive research about optical information security [2–25]. After double random phase encoding [3] was proposed, some researchers have further applied different wave propagation domains to enhance flexibility and security, such as the Fresnel transform [7], fractional Fourier transform [11] and gyrator transform [13].

A number of optical principles and structures [2–26], such as diffractive imaging [4], digital holography [24] and ghost imaging [26], have also been studied for securing information. It has been illustrated that plaintext can be flexibly encrypted, and different variables, such as phase and intensity, can be generated as ciphertexts. It has been demonstrated by researchers that in optical information security systems, stationary white noise can be generated as ciphertext, such as by using statistically-independent random phase-only patterns in the optical path [2–25]. However, it is found that optical security infrastructures based on double random phase encoding have a linear property [27, 28], and there is a vulnerability in the security of some established optical systems. Among the attack algorithms [27–29], the known plaintext attack and the chosen ciphertext attack have been widely studied to retrieve security keys from designed optical encoding systems.

In attack algorithms, there is a strong assumption that the setup parameters are known by attackers, and the same random phase-only patterns are employed to encode different plaintexts [27–29]. For example, in the known plaintext attack algorithm [28], the phase extraction approach can be used to retrieve or estimate a random phase-only pattern in the spatial frequency domain, and the same security keys are applied for encoding different input images. Since a strong assumption using the same security keys is used in some attack algorithms, updating random phase-only patterns can be directly employed to encode different plaintexts to withstand attacks [30]. Recently, novel attack algorithms [31, 32] have been studied to analyze the vulnerability of optical security systems. To enhance system security, various methods [21, 22, 30, 33–36], such as pixel scrambling and plaintext modifications [21, 22], can be employed.

Phase retrieval algorithms [37–45], such as using a computer-generated hologram, have also been investigated to retrieve noisy phase-only patterns as ciphertexts for security enhancement. Both iterative and non-iterative phase retrieval algorithms [37–45], such as the modified Gerchberg–Saxton algorithm [39, 46] and non-iterative interference-based phase retrieval algorithm [44], have been developed. In either iterative or non-iterative methods, the plaintext, such as data or images, can be flexibly encoded into noisy phase-only patterns. The objective of phase-retrieval-based optical encryption [37–45] is to find correct or approximate phase-only patterns using some constraints. In practice, the generated noisy phase-only patterns can be embedded into a spatial light modulator, therefore optical decryption can be easily implemented, compared to double random phase encoding [3]. For instance, Hwang *et al* [39] have studied a modified Gerchberg–Saxton algorithm to retrieve a noisy phase-only pattern and encode multiple input images. Xiao *et al* [42] proposed an improved phase-retrieval-based optical encoding method to generate noisy phase-only patterns as ciphertexts. The phase-retrieval-based optical encoding and decoding system can be applied as an effective and promising alternative for securing information, and may be used in many applications, such as passport checking and note verification.

In conventional attack algorithms, complete ciphertexts, such as simultaneous amplitude and phase, are assumed to be available to attackers. Recently, it has been found that some parameters in optical security setups, such as security keys and ciphertexts, can be flexibly compressed [47–51]. Due to data compression, the decoded data/images do not visually render information about the plaintext. Some researchers have further proposed to authenticate the decoded images instead of directly viewing them [47–52]. This authentication approach can open up a new research perspective for securing information, and can be applied in many optical encryption systems to enhance security. It has been illustrated [47–51] that a data compression operation can lead to the generation of a new security layer, since the decoded image should be nonlinearly correlated with the original information stored in a database for authentication. For instance, in the double random phase encoding system [3, 47], information obtained in the CCD plane can be flexibly compressed as ciphertext, and only sparse information (i.e. as ciphertext) is available to the receiver for decoding. The decoded images/data are noisy even when using the correct security keys due to the data compression operation in the CCD plane, and can be effectively authenticated without plaintext disclosure [47].

In recent years, single-pixel (ghost) imaging [26, 53–62] has been explored as an effective and promising approach for securing information. In a single-pixel imaging encoding system, a series of random phase-only patterns are sequentially embedded into a phase-only spatial light modulator, and a series of one-dimensional intensity points can be obtained as ciphertexts. In practice, various methods can be used to generate a series of random phase patterns in order to compress security keys and facilitate the transmission, such as by using chaotic maps [26]. Computational single-pixel (ghost) imaging has been widely used for securing information, since the patterns recorded at the reference beam arm can be computationally generated [63–66]. Although single-pixel imaging is promising for information security, there are several concerns as follows: (1) the values of the phase patterns (as keys) are relatively small which might affect the value range of the security keys; (2) the variety of security-key generation methods is low; (3) the number of varied single-pixel encoding/decoding structures is low; (4) attacks to single-pixel encoding systems have not been sufficiently designed and fully analyzed.

In this chapter, current developments of single-pixel optical information encoding and authentication are described. Single-pixel optical information encoding is presented by using an iterative phase extraction scheme, and a typical case is explained. Single-pixel optical information authentication is presented by using a signal binarization method, and a typical case is correspondingly illustrated. Subsequently, some discussions are given to describe the perspective and potential challenges of single-pixel optical information encoding and authentication methods. Finally, a conclusion is made to briefly summarize these approaches. It is believed that there will be a long-term benefit for the field of information security by optically encoding and authenticating information based on single-pixel imaging structures, and new single-pixel optical information encoding and authentication approaches can be continuously developed and integrated in a long-term study.

10.2 Single-pixel optical encoding

In a single-pixel optical information encoding system, it is well known that a series of noisy phase-only patterns can be generated and sequentially embedded into a phase-only spatial light modulator for encoding, and a series of one-dimensional intensity points can be correspondingly recorded by using a bucket detector (without spatial resolution). However, it is realized that the value range of the phase patterns is relatively small, which might affect the performance of the security keys. Recently, it has been illustrated that intensity modulation methods [58–60] can be developed and applied to generate a series of noisy phase-only patterns. Hence, the indirect phase space is dramatically enlarged, and system security can be correspondingly enhanced. For instance, a non-iterative interference-based phase retrieval algorithm [60] can be used to sequentially extract a series of noisy phase-only patterns from the pre-generated random intensity-only patterns.

In this chapter, a typical intensity-modulated single-pixel optical information encoding method is described. Figure 10.1 shows a schematic for single-pixel optical information encoding. A series of random intensity-only patterns are pre-generated, and an iterative phase retrieval algorithm [58] is developed to sequentially encode the series of random intensity-only patterns into a series of noisy phase-only patterns. For the sake of brevity, one random intensity-only pattern is encoded into two cascaded phase-only patterns, as shown in figure 10.2, and in practice it is straightforward to encode each pre-generated random intensity-only pattern into several phase-only patterns by using a computational approach [58]. In addition, different structures can also be designed and applied for encoding the random intensity-only pattern into phase-only patterns by using phase retrieval algorithms, as schematically illustrated in figure 10.3.

In figure 10.2, there are two types of extraction methods. For the sake of brevity, only the phase retrieval procedure for the first extraction type (i.e. M_e is placed before the M_f) is described, and it can be straightforward to extract a phase-only

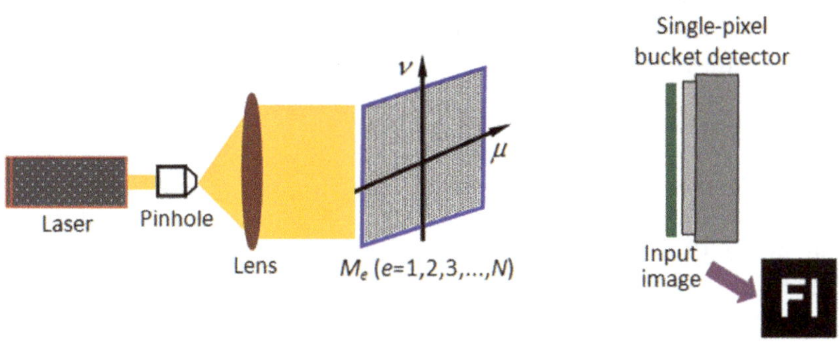

Figure 10.1. A schematic for single-pixel optical information encoding [58–62]: a series of phase-only patterns can be sequentially embedded into a phase-only spatial light modulator for encoding. A collecting lens can be placed between the input image plane and bucket detector plane.

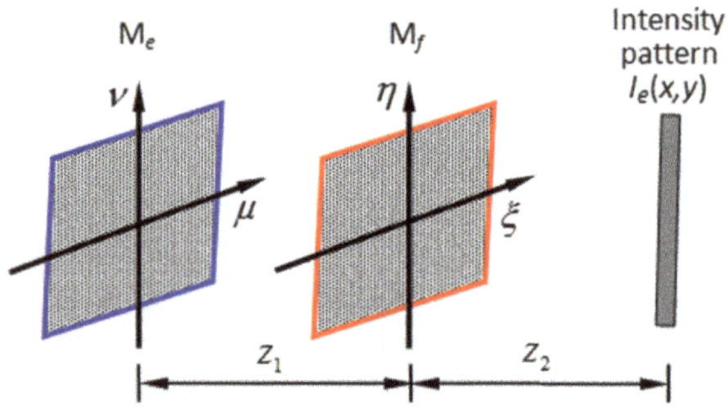

Figure 10.2. One pre-generated random intensity-only pattern is encoded into two cascaded phase-only patterns M_e and M_f: M_f, a fixed random phase-only pattern. The objective of the iterative phase retrieval algorithm is to extract one phase pattern M_e from each pre-generated intensity pattern $I_e(x,y)$. Hence, a series of phase-only patterns $[M_e(u,v), e=1,2,3,\ldots,N]$ can be generated for the subsequent encryption.

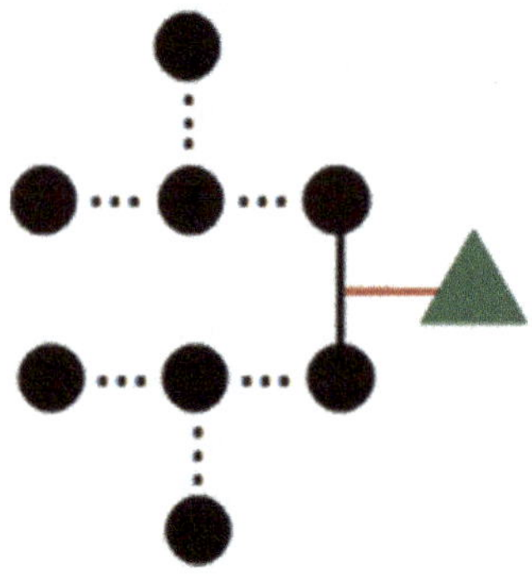

Figure 10.3. A typical structure for phase-pattern generation using a computational approach, such as a mixture of interference and cascaded: the triangle denotes a pre-generated random intensity pattern and the circles denote the retrieved noisy phase-only patterns.

pattern by using another extraction type (i.e. M_f is placed before the M_e). A modified Gerchberg–Saxton algorithm [46, 58] is developed and applied for the phase-pattern retrieval, and the extraction procedure consists of the following steps:

(1) From the pre-generated intensity-only pattern plane, the wave propagates back to the phase pattern $M_e(\mu, \nu)$ plane:

$$O^{(n)}(\mu, \nu) = \mathrm{FrT}_{\lambda,-z_1}\Big(\{\mathrm{FrT}_{\lambda,-z_2}[I_e(x, y)]\}\big[M_f(\xi, \eta)\big]^*\Big), \tag{10.1}$$

where $\mathrm{FrT}_{\lambda,-z}$ denotes the wave back-propagation [67–69], λ denotes the wavelength, n denotes the iterative number (integer 1, 2, 3,…), z_1 and z_2 denote axial distances as shown in figure 10.2, the asterisk denotes the complex conjugate, and $M_f(\xi, \eta)$ denotes the fixed random phase-only pattern.

(2) Apply a support constraint [58] to wavefront $O^{(n)}(\mu, \nu)$:

$$M_e^{(\mathrm{n})}(\mu, \nu) = O^{(n)}(\mu, \nu)/|O^{(n)}(\mu, \nu)|, \tag{10.2}$$

where | | represents a modulus operation.

(3) After the updated phase-only pattern is generated, wave propagation to the pre-generated intensity pattern plane is carried out:

$$O^{(n)}(x, y) = \mathrm{FrT}_{\lambda,z_2}\left(\{\mathrm{FrT}_{\lambda,z_1}[M_e^{(n)}(\mu, \nu)]\}\mathrm{M}_f(\xi, \eta)\right), \tag{10.3}$$

where $O^{(n)}(x, y)$ denotes the updated wavefront in the pre-generated intensity pattern plane. The correlation coefficient between $|O^{(n)}(x, y)|^2$ and the desired pattern $I_e(x, y)$ is calculated for monitoring the iterations.
(4) If the preset condition is not satisfied (i.e. the preset threshold), $O^{(n)}(x, y)$ is updated by [46, 58]:

$$\hat{O}^{(n)}(x, y) = [I_e(x, y)]^{1/2} O^{(n)}(x, y)/|O^{(n)}(x, y)|, \tag{10.4}$$

where $\hat{O}^{(n)}(x, y)$ denotes the updated wavefront in the pre-generated intensity pattern plane which can be further used for the next iteration ($n = n + 1$), i.e. starting from equation (10.1). When the preset condition is satisfied, $\mathrm{M}_e^{(n)}(\mu, \nu)$ generated in equation (10.2) can be considered as the finally-extracted phase pattern corresponding to the pre-generated intensity pattern $I_e(x, y)$. Similarly, other pre-generated random intensity-only patterns can be processed and encoded, and a series of noisy phase-only patterns $[\mathrm{M}_e(u, v), e = 1, 2, 3,\ldots, N]$ can be sequentially retrieved and used for subsequent encryption, as shown in figure 10.1.

Using the aforementioned steps, each pre-generated random intensity-only pattern $I_e(x, y)$ can be encoded, and one phase-only pattern can be correspondingly generated, i.e. $\mathrm{M}_e(u, v)$, $e = 1, 2, 3, \ldots, N$. During the encoding, a series of noisy phase patterns $[\mathrm{M}_e(u, v), e = 1, 2, 3, \ldots, N]$ is sequentially embedded into a phase-only spatial light modulator, and a series of one-dimensional intensity points $\{B_e\}$ ($e = 1, 2, 3, \ldots, N$) can be recorded by bucket detector as ciphertexts for the designed single-pixel information encoding system. A single-pixel bucket detector without spatial resolution is usually applied for the recording, and a 2D CCD camera might be used as an alternative in practical applications. As illustrated in figure 10.1, the aforementioned procedure can be further described by [58–60]

$$B_e = \iint |\{\mathrm{FrT}_{\lambda,d}[\mathrm{M}_e(\mu, \nu)]\}|^2 |p(k, l)|^2 \,\mathrm{d}k\mathrm{d}l, \tag{10.5}$$

where $p(k, l)$ denotes the plaintext to be encoded (64×64 pixels, a binary image shown in figure 10.1), and d denotes the axial distance between the phase-only pattern plane and the single-pixel bucket detector plane. A flow chart is given in figure 10.4 to show the single-pixel encoding procedure.

During the decoding, security keys, such as the series of random intensity-only patterns, should be available to the authorized receiver to extract phase-only patterns at first. In this case, the modified Gerchberg–Saxton algorithm should be correctly applied for phase extraction. Since computational single-pixel (ghost) imaging is employed, a series of reference intensity patterns $R_e(k, l)(e = 1, 2, 3, \ldots, N)$ can be numerically generated [58–60]. When the authorized receiver has the ciphertexts, a decoded image can be calculated by $[\langle BR(k, l)\rangle - \langle B\rangle\langle R(k, l)\rangle]$ (where $\langle . \rangle$ denotes the ensemble average) [58–60]. The quality of the decoded images

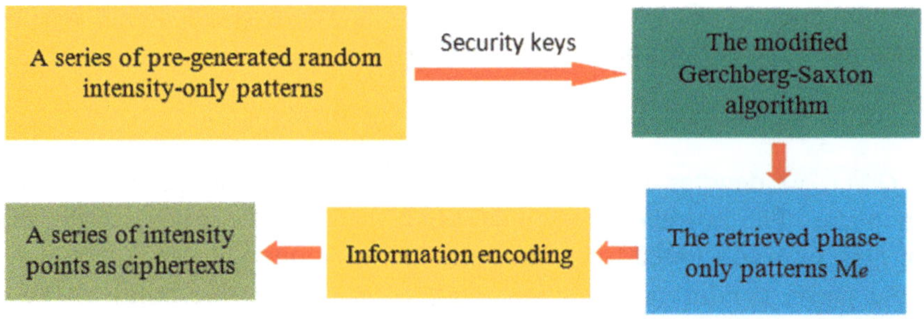

Figure 10.4. Flow chart to illustrate the single-pixel encoding procedure.

can be evaluated by using many parameters, such as the peak signal-to-noise ratio (PSNR) [58].

Results are obtained based on figures 10.1 and 10.2 via simulation to illustrate the single-pixel optical information encoding method. In figures 10.1 and 10.2, the plane wave is applied, and the laser wavelength is 630.0 nm. In figures 10.1 and 10.2, free space propagation [67, 68] is applied in the single-pixel optical information encoding process. Axial distances z_1, z_2 and d are 5.0 cm, 7.0 cm and 9.0 cm, respectively. In figure 10.2, a series of random intensity-only patterns $I_e(x, y)$, i.e. $N = 11\ 000$, are pre-generated, and the value range of each pre-generated random intensity pattern is (0, 450]. The random intensity-only patterns $I_e(x, y)$ are sequentially encoded, and a series of noisy phase-only patterns can be correspondingly generated. In the modified Gerchberg–Saxton algorithm, the correlation coefficient threshold is 0.92. In addition, the fixed random phase-only pattern M_f is distributed in the range of $[0, 2\pi]$. As seen in figures 10.1 and 10.2, when the phase-only patterns $[M_e(u, v), e = 1, 2, 3, \ldots, N]$ are sequentially embedded, the plaintext (64×64 pixels, a binary image shown in figure 10.1) can be encoded and a series of one-dimensional intensity points can be obtained as ciphertexts in the designed single-pixel optical information encoding system.

In figure 10.2, a fixed random phase-only pattern M_f is used, and the objective of the designed iterative phase retrieval algorithm is to extract a noisy phase-only pattern M_e from each pre-generated random intensity-only pattern. Figure 10.5 shows the fixed random phase-only pattern M_f. A series of random intensity-only patterns are pre-generated as principal security keys for the designed single-pixel optical encoding system, and figures 10.6(a)–(d) show four typical random intensity-only patterns. Using the modified Gerchberg–Saxton algorithm, a series of noisy phase-only patterns can be sequentially retrieved, and four typical phase-only patterns extracted respectively corresponding to figures 10.6(a)–(d) are shown in figures 10.7(a)–(d). To clearly illustrate the iterative phase extraction procedure, relationships between the number of iterations and correlation coefficient (CC) are shown in figures 10.8(a)–(d), i.e. respectively corresponding to figures 10.7(a)–(d). It can be seen in figures 10.8(a)–(d) that a rapid convergence rate is achieved, when the pre-generated random intensity-only pattern is encoded into phase-only patterns by using the modified Gerchberg–Saxton algorithm.

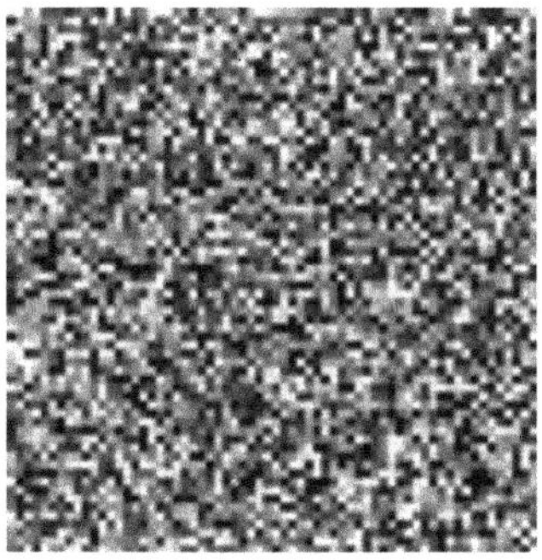

Figure 10.5. The fixed random phase-only pattern M_f.

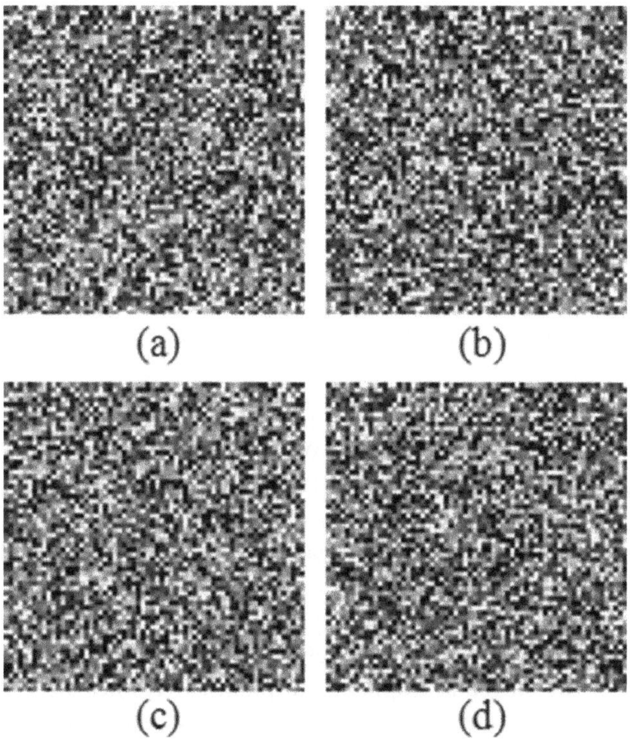

Figure 10.6. (a)–(d) Four typical random intensity-only patterns.

After the series of noisy phase-only patterns [$M_e(u, v)$, $e = 1, 2, 3, \ldots, 11\,000$] is extracted, they are sequentially embedded into a phase-only spatial light modulator for encoding, as shown in figure 10.1. Hence, a series of one-dimensional intensity points, i.e. ciphertexts, are obtained, such as by using a single-pixel bucket detector. Figure 10.9 shows the generated one-dimensional ciphertexts (intensity points). It can be seen in figure 10.9 that the one-dimensional ciphertexts are noisy, and the 2D input image (i.e. plaintext) has been effectively encoded. In practice, the generated

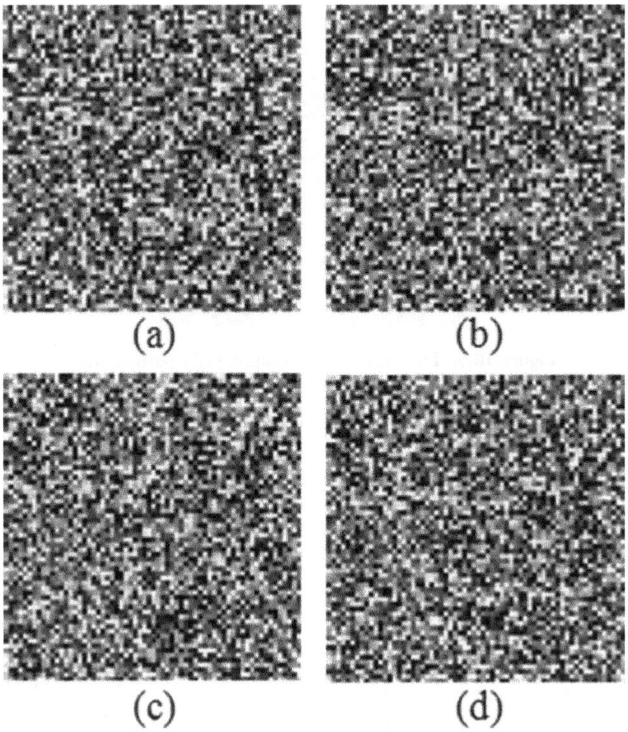

Figure 10.7. (a)–(d) Four typical phase-only patterns (M_e) extracted respectively corresponding to figures 10.6(a)–(d).

one-dimensional intensity points can be further modified or encoded, such as by using the compressed sensing approach. When the receiver has the correct ciphertexts and possesses the correct security keys, a decoded image is generated and shown in figure 10.10. The PSNR for figure 10.10 is 11.65 dB. It can be seen in figure 10.10 that the plaintext can be clearly observed, which satisfies the requirement of the designed single-pixel optical encoding system.

In the single-pixel optical encoding system, a series of pre-generated random intensity-only patterns and setup parameters are used as security keys. When the security keys are not correctly applied, the plaintext cannot be obtained. Figure 10.11(a) shows a decoded image, when only axial distance Z_1 contains an error of 0.5 cm during the decoding. The PSNR for figure 10.11(a) is 9.43 dB. Figure 10.11(b) shows a decoded image when only the wavelength contains an error of 10.0 nm during the decoding. The PSNR for figure 10.11(b) is 9.70 dB. Figure 10.11(c) shows a decoded image when only the random intensity-only patterns are incorrect during the decoding. The PSNR for figure 10.11(c) is 8.93 dB. For the sake of brevity, other setup parameters, such as *d*, are not analyzed and presented. It can be seen in figures 10.11(a)–(c) that when the security key is incorrect during decoding, the 2D input image (i.e. plaintext) cannot be retrieved or cannot be clearly observed. When several security keys are simultaneously incorrect, the decoded image will become fully noisy.

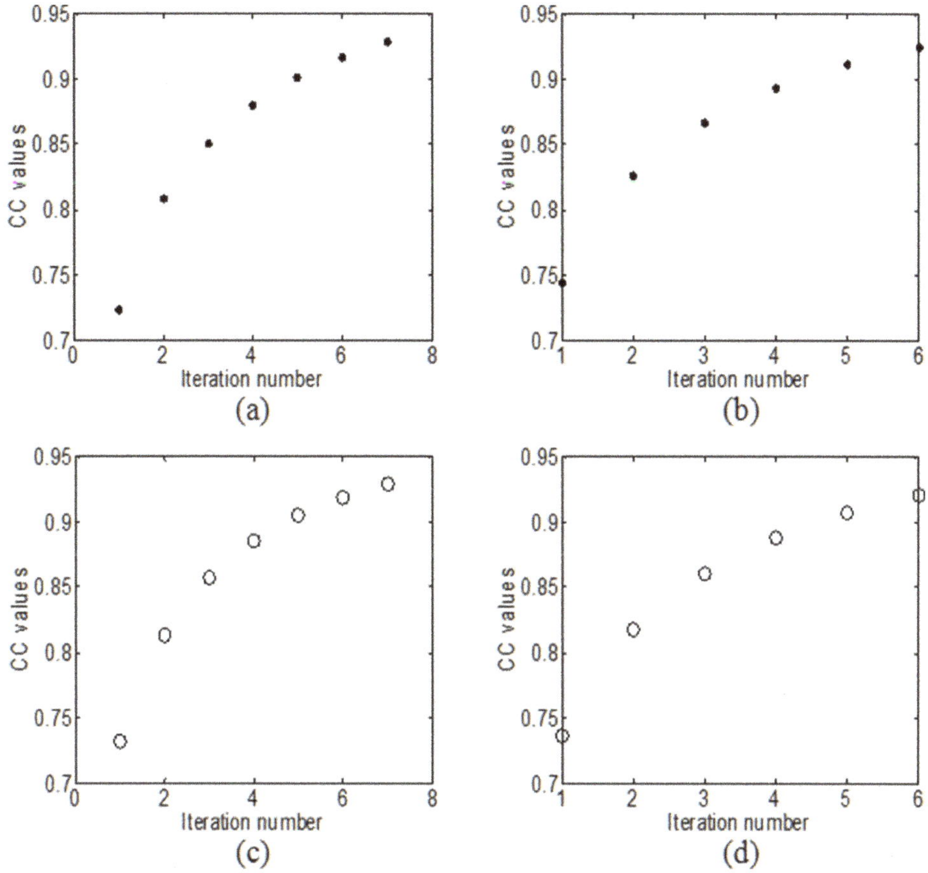

Figure 10.8. (a)–(d) Relationships between the number of iterations and correlation coefficient (CC), respectively, corresponding to figures 10.7(a)–(d).

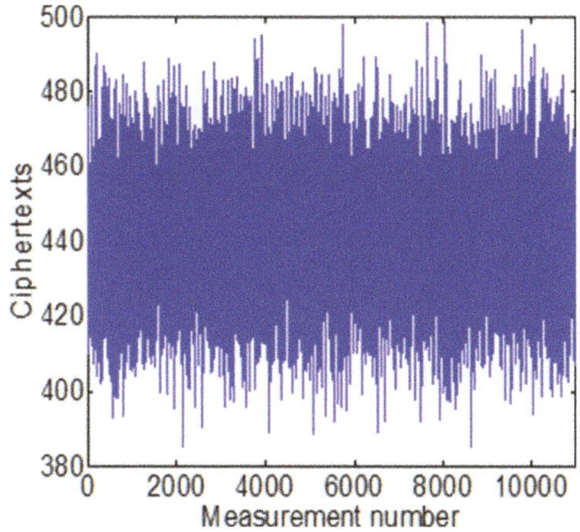

Figure 10.9. The one-dimensional ciphertexts (intensity points) obtained by using the designed single-pixel optical encoding approach.

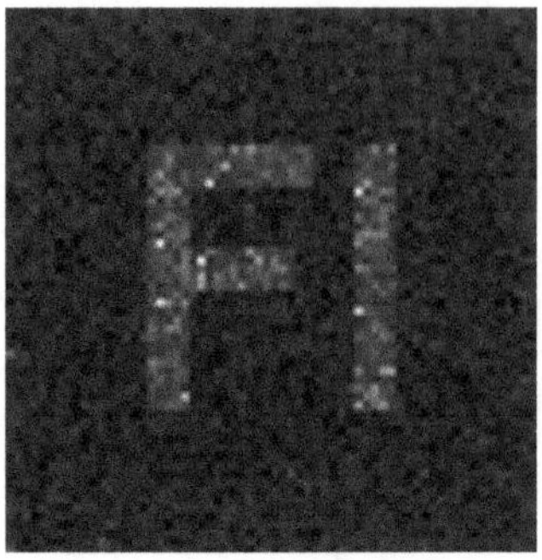

Figure 10.10. A decoded image obtained when the receiver has the correct ciphertexts and possesses the correct security keys.

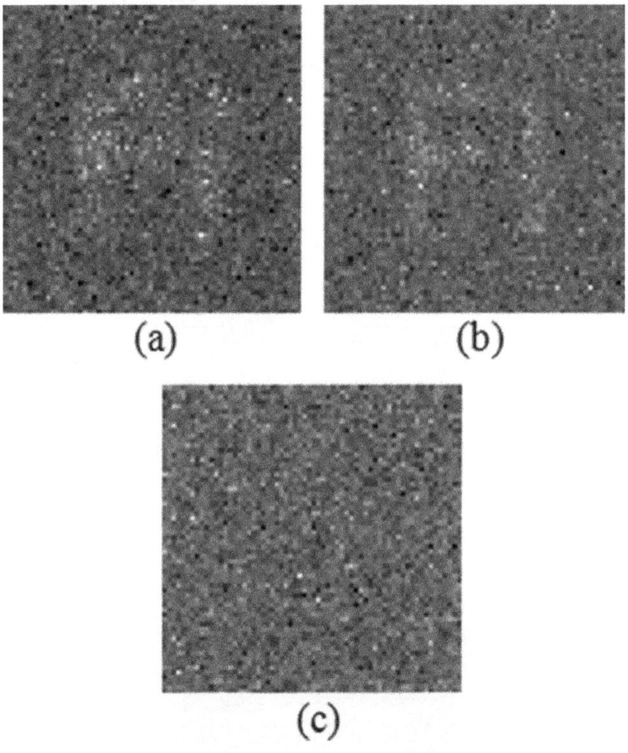

Figure 10.11. Decoded images obtained when (a) only the axial distance Z_1 contains an error of 0.5 cm during decoding, (b) only the wavelength contains an error of 10.0 nm during decoding, (c) only the random intensity-only patterns are incorrect during decoding.

In the intensity-modulated single-pixel optical encoding methods [58–60], a series of random intensity-only patterns can be pre-generated as principal security keys, and the value range of the security keys can be effectively enlarged. In addition, the indirect phase space is enlarged for single-pixel optical encoding.

Differing from the double random phase encoding method [2, 3], single-pixel detection is employed in a single-pixel optical encoding system for the recordings instead of a 2D CCD camera. In the intensity-modulated single-pixel optical encoding methods [58–60], either non-iterative or iterative phase extraction approaches can be designed and applied to extract noisy phase-only patterns for encoding, and high flexibility can be guaranteed for the intensity-to-phase conversion. In the intensity-modulated single-pixel optical encoding methods [58–60], the indirect phase space is large, and the ciphertexts (i.e. a series of one-dimensional intensity points) are not linearly related to the principal security keys (the random intensity-only patterns) which can enhance system security and also withstand attacks [58].

10.3 Single-pixel optical authentication

Although optical encryption techniques have been studied for securing information, it is still desirable to achieve a bigger variety and better security. In recent years, it has been found that optical information authentication strategies [47–51, 53–56, 62, 69–76] can be further developed and integrated into optical encoding setups. For instance, in the double random phase encoding setup [47], information recorded in the CCD plane can be flexibly compressed, and the generated sparse data are transmitted or stored as ciphertexts. This novel strategy can produce totally noisy decoded images. Although the decoded image cannot visually render plaintext information, it can be nonlinearly correlated with the plaintext for information authentication. In practice, original input data/images, i.e. plaintext, can be stored in a separate database for connection via a platform interface, and the plaintext cannot be directly observed by the receivers. Hence, a security layer can be further established over the optical encryption methods. It has been found that an optical information authentication strategy can be flexibly integrated into various optical encoding systems [18], and either security keys (such as setup parameters) or ciphertexts can be compressed or processed to become sparse.

In this chapter, a typical single-pixel optical information authentication method using binary signals [55] is introduced and described, and its application for securing information is also illustrated. In single-pixel (ghost) imaging [26, 53–66], it is well known that the object/information can be recovered by using intensity correlation algorithms. The propagating wave at the object beam arm is usually recorded via single-pixel detection (such as a bucket detector), and the propagating wave at the reference beam arm is usually recorded by using a 2D CCD camera. Applications using single-pixel (ghost) imaging have also been investigated [26, 53–66], such as high-contrast imaging via turbulent media and remote sensing [63]. However, a large number of measurements are usually requested for object/information recovery, which can increase the difficulty of data transmission or storage (especially in information security fields). A low-quantization level can be used as a promising approach in practical applications. In a single-pixel optical information authentication system, two quantization levels [55] can be flexibly used to compress the recorded information at the object beam arm or reference beam arm, and an optical nonlinear correlation approach [47–51, 53–56, 62, 69–72, 75, 76] can be further employed to authenticate the decoded object/information without disclosure of the plaintext information.

Figure 10.12 shows a schematic for the single-pixel optical information authentication method. It can be seen in figure 10.12 that the optical information authentication strategy is integrated into a single-pixel optical encoding setup, and the recording process is the same as that in the single-pixel optical encoding method. In a single-pixel optical information authentication system, the recordings at the reference beam arm can be numerically simulated (i.e. by using a computational approach [63–65]), and the CCD camera shown in figure 10.12 is not compulsory during the encoding process.

In figure 10.12, the plane wave is generated by the combination of a pinhole and a lens, and then reference beam and object beam can be generated by using a beam splitter cube. When a series of random phase-only patterns are sequentially used, intensity patterns at the reference beam arm can be correspondingly calculated by using the free-space wave propagation principle [67, 68] in a computational single-pixel (ghost) imaging system.

When the series of random phase-only patterns is sequentially embedded into the phase-only spatial light modulator, a series of one-dimensional intensity points can

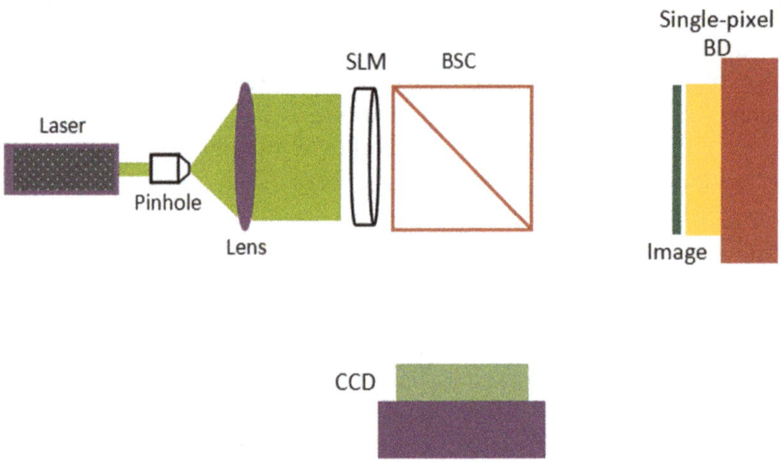

Figure 10.12. A schematic for the single-pixel optical information authentication method [53–56, 62]. A series of random phase-only patterns can be sequentially embedded into a phase-only spatial light modulator for encoding. A collecting lens can be placed between the input image plane and bucket detector. Note that the optical information authentication strategy is integrated into a single-pixel optical encoding system, and the recording process is the same as that in the single-pixel optical encoding method. CCD, charge-coupled device; BD, bucket detector; BSC, beam splitter cube; SLM, phase-only spatial light modulator. In this encoding setup, the axial distance between the SLM and the BD is the same as that between the SLM and the CCD.

be correspondingly recorded by using a single-pixel bucket detector at the object beam arm which can be expressed as [55]

$$B_i = \iint |p(k, l)O_i(k, l)|^2 \, \mathrm{d}k\mathrm{d}l, \tag{10.6}$$

where $p(k, l)$ denotes an input image (i.e. plaintext), $O_i(k, l)$ denotes the wavefront just before the image, and i denotes an integer (1, 2, 3,...). At the reference or object beam arm, binary signals (i.e. 0 and 1) can be generated by using a preset threshold [55], i.e. via post-processing. The threshold can be flexibly set and adjusted to satisfy different applications. In a single-pixel optical information authentication system, the recorded data are further compressed, and only two quantization levels (i.e., 0 and 1) are finally available at the reference or object beam arm. It is worth noting that the information binarization method may be simultaneously used at the reference and object beam arms. Reference intensity patterns and ciphertexts are transmitted to the receiver, and by using the classical optical coherence principle [65] the decoding process can be conducted via intensity correlation.

Since the information binarization operation is introduced and applied to compress the data, the decoded images cannot visually render plaintext information. A nonlinear correlation algorithm [47–51, 53–56, 62, 69–72, 75, 76] can be further used to authenticate the decoded image, which can be described by

$$X(k, l) = |\mathrm{IFT}(|\mathrm{FT}[p(k, l)]\{\mathrm{FT}[G(k, l)]\}^*|^{w-1}\, \mathrm{FT}[p(k, l)]\{\mathrm{FT}[G(k, l)]\}^*)|^2, \tag{10.7}$$

where w denotes the nonlinearity strength [47–51, 53–56, 62, 69–72], $G(k, l)$ denotes a decoded image, $X(k, l)$ denotes a correlation output, and FT and IFT denote the Fourier transform and inverse Fourier transform, respectively. In this chapter, the value w is set as 0.30. If only one remarkable peak is generated and the preset parameter matching is correct, it indicates that the receiver possesses the correct security keys or he/she is an authorized person. When only noisy correlation output is generated, this indicates that the receiver is not an authorized person or that the wrong security keys have been used for decoding. Since the plaintexts are further employed for information authentication, they should be separately or remotely stored for connection via a platform interface [55]. A flow chart is shown in figure 10.13 to illustrate the single-pixel optical information authentication process.

The schematic arrangement in figure 10.12 is computationally conducted to show the single-pixel optical information authentication method. For illumination, the plane wave (with wavelength of 630.0 nm) is generated by using a pinhole and a lens. The grayscale images are encoded in a single-pixel optical information authentication system. The image dimension is 64 × 64 pixels, and the image has 8 bits. The grayscale image to be encoded is the image ‘Baboon’ (http://sipi.usc.edu/database). In this encoding setup, the axial distance between the SLM and the detector is 10.0 cm. For simplicity, the threshold used in the information binarization method is set as an average value of the recorded intensity points (at the object beam arm) or each reference intensity pattern (at the reference beam arm). In practice, the threshold can be flexibly adjusted.

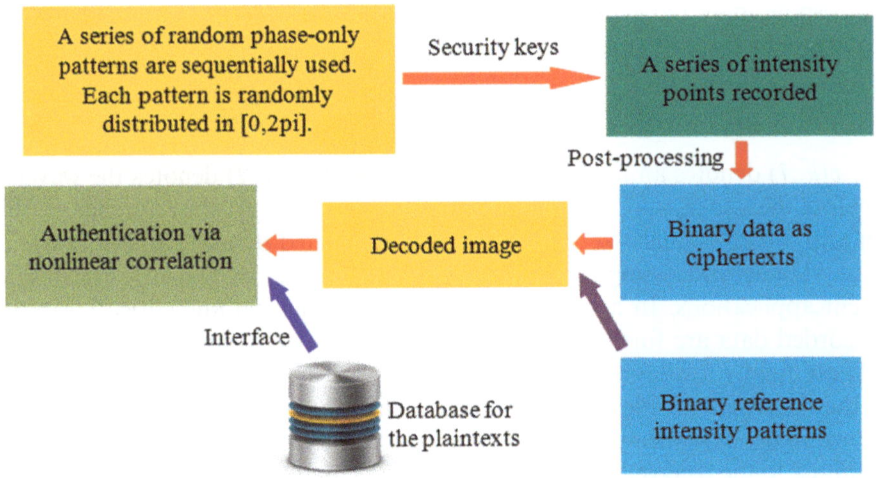

Figure 10.13. Flow chart to illustrate the single-pixel optical authentication process.

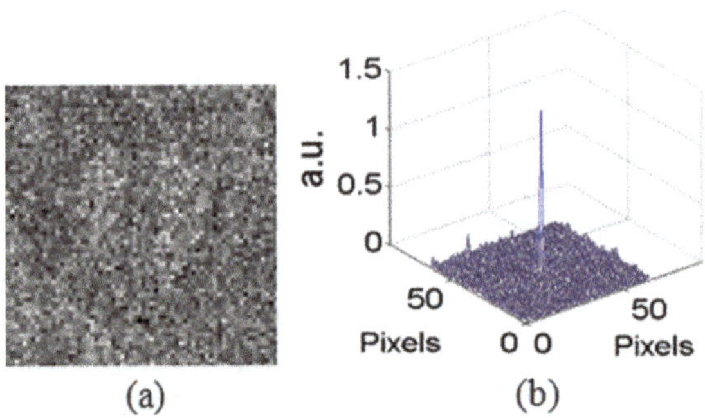

Figure 10.14. (a) A decoded image obtained when binary signals are used only at the object beam arm, and (b) a nonlinear correlation output obtained when the decoded image in (a) is nonlinearly correlated with the original input image (i.e. image 'Baboon').

Figure 10.14(a) shows a decoded image when the binary signals are used only at the object beam arm. All the security keys are correctly applied, and the PSNR for figure 10.14(a) is 3.73 dB. Only 730 measurements are employed, rather than a large number of measurements (such as 20 000 [58, 59, 65]). It can be seen that due to the information binarization operation, the decoded image cannot clearly render the plaintext information. Figure 10.14(b) shows a nonlinear correlation output, when the decoded image in figure 10.14(a) is nonlinearly correlated with the original input image (i.e. 'Baboon'). Only one remarkable peak is generated, indicating that the receiver is an authorized person or he/she possesses correct keys [55]. In practice, other parameters, such as the correlation peak value and peak-to-background ratio, can also be calculated and compared to those stored in a separate database for a matching operation. Figure 10.15(a) shows a decoded image when the binary signals are simultaneously employed at the reference and object beam arms. Only 1250 measurements are employed. All the security keys are correctly used, and the PSNR

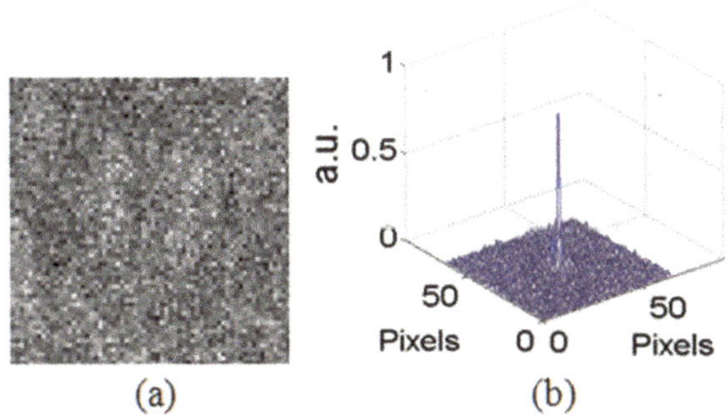

Figure 10.15. (a) A decoded image obtained when the binary signals are simultaneously generated and used at the reference and object beam arms, and (b) a nonlinear correlation output obtained when the decoded image in (a) is nonlinearly correlated with the original input image (i.e. image 'Baboon').

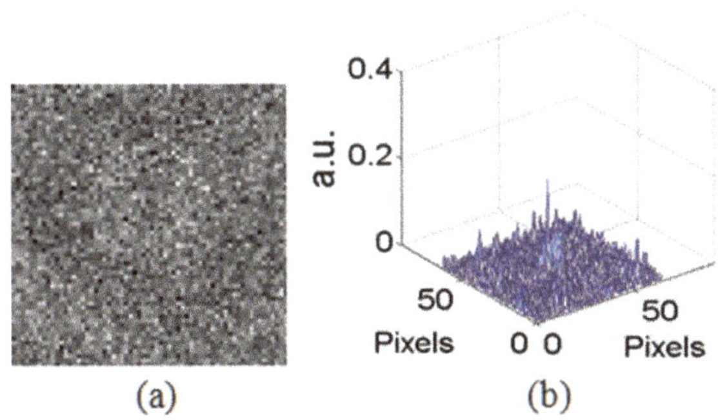

Figure 10.16. (a) A decoded image obtained when only the wavelength contains an error of 20.0 nm during the decoding, and (b) a nonlinear correlation output obtained when the decoded image in (a) is nonlinearly correlated with the original input image.

for figure 10.15(a) is 3.74 dB. Figure 10.15(b) shows a nonlinear correlation output, when the decoded image in figure 10.15(a) is nonlinearly correlated with the original input image (i.e. image 'Baboon'). It is illustrated that only one remarkable peak is generated over the backgrounds.

Since an optical information authentication strategy is integrated into the single-pixel optical encoding system, the setup parameters can also act as security keys for the single-pixel optical information authentication method. Figure 10.16(a) shows a decoded image when only the wavelength contains an error of 20.0 nm during the decoding. Binary signals are simultaneously employed at the reference and object beam arms. Only 1250 measurements are employed. The PSNR for figure 10.16(a) is 3.73 dB. Figure 10.16(b) shows a nonlinear correlation output, when the decoded image in figure 10.16(a) is nonlinearly correlated with the original input image (i.e. image 'Baboon'). It can be seen that only noisy correlation output is generated. When several parameters (i.e. different security keys) are simultaneously incorrect during the decoding, a totally noisy correlation output will be generated. For the sake of brevity, the other setup parameters are not analyzed and described.

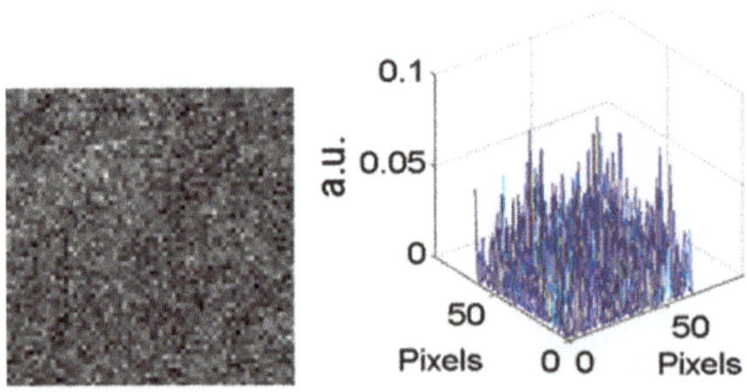

Figure 10.17. Discrimination test: (a) a decoded image obtained when another grayscale image (i.e. image 'Barbara') is encoded, and (b) a nonlinear correlation output obtained when the decoded image in (a) is nonlinearly correlated with original input image (i.e. image 'Baboon'). Binary signals are simultaneously employed at the reference beam arm and object beam arm. Only 1250 measurements are available and used.

To further demonstrate the performance of the single-pixel optical information authentication method, a discrimination test [55] is conducted. Figure 10.17(a) shows a decoded image, when another grayscale image (i.e. image 'Barbara' http://sipi.usc.edu/database) is encoded. Binary signals are simultaneously employed at the reference and object beam arms. The setup parameters are the same as those used for figure 10.15(a). Only 1250 measurements are employed. The PSNR for figure 10.17(a) is 5.12 dB. It can be seen in figure 10.17(a) that a noisy decoded image is also obtained due to the information binarization operation. Figure 10.17(b) shows a nonlinear correlation output when the decoded image in figure 10.17(a) is nonlinearly correlated with the original input image (i.e. image 'Baboon'). It can be seen in figure 10.17(b) that only noisy correlation output is generated. The test images used in this chapter can be found at http://sipi.usc.edu/database. Other images have also been encoded and tested, and similar authentication results can be obtained. The single-pixel optical information authentication method shows high discrimination capability, and can be suitable and useful for securing information in practical applications.

10.4 Discussion

In a single-pixel optical information encoding system, one concern is that the number of varied encoding structures is low. In addition, the direct phase space is not sufficiently large to guarantee system security. The intensity-modulated single-pixel optical information encoding method has been described in this chapter, and can be applied as a promising strategy for effectively enlarging the indirect phase space. It is desirable for researchers to continuously develop or design different encoding strategies and integrate them into single-pixel imaging systems to enrich this research field.

It is also desirable for nonlinear approaches to be studied and integrated into single-pixel optical information encoding systems. For instance, asymmetric methods and public keys can be studied for single-pixel optical information encoding. In recent years, it has been found that private keys [15, 23] can be generated and applied to optical security systems, and can be used by the receivers for decoding. It is believed that private keys can also be further generated and applied for single-pixel optical information encoding.

Nowadays, amplitude-only or phase-only spatial light modulators have been applied in various research fields, such as 3D display and optical information security. It is expected that with the rapid development of spatial light modulator techniques (such as higher diffraction efficiency), a higher decoding quality can be achieved and fewer measurements required for single-pixel optical information encoding.

In addition to high security, intensity-modulated single-pixel optical information encoding approaches also possess high flexibility. Various optical imaging setups can be flexibly designed and applied to generate a series of phase-only patterns which are further embedded into phase-only spatial light modulators for encoding. For instance, many iterative and non-iterative phase retrieval algorithms can be used, and different phase retrieval structures (such as hierarchical or cascaded) can be designed and applied.

The series of one-dimensional intensity points (i.e. as ciphertexts) generated in single-pixel optical information encoding systems can be further processed and then transmitted by using channels, such as optical fiber communication. Practical application cases can be continuously explored by using single-pixel optical encoding methods for securing information.

In single-pixel optical information authentication methods, the number of measurements can be flexibly adjusted to satisfy the requirements of different application cases. Advanced methods might be further investigated to effectively determine how many measurements are suitable for encoding and authenticating different input images. For instance, in the single-pixel information authentication approach with binary signals as illustrated in section 10.3, only 730 measurements have been used. However, determining the optimal value remains unclear for authenticating different input images, which can be further studied.

In single-pixel optical information authentication methods, there are no sufficient and varied strategies to establish encoding and authentication infrastructures. It is desirable for the varied single-pixel optical encoding structures to be continuously studied and applied for information authentication. The simple usage of random phase-only patterns should be modified or replaced in single-pixel optical information authentication systems. Since single-pixel optical information authentication is established based on single-pixel optical information encoding setups, security keys generated in a single-pixel optical encoding method can also be considered as keys for a single-pixel optical information authentication system. In addition, the question of how to effectively distribute security keys also becomes an important topic in these systems.

In single-pixel optical information authentication methods, nonlinear correlation algorithms [47–51, 53–56, 62, 69–72, 75, 76] have already been used to verify the decoded images. It is desirable that other correlation algorithms are further studied for single-pixel optical information authentication. When the number of optical correlation algorithms that are available in the system is larger, the designed single-pixel optical information authentication system can possess higher security in practical applications.

In single-pixel optical information encoding or authentication systems, many image/signal processing algorithms [21, 22] can be further studied and integrated. For instance, compressed sensing methods and various transforms can be studied, which can be further integrated into designed single-pixel optical information encoding or authentication systems for securing information.

In single-pixel optical information encoding or authentication systems, real-time information processing may be a challenge. The combination of electronic and optical principles can facilitate applications of single-pixel optical information encoding or authentication methods. When electronic cryptographic strategies are employed or integrated, it is expected that more single-pixel imaging setups and systems could be established for securing information.

10.5 Conclusions

This chapter presents the development of single-pixel optical information encoding and authentication methods for information security. In the single-pixel optical information encoding method, intensity modulation strategies have been introduced to conduct intensity-to-phase conversion, and its validity has been illustrated. Intensity modulation strategies can effectively enlarge the indirect phase space for single-pixel optical information encoding methods, and high security and high flexibility can be guaranteed. In the single-pixel optical information authentication method, binary signals have been used, and validity has also been illustrated. Since the decoded image cannot clearly render plaintext information, authentication is conducted to effectively verify the decoded images without disclosure of the plaintexts. In practice, the plaintexts may be stored in a separate or remote database for connection by using a platform interface. It is believed that single-pixel optical information encoding and authentication can provide a promising strategy for optically securing information, and that practical application cases could be explored in the future. It is expected that more researchers will be inspired to study this rich research area, i.e. single-pixel optical information encoding and authentication.

Acknowledgements

Support from the National Natural Science Foundation of China (NSFC) under Grant No. 61605165; the Hong Kong Research Grants Council Early Career Scheme under Grant No. 25201416 and Shenzhen Science and Technology Innovation Commission (Basic Research Program) under Grant No. JCYJ20160531184426473 are acknowledged.

References

[1] Petitcolas F A P, Anderson R J and Kuhn M G 1999 Information hiding—a survey *Proc. IEEE* **87** 1062–78

[2] Javidi B 1997 Securing information with optical technologies *Phys. Today* **50** 27–32

[3] Refregier P and Javidi B 1995 Optical image encryption based on input plane and Fourier plane random encoding *Opt. Lett.* **20** 767–9
[4] Chen W, Chen X and Sheppard C J R 2010 Optical image encryption based on diffractive imaging *Opt. Lett.* **35** 3817–9
[5] Chen W and Chen X 2011 Optical cryptography topology based on a three-dimensional particle-like distribution and diffractive imaging *Opt. Express* **19** 9008–19
[6] Chen W, Situ G and Chen X 2013 High-flexibility optical encryption via aperture movement *Opt. Express* **21** 24680–91
[7] Matoba O and Javidi B 1999 Encrypted optical memory system using three-dimensional keys in the Fresnel domain *Opt. Lett.* **24** 762–4
[8] Situ G and Zhang J 2005 Multiple-image encryption by wavelength multiplexing *Opt. Lett.* **30** 1306–8
[9] Kishk S and Javidi B 2003 Watermarking of three-dimensional objects by digital holography *Opt. Lett.* **28** 167–9
[10] Wang X, Zhao D, Jing F and Wei X 2006 Information synthesis (complex amplitude addition and subtraction) and encryption with digital holography and virtual optics *Opt. Express* **14** 1476–86
[11] Unnikrishnan G, Joseph J and Singh K 2000 Optical encryption by double-random phase encoding in the fractional Fourier domain *Opt. Lett.* **25** 887–9
[12] Chen W, Chen X, Anand A and Javidi B 2013 Optical encryption using multiple intensity samplings in the axial domain *J. Opt. Soc. Am.* A **30** 806–12
[13] Liu Z, Xu L, Lin C and Liu S 2010 Image encryption by encoding with a nonuniform optical beam in gyrator transform domains *Appl. Opt.* **49** 5632–7
[14] Meng X F, Cai L Z, Xu X F, Yang X L, Shen X X, Dong G Y and Wang Y R 2006 Two-step phase-shifting interferometry and its application in image encryption *Opt. Lett.* **31** 1414–6
[15] Qin W and Peng X 2010 Asymmetric cryptosystem based on phase-truncated Fourier transforms *Opt. Lett.* **35** 118–20
[16] He W, Peng X, Meng X and Liu X 2012 Optical hierarchical authentication based on interference and hash function *Appl. Opt.* **51** 7750–7
[17] Alfalou A and Brosseau C 2009 Optical image compression and encryption methods *Adv. Opt. Photon.* **1** 589–636
[18] Chen W, Javidi B and Chen X 2014 Advances in optical security systems *Adv. Opt. Photon.* **6** 120–55
[19] Alfalou A and Brosseau C 2010 Dual encryption scheme of images using polarized light *Opt. Lett.* **35** 2185–7
[20] Alfalou A, Brosseau C, Abdallah N and Jridi M 2011 Simultaneous fusion, compression, and encryption of multiple images *Opt. Express* **19** 24023–9
[21] Hennelly B and Sheridan J T 2003 Optical image encryption by random shifting in fractional Fourier domains *Opt. Lett.* **28** 269–71
[22] Naughton T J, Hennelly B M and Dowling T 2008 Introducing secure modes of operation for optical encryption *J. Opt. Soc. Am.* A **25** 2608–17
[23] Liu W, Liu Z and Liu S 2013 Asymmetric cryptosystem using random binary phase modulation based on mixture retrieval type of Yang–Gu algorithm *Opt. Lett.* **38** 1651–3
[24] Matoba O, Nomura T, Cabré E P, Millán M S and Javidi B 2009 Optical techniques for information security *Proc. IEEE* **97** 1128–48

[25] Shi Y, Situ G and Zhang J 2008 Multiple-image hiding by information prechoosing *Opt. Lett.* **33** 542–4
[26] Clemente P, Durán V, Torres-Company V, Tajahuerce E and Lancis J 2010 Optical encryption based on computational ghost imaging *Opt. Lett.* **35** 2391–3
[27] Carnicer A, Usategui M M, Arcos S and Juvells I 2005 Vulnerability to chosen-cyphertext attacks of optical encryption schemes based on double random phase keys *Opt. Lett.* **30** 1644–6
[28] Peng X, Zhang P, Wei H and Yu B 2006 Known-plaintext attack on optical encryption based on double random phase keys *Opt. Lett.* **31** 1044–6
[29] Barrera J F, Vargas C, Tebaldi M, Torroba R and Bolognini N 2010 Known-plaintext attack on a joint transform correlator encrypting system *Opt. Lett.* **35** 3553–5
[30] Frauel Y, Castro A, Naughton T J and Javidi B 2007 Resistance of the double random phase encryption against various attacks *Opt. Express* **15** 10253–65
[31] Liao M, He W, Lu D and Peng X 2017 Ciphertext-only attack on optical cryptosystem with spatially incoherent illumination: from the view of imaging through scattering medium *Sci. Rep.* **7** 41789
[32] Li G, Yang W, Li D and Situ G 2017 Cyphertext-only attack on the double random-phase encryption: experimental demonstration *Opt. Express* **25** 8690–7
[33] He M, Tan Q, Cao L, He Q and Jin G 2009 Security enhanced optical encryption system by random phase key and permutation key *Opt. Express* **17** 22462–73
[34] Kumar P, Kumar A, Joseph J and Singh K 2009 Impulse attack free double-random-phase encryption scheme with randomized lens-phase functions *Opt. Lett.* **34** 331–3
[35] Peng X, Wei H and Zhang P 2006 Asymmetric cryptography based on wavefront sensing *Opt. Lett.* **31** 3579–81
[36] Chen W and Chen X 2014 Double random phase encoding using phase reservation and compression *J. Opt.* **16** 025402
[37] Wang R K, Watson I A and Chatwin C 1996 Random phase encoding for optical security *Opt. Eng.* **35** 2464–9
[38] Li Y, Kreske K and Rosen J 2000 Security and encryption optical systems based on a correlator with significant output images *Appl. Opt.* **39** 5295–301
[39] Hwang H E, Chang H T and Lie W N 2009 Multiple-image encryption and multiplexing using a modified Gerchberg–Saxton algorithm and phase modulation in Fresnel-transform domain *Opt. Lett.* **34** 3917–9
[40] Johnson E G and Brasher J D 1996 Phase encryption of biometrics in diffractive optical elements *Opt. Lett.* **21** 1271–3
[41] Chen W, Chen X, Stern A and Javidi B 2013 Phase-modulated optical system with sparse representation for information encoding and authentication *IEEE Photon. J* **5** 6900113
[42] Xiao Y L, Zhou X, Yuan S, Liu Q and Li Y C 2009 Multiple-image optical encryption: an improved encoding approach *Appl. Opt.* **48** 2686–92
[43] Liu Z, Guo Q, Xu L, Ahmad M A and Liu S 2010 Double image encryption by using iterative random binary encoding in gyrator domains *Opt. Express* **18** 12033–43
[44] Zhang Y and Wang B 2008 Optical image encryption based on interference *Opt. Lett.* **33** 2443–5
[45] Shao Z, Shu H, Wu J, Dong Z, Coatrieux G and Coatrieux J L 2014 Double color image encryption using iterative phase retrieval algorithm in quaternion gyrator domain *Opt. Express* **22** 4932–43

[46] Gerchberg R W and Saxton W O 1972 A practical algorithm for the determination of phase from image and diffraction plane pictures *Optik* **35** 237–46

[47] Pérez-Cabré E, Cho M and Javidi B 2011 Information authentication using photon-counting double-random-phase encrypted images *Opt. Lett.* **36** 22–4

[48] Chen W 2016 Single-shot imaging without reference wave using binary intensity pattern for optically-secured-based correlation *IEEE Photon. J* **8** 6900209

[49] Chen W and Chen X 2016 Digital holography-secured scheme using only binary phase or amplitude as ciphertext *Appl. Opt.* **55** 6740–6

[50] Chen W 2015 Multiple-wavelength double random phase encoding with CCD-plane sparse-phase multiplexing for optical information verification *Appl. Opt.* **54** 10711–6

[51] Wang X G, Chen W and Chen X 2015 Optical information authentication using compressed double-random-phase-encoded images and quick-response codes *Opt. Express* **23** 6239–53

[52] Sadjadi F and Javidi B 2007 *Physics of the Automatic Target Recognition* (Berlin: Springer)

[53] Chen W and Chen X 2013 Object authentication in computational ghost imaging with the realizations less than 5% of Nyquist limit *Opt. Lett.* **38** 546–8

[54] Chen W and Chen X 2013 Ghost imaging for three-dimensional optical security *Appl. Phys. Lett.* **103** 221106

[55] Chen W and Chen X 2015 Grayscale object authentication based on ghost imaging using binary signals *EPL* **110** 44002

[56] Chen W and Chen X 2014 Marked ghost imaging *Appl. Phys.Lett.* **104** 251109

[57] Tanha M, Kheradmand R and Ahmadi-Kandjani S 2012 Gray-scale and color optical encryption based on computational ghost imaging *Appl. Phys. Lett.* **101** 101108

[58] Chen W 2016 Optical cryptosystem based on single-pixel encoding using the modified Gerchberg–Saxton algorithm with a cascaded structure *J. Opt. Soc. Am.* A **33** 2305–11

[59] Chen W 2016 Correlated-photon secured imaging by iterative phase retrieval using axially-varying distances *IEEE Photon. Technol. Lett.* **28** 1932–5

[60] Chen W 2016 Optical data security system using phase extraction scheme via single-pixel detection *IEEE Photon. J.* **8** 7801507

[61] Chen W 2016 Modulating phase via rotation for optical encoding based on correlated photon imaging *IEEE Photon. Technol. Lett.* **28** 540–3

[62] Chen W and Chen X 2015 Optical authentication via photon-synthesized ghost imaging using optical nonlinear correlation *Opt. Lasers Eng.* **73** 123–7

[63] Erkmen B I 2012 Computational ghost imaging for remote sensing *J. Opt. Soc. Am.* A **29** 782–9

[64] Shapiro J H 2008 Computational ghost imaging *Phys. Rev.* A **78** 061802(R)

[65] Erkmen B I and Shapiro J H 2010 Ghost imaging: from quantum to classical to computational *Adv. Opt. Photon.* **2** 405–50

[66] Wu J, Haobogedewude B, Liu Z and Liu S 2017 Optical secure image verification system based on ghost imaging *Opt. Commun.* **399** 98–103

[67] Goodman J W 1996 *Introduction to Fourier Optics* 2nd edn (New York: McGraw-Hill)

[68] Schnars U and Jueptner W 2005 *Digital Holography: Digital Hologram Recording, Numerical Reconstruction, and Related Techniques* (Berlin: Springer)

[69] Javidi B *et al* 2016 Roadmap on optical security *J. Opt.* **18** 083001

[70] Pérez-Cabré E, Abril H C, Millan M S and Javidi B 2012 Photon-counting double-random-phase encoding for secure image verification and retrieval *J. Opt.* **14** 094001

[71] Chen W and Chen X 2014 Optical multiple-image authentication based on modified Gerchberg–Saxton algorithm with random sampling *Opt. Commun.* **318** 128–32

[72] Chen J X, Zhu Z L, Fu C, Yu H and Zhang L B 2015 Gyrator transform based double random phase encoding with sparse representation for information authentication *Opt. Laser Technol.* **70** 50–8

[73] Carnicer A and Javidi B 2017 Optical security and authentication using nanoscale and thin-film structures *Adv. Opt. Photon.* **9** 218–56

[74] Wang Q, Alfalou A and Brosseau C 2017 New perspectives in face correlation research: a tutorial *Adv. Opt. Photon.* **9** 1–78

[75] Yi F, Jeoung Y and moon I 2017 Three-dimensional image authentication scheme using sparse phase information in double random phase encoded integral imaging *Appl. Opt.* **56** 4381–7

[76] Chen W 2017 Ghost identification based on single-pixel imaging in big data environment *Opt. Express* **25** 16509–16

IOP Publishing

Advanced Secure Optical Image Processing for Communications

Ayman Al Falou

Chapter 11

Compressive optical image security

Jun Li, Miao Lei, Xiaofang Dai, Jiaosheng Li, Ting Zhong, Hongbing Li Bangwen Jia, Chanxia Yang, Shangyuan Wang and Chen Wang

11.1 Compressive sensing

The traditional method of signal acquisition and processing consists of four parts: sampling, compression, transmission, and decompression. The signal acquisition must satisfy the requirement of the Nyquist sampling theorem in the sense that the sampling rate must be larger than twice the maximum frequency of the signal. After the signal system collects large amounts of data, in order to reduce the storage, processing and transmission costs, the acquired data is compressed by only keeping a small portion of the data that is sufficient to meet the requirements of human visual perception. This high-speed sampling and recompression process results in a wastage of sampling resources. In 2006, Donoho, Candes and Tao proposed a novel compressive sensing (CS) theory [1], which uses a new method of signal sampling and reconstruction based on sparsity or compressibility of the signal. One of the important features of CS theory is that it can make full use of the sparsity of the input signal to realize the high-dimensional sparse signal perception through a lower-dimensional uncorrelated measurement. Unlike the conventional Nyquist sampling theorem, the sampling rate in CS theory is not limited by the signal bandwidth, but only by sparsity of the signal. The higher the sparsity, the lower the sampling rate can be. The core idea of CS is to sample and compress the signal at the same time. This can further improve the sampling rate under the same hardware constraints. The original signal is reconstructed by a corresponding reconstruction algorithm.

11.1.1 The basic framework of compressive sensing

The premise of CS theory is that the input signal is sparse or compressible. If the signal satisfies these requirements, we can reconstruct the signal by making fewer

doi:10.1088/978-0-7503-1457-2ch11

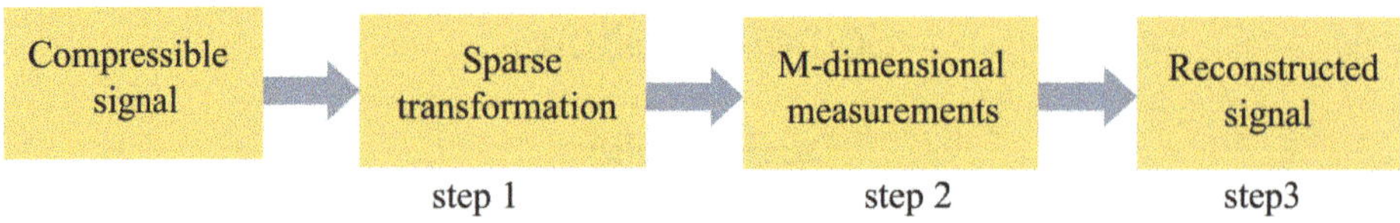

Figure 11.1. The method of signal acquisition and coding process using compressive sensing theory.

measurements than required by a conventional sampling method. Figure 11.1 shows the method of signal acquisition and coding process using CS theory. This process consists of three parts: sparse representation, random measurement projection, and original signal reconstruction.

Let $x \in R^{N\times 1}$, $x = (x_1, x_2, \ldots, x_N)^T$ denote a real valued one-dimensional finite length discrete-time signal, and let $\Phi = [\varphi_1, \varphi_2, \ldots, \varphi_N] \in R^{N\times N}$ denote a basis matrix composed of column vectors φ_i. The sparse signal x is represented using the basis matrix Φ as

$$x = \Phi\alpha \tag{11.1}$$

where α is the coefficient (weight) vector, and which can be expressed as $\alpha = \Phi^T x$. If only K non-zero values exist in α, α is the K-sparse representation of the signal x using the basis matrix Φ. The essence of CS theory is to obtain the observation vector by random projection, i.e., the observed value y:

$$y = \Psi x \tag{11.2}$$

where y is an $M \times 1$ column vector, $\Psi \in R^{M\times N}$ is a measurement matrix that is uncorrelated with the coefficient basis Φ, which can be considered as an encryption key to extend the key space. Since $M \ll N$, the number of equations in (11.2) is far fewer than the number of unknowns. Hence, the system of equations cannot be solved. Therefore, another approach to solve for x is by using an l_1 norm optimization problem:

$$\hat{x} = \text{argmin} \, ||x||_1 \quad \text{s.t.} \quad y = \Psi x. \tag{11.3}$$

In summary, CS theory mainly includes sparse signal representation, a measurement matrix and signal reconstruction.

11.1.2 The core problem of compressive sensing

- **Sparse representation**

  The sparse signal representation is an *a priori* condition for CS, that is, a signal can be converted to a sparse signal representation using a transformation. As shown in figure 11.2, the known discrete-time signal x is represented using the basis matrix Φ using (11.1).

  As can be seen from the illustration in figure 11.2, there are more non-zero entries in x, but only K non-zero entries in α, and all other $N-K$ coefficients are either zero or approximately zero. In this case, the signal α is the K-sparse representation of the signal x on the sparse basis Φ. Presently, the commonly

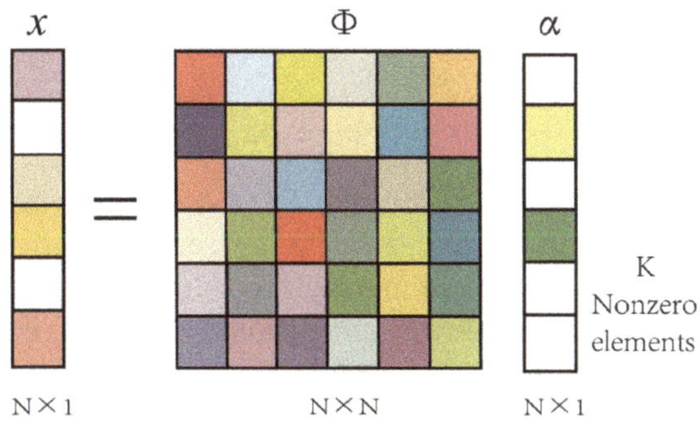

Figure 11.2. Sparse representation by a sparse basis Φ.

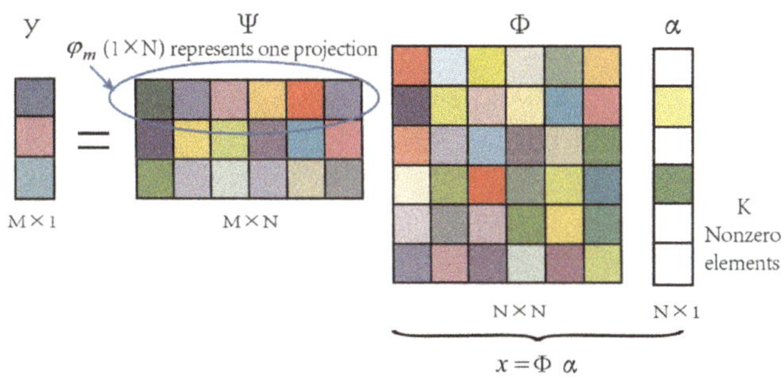

Figure 11.3. Compressed sensing projection measurement procedure.

used sparse transformations in CS are the wavelet transform, Fourier transform and discrete cosine transform.

Typically, all signals are not sparse in nature. The sparsity of the signal is proportional to the number of coefficients that are either zero or close to zero. In general, the sparseness of a signal is measured by transforming the sparse decay rate. Let us consider $x = \Phi\alpha$ as an example, and sort the sparse coefficients α_n such that, $|\alpha_1| \geqslant |\alpha_2| \geqslant |\alpha_3| \geqslant \cdots \geqslant |\alpha_n|$. If there exists a constant P, $c \geqslant 1$, and satisfy

$$|x_n| \leqslant P \cdot n^{-c} \tag{11.4}$$

the sparse coefficient vector is said to follow the law of power decay, while the signal x satisfies the sparsity requirement on the sparse basis Φ. As the value of c increases, the decay rate is faster, leading to higher compressibility of the signal.

- **Measurement matrix**

  In the CS measurement process, the signal x is projected onto the measurement matrix Ψ that is uncorrelated with the sparse basis matrix Φ to affect the projection measurement. The measurement matrix Ψ can be a random matrix, and can be considered as an encryption key to extend the key space. As shown in figure 11.3, the M-length measured value is obtained by the inner product of the measurement matrix Ψ and the signal x, which can be expressed as:

$$y = \Psi x = \Psi\Phi\alpha = \theta\alpha. \tag{11.5}$$

The measurement matrix Ψ is an $M \times N(M \ll N)$ matrix, and the measured value y is an $M \times 1$ vector and $\theta = \Psi\Phi$ is a sensing matrix.

The role of the measurement matrix is to transform the higher-dimensional original sparse signals into lower-dimensional subspace. Finally, the original signal is reconstructed by using special reconstruction algorithms. In order to convert as much useful information as possible into the measured values, it is required that the projection matrix satisfies the restricted isometry property (RIP). In other words, the measurement basis in the measurement matrix is uncorrelated with the signal sparse basis, and it can be expressed using the RIP condition as:

$$1 - \delta \leqslant \frac{||\theta x||}{||x||_2^2} \leqslant 1 + \delta \text{ and } \delta > 0. \tag{11.6}$$

- **Signal reconstruction**
  Signal reconstruction refers to reconstructing the original N-length signal σ^2 from the low dimension M-length measured value y. The original signal can be reconstructed by solving the following l_0 norm optimization problem:

$$\hat{\alpha} = \underset{\alpha}{\text{argmin}}\ ||\alpha||_0 \quad \text{s.t.}\ \ \theta\alpha = y \tag{11.7}$$

where $||\alpha||_0$ represents the number of the non-zero elements in α. If the measurement number M satisfies the requirement that $M = O(K\ \log(N))$, the original sparse signal x can be reconstructed accurately.

However, if the measurement number M satisfies $M \geqslant \mathrm{c} \cdot K \cdot\ \log(N/K)$, the $\Psi x + e = y$ norm optimization problem in (11.7) can be converted into an l_1 norm convex optimization problem to rebuild the K sparse vector, which can be transformed into a linear programming problem as in (11.8), and solved by second-order cone programming:

$$\hat{\alpha} = \underset{\alpha}{\text{argmin}}||\alpha||_1 \ \text{ s. t. }\ \ \theta\alpha = y. \tag{11.8}$$

The optimization problem under the minimum l_1 norm is also called basis pursuit (BP). Algorithms that are commonly used to solve the problem in (11.8) include the classical algorithms for convex optimization problems such as the gradient projection method, BP algorithm, interior point method, conjugate gradient projection method and iterative shrinkage/threshold (IST) algorithm. Typically, the BP algorithm is more complex and the interior point method is slow but highly accurate. The gradient projection method has the advantage of good operation speed. The iterative threshold method is relatively simple and widely used in image processing, and it can reconstruct the original signal by a fixed iterative formula. Because of its strong dependence on the observation operator, the convergence rate of the IST algorithm is not ideal for finding a solution. In order to further accelerate the convergence rate of this algorithm, a two-step iterative shrinkage/threshold algorithm (TwIST) [2] was proposed. The TwIST algorithm preserves the denoising advantage of the IST algorithm and can deal with ill-conditioned problems efficiently.

11.2 Compressive optical image security

Image information security has become increasingly important in many image applications due to the continuous growth of transmission demand and image information sharing. It is important to consider the security of these systems to protect information resources against human error and malicious attacks. In recent years, encryption and decryption methods based on optoelectronics have proven to be highly accurate for the security of optical systems. Following the invention of the optical image encryption method based on double random phase encoding (DRPE) [3] technology, many research scholars have carried out research on this topic, and this has led to an emerging cross-field research topic known as optical image security. Optical image security using holographic technology is one of the mainstream programs [4–9].

Research related to image security based on optical information processing includes optical image encryption, optical image hiding (or watermarking), and optical image encryption and hiding. The relevant optical safety research can be found in the previous chapter of this book.

11.2.1 The challenge in optical image security

Large volume of data has been the main limiting factor for optical image security due to the requirements of storing or transmission of holograms. In order to address this issue, many compression techniques have been proposed in recent years [10–13]. However, the holographic compression effect is not ideal due to the implementation of hologram compression using electronic means, and the holographic speckle effect. The CS theory is applicable on the optical domain data and provides a powerful tool for hologram compression.

The joint approach of combining image security with CS theory offers several benefits. This approach can not only reduce the effort required for encrypted image acquisition and ciphering text size but also increase the encryption layer and pre-process the input signal for the optical encryption system [14]. Research on digital domain image security based on compression perception has been performed by many scholars [15–20]. However, previous research work mostly adopted the charge coupled device (CCD) surface array imaging, and achieved image security via electronic means. Although these methods can perform well to compress a large volume of data, they cannot be applied to pure light systems, such as applications using an all-optical network. Moreover, the previous work did not make full use of the high speed and high parallelism characteristics of an optical system. These drawbacks seriously restrict the application of existing methods, so they are less likely to perform the function of optical image sensing, optical image security processing and optical image compression simultaneously in an optical imaging system.

Therefore, if the hologram obtained by optical image security processing in the optical domain can be compressed, it will effectively reduce the amount of information needed for optical image security. In addition, this process will solve the requirement of storage and transmission of mass data. Moreover, this approach will also provide an effective solution to the processing of large amounts of data in other technical fields such as the Internet of Things (IoT), all-optical network, energy source and information security, as well as the secure acquisition of 2D or 3D scene information, which has great theoretical and practical significance. In contrast to traditional holographic imaging, when acquiring scene imaging in an optical system, the process also performs optical image encryption, optical image hiding (or watermarking) and optical compression of the scene. This approach will not only improve the information security in a system, but also reduce the amount of data storage and transmission requirements significantly. Moreover, this approach also provides the theoretical and technical basis for solving the problems of the holographic imaging process.

11.2.2 The solution for the challenge of optical image security

An optical image security diagram based on single-pixel compression imaging performed in the optical domain is shown in figure 11.4.

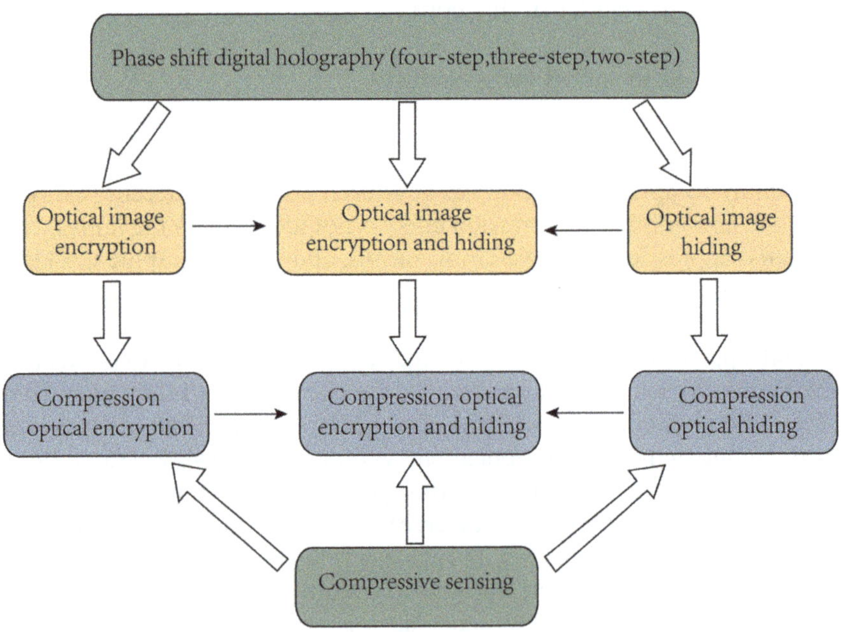

Figure 11.4. Optical image security schematic based on single-pixel compression imaging performed in the optical domain.

Optical image encryption and optical image hiding are studied in combination with a two or three or four-step phase-shifting holography. Using this method, optical image encryption and optical image hiding techniques are combined to carry out optical image encryption and hiding. Hence, the system is more secure, and does not allow access to unauthorized users. Furthermore, we apply the CS technique in a purely optical domain to perform a compressive optical image security study based on the single-pixel compression imaging technique. If the measurement matrix Ψ is safe, then the CS theory provides an additional encryption layer into optical image security processing, thereby improving the overall security of the system. The methods for compressive optical image security include compressive optical image encryption, compressive optical image hiding, and compressive optical image encryption and hiding. By adopting the CS technology, these methods will effectively reduce the amount of information recording needed for providing optical information security. Furthermore, this approach can also address the information security issues in the areas of the IoT, all-optical network and 2D or 3D scene information security access etc.

11.3 Compressive optical image encryption

From the time that DRPE technology was invented, optical image encryption technology has attracted the attention of many research scholars. By combining the optical image encryption technique with the CS technique, the encrypted image is compressed and sampled at the same time. This approach can make full use of the high parallelism and high rate optical image processing, and can also fully utilize the advantages of compressed sampling to reduce the volume of data. Storage and transmission of data will be further reduced, and system security is also improved.

We propose two compression optical image encryption architectures based on a Mach–Zehnder interferometer (MZI) and joint transform correlator (JTC), respectively.

11.3.1 Compressive optical image encryption based on MZI

In the MZI, an encrypted hologram is obtained by encrypting the image using the DRPE technique in the optical path of the object. Next, the encrypted hologram is compressed and sampled in the sparse domain using the optimized selection of the measurement matrix. Finally, the reconstruction algorithm is used to reconstruct the hologram, from which the original encrypted object can be obtained. A schematic diagram of compressive optical image encryption based on MZI is shown in figure 11.5.

The system shown in figure 11.5 comprises of two parts: optical image encryption and single-pixel compression imaging. A laser beam is split into two beams: an object light and a reference light. First, the object light illuminates the object image for encryption and then passes through two random phase masks to implement the encryption using the DRPE technique. On the reference optical path, the reference

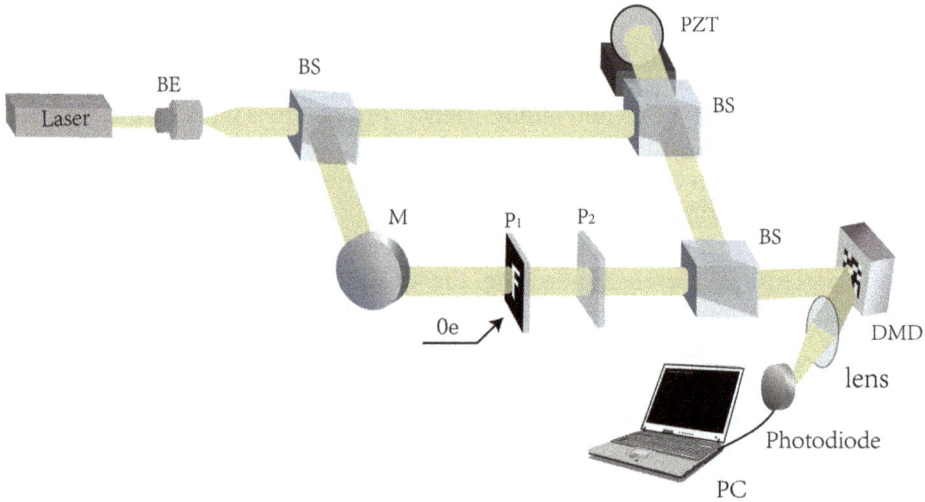

Figure 11.5. A schematic diagram of a compressive optical image encryption system based on MZI. BE, beam expander; L, lens; BS, beam splitter; M, mirror; P, random phase plate; PZT, piezoelectric transducer mirror; Oe, object image.

light passes through the piezoelectric transducer mirror (PZT) to realize the phase-shift. Next, the two beams superimpose on another beam splitter, where they will overlap to form an interference pattern on the surface of the DMD. Compressive sampling values are obtained by calculating the random linear measurements of the measurement matrix loaded on the DMD plane and the hologram. Following this, the measurements are collected by using a high-sensitivity photodiode. Next, the encrypted hologram is obtained through the traditional communication channel, and the encrypted hologram on the DMD plane is reconstructed using a CS algorithm. Finally, the original object can be decrypted using the Fresnel inverse transform, the appropriate parameters of the optical system and the decryption key. A description of the principle is given below.

Suppose that the reference wave is expressed by the real amplitude R. $U_0(x_0, y_0)$ is a complex object field in the plane P_1. The complex amplitude transmittances of P_1 and P_2 are $\exp[i2\pi \cdot p(x_0, y_0)]$ and $\exp[i2\pi \cdot q(x_0, y_0)]$, respectively, where $p(x_0, y_0)$ and $q(x_0, y_0)$ are two independent white noises uniformly distributed in [0, 1]. z_1 is the distance between planes P_1 and P_2, and z_2 is the distance between P_2 and DMD. The complex object field $U(x, y)$ on the DMD plane can be written as

$$U(x, y) = FR_{Z2}\left\{FR_{Z1}\left\{U_0(x_0, y_0)\exp\left[i2\pi \cdot p(x_0, y_0)\right]\right\} \times \exp\left[i2\pi \cdot q(x_0, y_0)\right]\right\}, \tag{11.9}$$

where FR_Z denotes the Fresnel transform of distance z. Based on the traditional three-step phase-shifting holography, we set the phases of the reference wave by PZT to 0, $\pi/2$ and π, respectively, then three interference intensities of the complex amplitude field are expressed as

$$I_1(x, y; 0) = | R + U(x, y) |^2 = I_0(x, y) + 2\,\mathrm{Re}\,[U(x, y)] \cdot R, \tag{11.10}$$

$$\begin{aligned} I_2(x, y; \pi/2) &= | R \cdot \exp(\mathrm{i} \cdot \pi/2) + U(x, y) |^2 \\ &= I_0(x, y) + 2\mathrm{Im}[U(x, y)] \cdot R, \end{aligned} \tag{11.11}$$

$$\begin{aligned} I_3(x, y; \pi) &= | R \cdot \exp(\mathrm{i} \cdot \pi) + U(x, y) |^2 \\ &= I_0(x, y) - 2\,\mathrm{Re}\,[U(x, y)] \cdot R, \end{aligned} \tag{11.12}$$

where I_0 is the zero-order light given by

$$I_0(x, y) = R^2 + | U(x, y) |^2. \tag{11.13}$$

DMD is actually a spatial light modulator (SLM), which corresponds to a random measurement matrix. On the basis of the restricted isometry property (RIP) [21] in CS theory, the mirrors on DMD were in a certain pseudo-random condition when the hologram I formed on the DMD plane. With the control of a pseudo-random number generator (RNG), only the light reflected by the mirrors in orientation towards the lens can be summed at the photodiode to compute the measurement as its output voltage.

$$Y(\mathrm{m}) = \{y_{1m}, y_{2m}, y_{3m}\} = \varphi_m[I_1, I_2, I_3], \tag{11.14}$$

where φ_m is the mth pseudo-random matrix on the DMD plane, $m = \{1, 2, \ldots, M\}$. The pseudo-random number generator is generated with the TwIST algorithm package by computer. Through the production of M measurement matrixes uploaded into DMD device, we can perform M times compressive measurements to achieve the output value Y on the single-pixel photodiode.

$$Y = [y_1, y_2, y_3] = \psi[I_1, I_2, I_3], \tag{11.15}$$

where $(I_1, I_2, I_3) = \begin{bmatrix} I_{11} & I_{21} & I_{31} \\ I_{12} & I_{22} & I_{32} \\ \vdots & \vdots & \vdots \\ I_{1M} & I_{2M} & I_{3M} \end{bmatrix}$, I_{1M} represents the hologram I_1 in the Mth measurement; $\psi \in R^{M\times N}$ is the measurement matrixes uploaded into DMD device; $Y \in R^{M\times 3}$ is the measurement data on the photodiode; and $y_k \in R^{M\times 1}$, $I_k \in R^{N\times 1}$, $k = 1, 2, 3$.

The optical signal is converted into an electrical signal using a floating point type photodiode. And through an AD converter the electrical signal is discretized to the digital signal according the quantization bits of the AD converter. In the process, we will introduce a quantization error to influence the quality of the reconstruction image, but we can handle the quantization noise through adopting a higher quantization bits-based AD converter and image filtering technology according to the actual system needs. Finally, these digital signals can be transmitted via a conventional channel to the computer where we can reconstruct and decrypt the image.

The above description is related to compressive optical image encryption by using three-step phase-shifting holography. Further, in the case of the two-step algorithm based on the traditional two-step quadrature phase-shifting holography [22, 23], when we set the phases of the reference wave in the first and second exposure to 0 and $\pi/2$ respectively, then two on-axis quadrature-phases (as in (11.10) and (11.11)) cause a larger decrease in the holograms data volume than the three-step phase-shifting holography algorithm.

The following shows the process of image reconstruction and decryption. We can retrieve the interference patterns $\hat{I}_k$ by adopting the TwIST algorithm to solve the optimal problem below:

$$\min_{I_k} \frac{\mu}{2} \| Y_k - \Psi \hat{I}_k \|_2^2 + TV(\hat{I}_k), \quad \text{s.t.} \quad Y_k = \Psi I_k \tag{11.16}$$

where $\| \cdot \|_2$ stands for the l_2 norm of a vector and μ is a constant. The first penalty is a least-squares term that is very small when $\hat{I}_k$ is consistent with the correlation vector Y_k. The second penalty $TV(\hat{I}_k)$ is the signal's total variation, which can be described as

$$TV(\hat{I}_k) = \sum_{i,j} \sqrt{\left(\hat{I}_{k(i+1),j} - \hat{I}_{ki,j}\right)^2 + \left(\hat{I}_{ki,j+1} - \hat{I}_{ki,j}\right)^2}, \tag{11.17}$$

where indices i, j run over all pairs of the adjacent pixels in $\hat{I}_k$. Due to the non-differentiability of the Euclidean norm at the origin, we will take an approximated TV as follows:

$$TV(\hat{I}_k) = \sum_{i,j} \sqrt{\left(\hat{I}_{k(i+1),j} - \hat{I}_{ki,j}\right)^2 + \left(\hat{I}_{ki,j+1} - \hat{I}_{ki,j}\right)^2 + \xi^2}, \tag{11.18}$$

here ξ is a small positive parameter. Apart from the consideration of sparsity and fidelity for a given reconstruction, TV regularization preserves the edges of the recovered image while suppressing any spurious high-frequency components. After setting up the cost function, fixed-point iteration is employed to extract a solution for the problem.

By using the three-step phase-shifting holography technique, the reconstruction process will be described below. For the reconstruction process for the two-step algorithm one can refer to [24, 25], in which the two-step phase-shifting holography technique is described based on the calculated intensity.

When the intensity patterns $\hat{I}_1$, $\hat{I}_2$, $\hat{I}_3$ are reconstructed with the TwIST algorithm, we can calculate the complex amplitude on the DMD plane

$$U(x, y) = \frac{\hat{I}_1 - \hat{I}_3 + i(2\hat{I}_2 - \hat{I}_1 - \hat{I}_3)}{4}. \tag{11.19}$$

Once the complex amplitude $U(x, y)$ is known, in addition to random phase masks P_1 and P_2, z_1, z_2 and λ, we can digitally or optically retrieve the original object image from the encrypted image as

$$O'(x_0, y_0) = IFR_{Z1}\{IFR_{Z2}\{U(x, y)\exp[-i2\pi \cdot q(x_0, y_0)]\} \times \exp[-i2\pi \cdot p(x_0, y_0)]\}. \tag{11.20}$$

where IFR_Z denotes the inverse Fresnel transformation of the distance z. Specifically, the original image has been perfectly reconstructed and decrypted.

11.3.2 Compressive optical image encryption based on JTC

In the JTC-based optical image encryption architecture shown in figure 11.6, the object image is permutated using a binary scrambling method. Next, the permutated object is encrypted and registered as holograms by phase-shifting interferometry (PSI). Subsequently, the encrypted images and the key are compressed using single-pixel compressive imaging. Finally, the original image can be reconstructed and decrypted using a specified algorithm. This method presents an effective scheme to overcome the alignment drawbacks existing in the systems based on MZI architecture [26–28].

The object image $f(x, y)$ is permutated using the binary scrambling method first. The steps of permutation are as follows. Make the logic operation of XOR between the scrambling keys (key_1 and key_2) and coordinates of each pixel address of the image. That is $(f(x, y) \to f_0(x \oplus key_1, y \oplus key_2) \to f_0(x_0, y_0))$. Where $\oplus$ denotes the operation of XOR; x, y, and x_0, y_0 denote location coordinates of the object image and permutated image, respectively.

The collimated linear laser beam illuminates a SLM controlled by a computer, and displays the permutated image $f_0(x_0, y_0)$, and a random phase mask (RPM) $n_1(x_0, y_0)$ on its right side and the other random phase mask (RPM) $n_2(x_0, y_0)$ on its left side. The right and the left half waves passing through the SLM correspond to the object wave and the reference wave in the PSI, respectively. Next, the two waves overlap to form the interferogram on a DMD plane. After this, the compressive sampling data is obtained by using random linear measurements of the interferogram I and the measurement matrix Φ on the DMD plane. The data is then collected by a highly sensitive photodiode. Finally, the compressed hologram image is acquired by a traditional communication channel and then retrieved by a specified algorithm. Finally, the object image is decrypted using an inverse Fresnel transformation with reconstructed holograms and the correct decryption keys. A description of the principle is given below.

Let us assume that the permutated image $f_0(x_0, y_0)$ has been normalized into the range of $[0, \frac{1}{2}]$, $n_1(x_0, y_0)$ and $n_2(x_0, y_0)$ are two independent and random white noises uniformly distributed in [0, 1]. The distance between the centers of the object scene and the reference scene is $2c$, the wavelength of the input plane wave is λ, and the diffraction distance is d. In combination with the adequate state of light polarization, the SLM can reach a phase modulation in $[0, 2\pi]$. The object wave and the reference wave on the back side of SLM can be described, respectively, as

$$o(x_0 - c, y_0) = \exp[i2\pi f_0(x_0 - c, y_0)] \times \exp[i2\pi n_1(x_0 - c, y_0)], \tag{11.21}$$

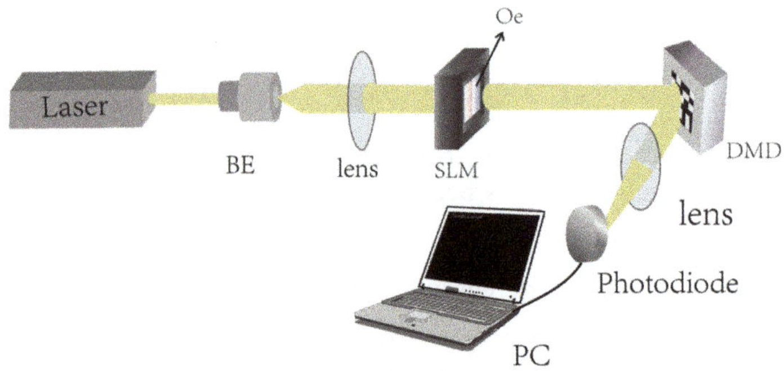

Figure 11.6. A schematic diagram of compressive optical image encryption system based on JTC. BE, beam expander; Oe, object image.

$$r(x_0 + c, y_0) = \exp\left[i2\pi n_2(x_0 + c, y_0)\right], \tag{11.22}$$

where $o(x_0 - c, y_0)$ denotes the object wave, and $r(x_0 + c, y_0)$ denotes the reference wave on the input plane.

After the Fresnel diffraction, the two complex waves on the DMD plane become

$$O(x, y) = FR_d\left[o(x_0 - c, y_0)\right] = A_0 \exp\left[i\varphi_0(x, y)\right], \tag{11.23}$$

$$R(x, y) = FR_d\left[r(x_0 + c, y_0)\right] = A_r \exp\left[i\varphi_r(x, y)\right], \tag{11.24}$$

where FR_d represents the Fresnel diffraction of the distance d. $O(x, y)$ and $R(x, y)$, respectively, represent the diffracted fields of the object wave $o(x_0 - c, y_0)$ and the reference wave $r(x_0 + c, y_0)$ on the DMD. A_0, A_r, Φ_0, Φ_r denote the amplitude and phase of the object wave and the reference wave on the DMD plane, respectively.

Using a standard three-step PSI on the JTC with phase shifting angles of 0, $\frac{\pi}{2}$ and π, then the three interferograms I_k on the DMD plane are sampled sequentially and expressed as

$$I_1(x, y) = A_0^2 + A_r^2 + 2A_0A_r \cos\left[\varphi_0(x, y) - \varphi_r(x, y) + 2cx\right], \tag{11.25}$$

$$I_2(x, y) = A_0^2 + A_r^2 + 2A_0A_r \cos\left[\varphi_0(x, y) - \varphi_r(x, y) + 2cx + \frac{\pi}{2}\right], \tag{11.26}$$

$$I_3(x, y) = A_0^2 + A_r^2 + 2A_0A_r \cos\left[\varphi_0(x, y) - \varphi_r(x, y) + 2cx + \pi\right]. \tag{11.27}$$

In the same way as in the MZI architecture, the DMD is used to load the measurement matrix and generate the hologram. In the receiving terminal, the method used in JTC can be seen in the MZI, so we will not repeat them.

The following shows the process of image reconstruction and decryption. We also adopt the TwIST algorithm to reconstruct the interference wave intensity $\hat{I}_k$. The process of reconstructing the interference wave intensity $\hat{I}_k$ is the same as in the previous section, which will not repeat here. When the intensity patterns $\hat{I}_1$, $\hat{I}_2$, $\hat{I}_3$ are reconstructed with the TwIST algorithm, we can get the complex amplitude on the DMD plane

$$O_e(x, y) = \frac{\hat{I}_1 - \hat{I}_3 + i(2\hat{I}_2 - \hat{I}_1 - \hat{I}_3)}{4}. \tag{11.28}$$

In order to decrypt $O(x, y)$, we must obtain the principle key $R'(x, y)$ which is obtained from the diffracted field $R(x, y)$ with a phase deviation. The method of the principle key encryption is basically the same as the method of the permutated image. So, we carry out the three-step PSI process again, but now the RPM $n_2(x, y)$ is displayed on the left half of the SLM, zero on its right half. Next the diffracted by the object wave, on the DMD plane, becomes a plane wave represented as $C(x, y) = \exp(i\varphi_c)$, where φ_c is a constant. Then through the CS, we can get the intensity patterns $\hat{I}'_k$ (k=1, 2, 3). So, we can calculate the complex amplitude on the DMD plane

$$R'(x, y) = A_r \exp\left[i(\phi_c - \phi_r + 2cx)\right] = \frac{\hat{I}'_1 - \hat{I}'_3 + i(2\hat{I}'_2 - \hat{I}'_1 - \hat{I}'_3)}{4}, \tag{11.29}$$

and further obtain $O'(x, y)$, which represents the product of the object wave $O_0(x, y)$ and a constant phase term, by

$$O'(x, y) = O_e(x, y)/R'(x, y) = A_0 \exp\left[i(\phi_0(x, y) - \phi_c)\right]. \tag{11.30}$$

Then the object wave $o(x_0 - c, y_0)$ can be reconstructed from

$$o(x_0 - c, y_0) = IFR_d[O'(x, y)], \tag{11.31}$$

where IFR_d stands for the inverse Fresnel transform of the distance d. Finally, after multiplying $o(x_0 - c, y_0)$ by $\exp[-i2\pi n_1(x_0, y_0)]$ and extracting the phase of the result and dividing it by π, we can obtain the permutated image $f_0(x_0, y_0)$. Finally, we can get the original image $f(x, y)$ by using the binary scrambling method again.

Once $\hat{I}_k$ and $\hat{I}'_k$ are known, in addition to the scrambling keys of decryption (key_1 and key_2), λ, d and the random mask $n_1(x_0, y_0)$, we can digitally or optically retrieve the original object image from the encrypted image.

11.3.3 Computer simulations

A series of numerical simulations have been performed to verify the effectiveness of our proposed method. In the MZI, a HeNe laser (λ = 632.8 nm) is used, and the other parameters are z_1 = 0.125 m, z_2 = 0.125 m, R = 1.5, respectively. The measurement matrix size of the DMD used was 256 × 256 pixels. The measurement matrix generated by the DMD has random sequences of 0/1. The original object images used in the simulation are shown in figures 11.7(a) and (c), all 256 × 256 pixels in size. Figure 11.7(b) shows the recovered object images from 256 × 256 × 18.7% measurements by using a three-step phase-shifting digital holography method. Figure 11.7(d) shows the recovered object images from

256 × 256 × 10.6% measurements using a two-step-only quadrature phase-shifting digital holography method. The computer simulations show that this compressive optical image encryption method based on MZI is performed using a completely optical scheme in the Fresnel domain. Moreover, two-step-only quadrature phase-shifting digital holography has fewer numbers of holograms than three-step or four-step phase-shifting digital holography.

The parameters used on the JTC setup are λ = 632.8 nm, d = 0.5 m, where we adopt a three-step phase-shifting digital holography method to demonstrate the effectiveness of this method. All the frames are digitized to 256 gray levels with the size of 256 × 256 pixels. $key_1 = key_2 = 85$. Figure 11.8(a) is the original image, figure 11.8(b) shows the recovered object image from 256 × 256 × 51.8% measurements using our method. The computer simulations show that this compressive optical image encryption method based on JTC is performed to demonstrate feasibility and validity.

To further investigate the security of this compressive optical image encryption method, we reconstruct the original grayscale image with 256 × 256 × 51.8%

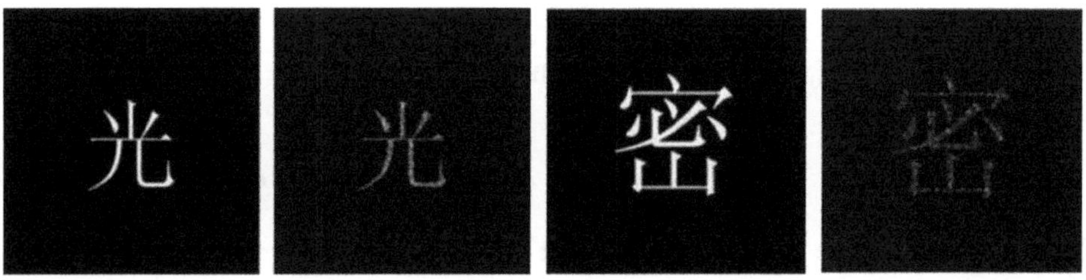

(a) (b) (c) (d)

Figure 11.7. Computer simulation results. (a) Binary image; (b) retrieved image from 256 × 256 × 18.7% measurements using a three-step phase-shifting digital holography method; (c) binary image; (d) retrieved image from 256 × 256 × 10.6% measurements using a two-step-only quadrature phase-shifting digital holography method.

(a) (b)

Figure 11.8. Computer simulation results. (a) Grayscale image; (b) retrieved image from 256 × 256 × 51.8% measurements using our method.

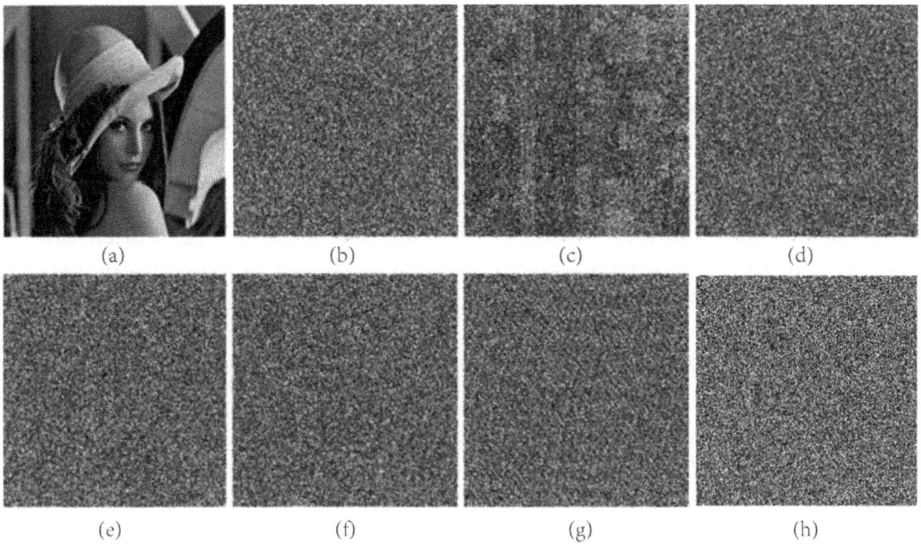

Figure 11.9. The retrieved images with one incorrect key in the decryption process. (a) Original image; (b) when the principal key $n_2(x_0, y_0)$ is not used; (c) when one does not use the scrambling key; (d) when the scrambling keys are incorrect; (e) when d is an error; (f) when λ is an error; (g) when $n_1(x_0, y_0)$ is not eliminated; (h) when the measurement matrix φ is incorrect.

measurements when one of the keys is incorrect, as shown in figure 11.9. If one of the keys is incorrect, the retrieved image will be greatly affected. Among these keys, the principal key $n_2(x_0,y_0)$ plays a critical role in the security system, as shown in figure 11.9(b), in which the retrieved image is the same as the noise and fully unrecognizable when the scrambling keys are not used or are incorrect, the reconstruction images are shown in figures 11.9(c) and (d). When the diffraction distance of the object image and the wavelength of the He–Ne laser exhibit a relative error, there are no correct reconstruction images, as shown in figures 11.9(e) and (f). When $n_1(x_0, y_0)$ is not eliminated, the reconstruction image is shown in figure 11.9(g). When the measurement matrix φ is incorrect, the reconstruction image is shown in figure 11.9(h). We have also performed some simulations with other binary images and obtained satisfactory retrieval results, which are similar to those obtained with the gray-level images. Therefore, they are not considered here.

11.4 Compressive optical image hiding

The principle behind optical image hiding technology is that the human sensory organs are not sensitive to some of the redundant information in the image signal. In an optical image hiding system, the information of a secret image is embedded in the host image for the purpose of hiding secret image data. This is achieved while leaving the original visual information of the host image roughly unchanged. This can also achieve safe transmission and storage of information, and also avoid the attention of thieves.

11.4.1 Compressive optical image hiding based on MZI

A schematic diagram of the compressed optical image hiding system based on a modified MZI is shown in figure 11.10. In this system, the spatial distribution of a two-dimensional reference light is generated by introducing the host object in the reference optical path such that concealment of the image is realized. Next, the hidden hologram is obtained using sparse domain sampling and the measurement matrix. After this, it is reconstructed using TwIST. Finally, the hidden original object is reconstructed using four-step phase shift digital holography or two-step phase shift digital holography. Below, we discuss the two-step phase shift digital holography method.

The linearly polarized laser beam is divided into a reference beam and an object beam. In the object arm, the object beam illuminates the object image that will be hidden and displayed on SLM. In the other arm, the phase of the reference beam is modulated first by a $\lambda/2$ wave plate and then by a $\lambda/4$ wave plate to realize the phase-shift. The phase-shifted reference beam then illuminates the host image. The object wave and the reference wave from the two beams overlap to form the phase-shifted interferogram on the plane of a DMD. After this, the compressive sampling data is obtained by computing the measurements of the interferogram H and the measurement matrix Ψ in DMD plane. The samples are then gathered by a highly sensitive photodiode. Finally, the compressed hologram image is obtained by a traditional communication signal path, and the object image is reconstructed using a certain signal recovery algorithm of the CS. Finally, the original object can be reconstructed via inverse Fresnel-transform with only two holograms.

According to the Fresnel diffraction formula, the complex hologram coming from the object beam on the DMD plane is represented by the complex amplitude distribution as

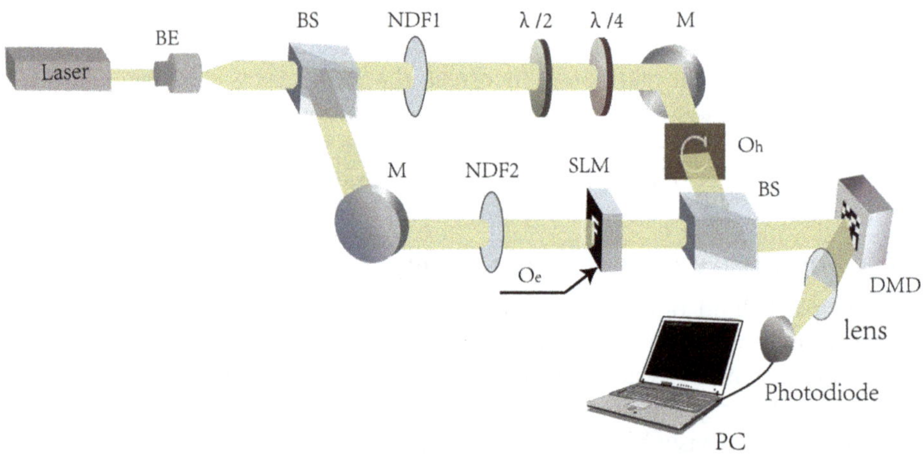

Figure 11.10. A schematic diagram of a compressed optical image hiding system based on a modified MZI. BE, beam expander; BS, beam splitter; NDF, neutral density filters; M, mirror; Oe, object image; Oh, host image.

$$U(x, y) = Frt\left[O(x_0, y_0)\right] = A_0(x, y)\exp\left[i\phi_0(x, y)\right] \tag{11.32}$$

where $O(x_0, y_0)$ represent the transmittance of the object image, *Frt* denotes the Fresnel transform. The amplitude of an on-axis reference plane wave in this plane is $A_r(x, y)$, and the complex distribution diffracted field $R(x, y)$ by the host image on the DMD plane can be expressed as

$$R(x, y) = A_r(x, y)\exp\left[i\phi_r(x, y)\right]\exp(i\delta) \tag{11.33}$$

where δ represents the stepped phase. The reference light beam carrying the host image and the object light beam carrying the object image form the coaxial interference light and superimpose to generate the interference hologram on the DMD plane, which contains the hidden object image information, and it can be described as

$$\begin{aligned} H(x, y, \delta) = |U(x, y) + R(x, y)|^2 &= A_0^2(x, y) + A_r^2(x, y) \\ &+ 2A_0(x, y)A_r(x, y) \times \cos[\varphi_0(x, y) - \varphi_r(x, y) - \delta] \\ &\times \left(\delta = 0, \frac{\pi}{2}\right). \end{aligned} \tag{11.34}$$

The object image is embedded in the diffraction field of the host image by the image hiding process. By adjusting the intensity ratio of two paths via NDF in the modified MZI, different embedded levels for image hiding can be implemented. The purpose is to acquire interferograms that look like the Fresnel diffraction image of the host image to achieve image hiding in the Fresnel transform domain. When the light is reflected by the mirrors of the DMD plane in the direction towards the lens and the light is summed at the photodiode to calculate the measurement as its output voltage value

$$Y(m) = \{y_{1m}, y_{2m}\} = \phi_m[H_1, H_2]. \tag{11.35}$$

Here, ϕ_m is the mth pseudo random matrix on the DMD, $m \in \{1, 2, \cdots, M\}$. When we repeat the process M times, we can obtain the output value Y on the photodiode.

$$\mathrm{Y} = [\mathbf{y}_1, \mathbf{y}_2] = \Psi[\mathrm{H}_1, \mathrm{H}_2] \tag{11.36}$$

where $[H_1, H_2] = \begin{bmatrix} H_{11} & H_{21} \\ H_{12} & H_{22} \\ \vdots & \vdots \\ H_{1M} & H_{2M} \end{bmatrix}$, $\Psi \in R^{M\times N}$ is the measurement matrix uploaded into the DMD and $Y \in R^{M\times 2}$ is the measurement data; $\mathbf{y}_\mathbf{k} \in R^{M\times 1}$ and $\mathrm{H}_\mathbf{k} \in R^{N\times 1}$, $k = 1,2$.

The restoring process is an inverse problem to estimate the original image from the observations. Hence, this problem can be converted into the optimal problem, where we can retrieve the interference patterns $\hat{H}_k$ by adopting the TwIST algorithm.

We get two on-axis quadrature-phase intensity patterns $\hat{H}_1$ and $\hat{H}_2$ with the phases of the reference wave of 0 and $\pi/2$, respectively, and the clear reconstructed image without a twin-image and zero-order image can be obtained with only two phase-shifting holograms by a certain algorithm.

When we set the phases of the reference wave in the first and second exposure to 0 and $\pi/2$, respectively, the intensity distributions of the two interferograms are

$$\hat{H}_1(x, y) = A_0^2(x, y) + A_r^2(x, y) + 2A_0(x, y)A_r(x, y)\cos\varphi(x, y) \tag{11.37}$$

$$\begin{aligned}\hat{H}_2(x, y) &= A_0^2(x, y) + A_r^2(x, y) + 2A_0(x, y)A_r(x, y) \\ &\quad\times \cos\left[\varphi(x, y) - \frac{\pi}{2}\right] \\ &= A_0^2(x, y) + A_r^2(x, y) + 2A_0(x, y)A_r(x, y) \\ &\quad\times\left[\cos\varphi(x, y)\cos\frac{\pi}{2} + \sin\varphi(x, y)\sin\frac{\pi}{2}\right]\end{aligned} \tag{11.38}$$

where we can designate $\varphi(x, y) = \varphi_0(x, y) - \varphi_r(x, y)$, and $p(x, y) = A_0^2(x, y) + A_r^2(x, y)$, from (11.37) and (11.38), we can obtain

$$A_0(x, y)\cos\varphi(x, y) = \frac{\hat{H}_1 - p}{2A_r(x, y)} \tag{11.39}$$

$$A_0(x, y)\sin\varphi(x, y) = \frac{\hat{H}_2 - \hat{H}_1\cos\frac{\pi}{2} - \left(1 - \cos\frac{\pi}{2}\right)p}{2A_r(x, y)\sin\frac{\pi}{2}}. \tag{11.40}$$

Under the relation $\sin^2\varphi + \cos^2\varphi = 1$, and if $\hat{H}_1$, $\hat{H}_2$, $A_r(x, y)$ are known, we can easily calculate $p(x, y)$.

$$p(x, y) = \frac{-v - \sqrt{v^2 - 4uw}}{2w} \tag{11.41}$$

$$\text{where } w = 2\left(1 - \cos\frac{\pi}{2}\right) \tag{11.42}$$

$$v = -2\left(1 - \cos\frac{\pi}{2}\right)(\hat{H}_1 + \hat{H}_2) - 4A_r^2(x, y)\sin^2\frac{\pi}{2} \tag{11.43}$$

$$u = \left(\hat{H}_1 - \hat{H}_2\cos\frac{\pi}{2}\right)^2 + \hat{H}_2^2\sin^2\frac{\pi}{2} + 4A_r^4(x, y)\sin^2\frac{\pi}{2}. \tag{11.44}$$

When $p(x, y)$ is calculated, we can also calculate the amplitude $A_0(x, y)$ and phase $\varphi_0(x, y)$ of the hidden image as

$$A_0(x, y) = \sqrt{p(x, y) - A_r^2(x, y)} \tag{11.45}$$

$$\varphi_0(x, y) = \tan^{-1}\left[\frac{\hat{H}_1 - p}{2A_r(x, y)} + i\frac{\hat{H}_2 - \hat{H}_1\cos\frac{\pi}{2} - \left(1 - \cos\frac{\pi}{2}\right)p}{2A_r(x, y)\sin\frac{\pi}{2}}\right] + \varphi_r(x, y). \quad (11.46)$$

When we calculate the amplitude $A_0(x, y)$ and the phase $\varphi_0(x, y)$, we can then perform inverse Fresnel propagation to recover the original object image through digital or optical means. This means that the complex object wave field in the recording plane can be retrieved by using only two interferograms.

11.4.2 Compressive optical image hiding based on JTC

The compressive optical image hiding system based on JTC is similar to the compressive optical image encryption system. The only difference is the host image on the left half of SLM. Hence, we will not discuss this topic in detail.

11.4.3 Experimental/simulation results

The wavelength of the He–Ne laser is 632.8 nm and the ratio of light amplitude between the object beam and reference beam is 0.1:1, the largest number of pixels of a HOLOEYE LC 2002 type SLM is 800 (H) by 600 (V) with a pixel pitch of 32 μm. The size of the host image 'C' is 0.4 cm × 0.3 cm, which is shown in figure 11.11(a). The size of the object image 'F' used for hiding is 896 μm × 448 μm, which is shown in figure 11.11(b). We took 1024 × 1024 × 18.8% image data from the hidden image in our compressive image hiding system, and then once we received the compressed data of the two hidden interferograms, the two compressive hidden holograms could be exactly reconstructed from much less than the Nyquist sampling number of observations by solving an optimization. The experimental results for compressive optical image hiding in a modified MZI are shown in figures 11.11(c) and (d). Figure 11.11(c) shows the retrieved original image without compression. Figure 11.11(d) shows the reconstructed original image from 1024 ×1024 × 18.8% image data using our method. The reconstruction results employing 18.8% and 100% measurements are almost the same, which indicates that this compressive optical image hiding method is performed by using a completely optical scheme in the Fresnel domain, and the hidden image can be reconstructed only with two interference holograms.

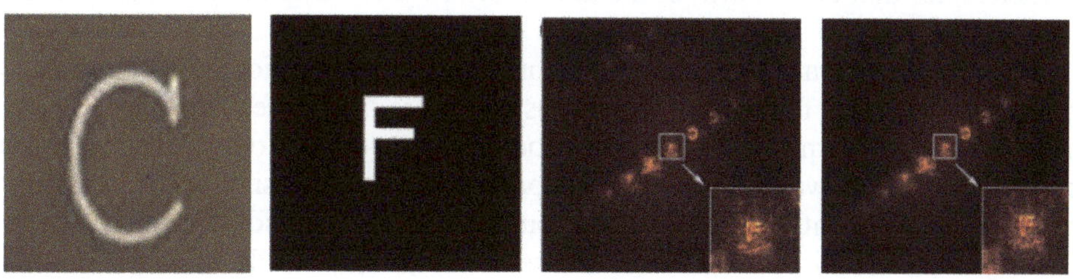

(a) (b) (c) (d)

Figure 11.11. Experimental results. (a) The host image 'C'; (b) the object image 'F'; (c) retrieved original image without compression; (d) reconstructed image from 1024 × 1024 × 18.8% measurements using our method.

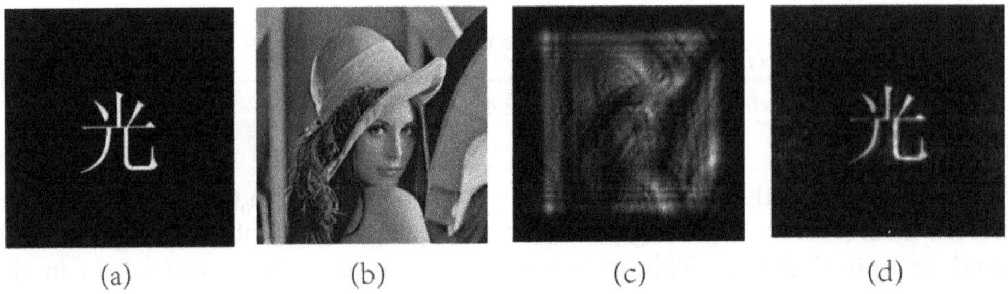

Figure 11.12. The simulation results. (a) Object image; (b) host image; (c) one of the interferograms from when the watermark image has been embedded into the host image; (d) retrieved image from 256 × 253 × 23% measurements using our method.

On the JTC setup, the wavelength of the He–Ne laser and the light intensity ratio between the object beam and reference beam is the same as the value in MZI. The recording distances of the object image and host image are both 0.2 m. The object image and the host image are both 256 × 256 pixels. The simulation results for compressive optical image hiding on a JTC are shown in figure 11.12. Figure 11.12(a) is the object image, figure 11.12(b) is the host image. One of these interferograms of the object image embedding into the host image is shown in figure 11.12(c). Figure 11.12(d) shows the recovered object image from 256 × 253 × 23% measurements using this method. The computer simulations show that this method can be performed using a fully optical scheme in the Fresnel domain and it can be used to reconstruct the original image well with only 23% of the number of measurements established by the Nyquist criterion. We have also made some simulations with other images and obtained satisfactory retrieval results, which are similar to those obtained with the use of grayscale images, therefore they are ignored here.

11.5 Compressive optical image encryption and hiding

By integrating the methods described in the previous sections, the encryption and hiding techniques are further combined to carry out an optical image encryption and hiding technique. This technique can simultaneously perform the protection and compression of the sensing scene information in the imaging process. It has characteristics that do not attract the unauthorized user's attention, so the system is more secure.

A schematic diagram of compressed optical image encryption and hiding based on MZI is shown in figure 11.13. The host object is introduced into the reference path of the interferometer branches to produce a two-dimensional spatial distribution of the reference wave. Through the modified MZI, the sensing object is not only imaged via a coherent imaging method, but also encrypted and embedded into the host object information, to perform image security; Meanwhile, the secure object holograms are further compressed by sparse sampling to reduce the data recorded by a single-pixel detector to represent the original imaging object.

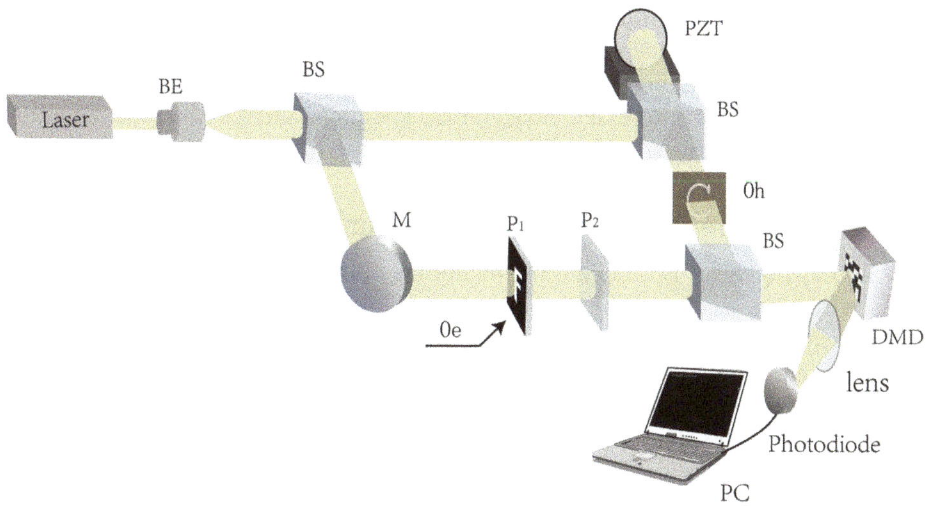

Figure 11.13. A schematic diagram of compressed optical image encryption and hiding based on MZI. BE, beam expander; BS, beam splitter; M, mirror; P, random phase plate; PZT, piezoelectric transducer mirror; Oe, object image; O_h, host image.

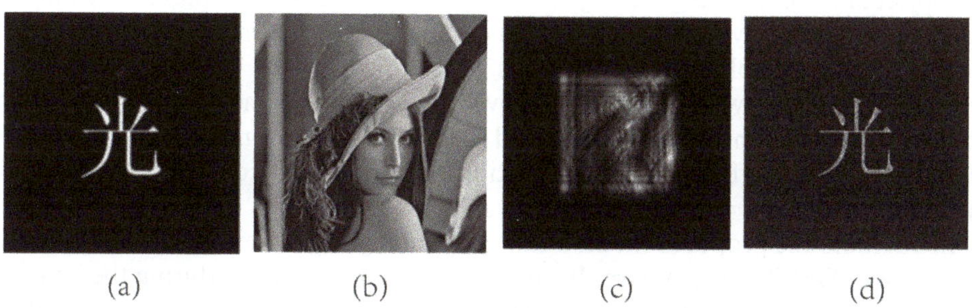

Figure 11.14. The simulation results. (a) Object image; (b) host image; (c) one of the interferograms from when the watermark image has been embedded into the host image; (d) retrieved image from 256 × 256 × 54.2% measurements using our method.

We used computer simulations to verify the feasibility of this method with functions of encryption and hiding. The original image object that needed to be encrypted and hidden is shown in figure 11.14(a); the host image is shown in figure 11.14(b); all images are 256 × 256 pixels in size. The wavelength of the He–Ne laser we used in this compressive holographic imaging system with security is 632.8 nm. The light amplitude ratio between the object beam and the reference beam is 0.000 001:1. In the optical path, the first and second diffraction distance of the object image denoted by z_1 and z_2 are 0.1 m and 0.2 m, respectively. The diffraction distance of the host image z_3 is 0.3 m. We took 256 × 253 × 54.2% measurements from the encrypted and hidden image in our compressive holographic imaging system with security features. Once we received the compressed data of the encrypted and hidden image, we reconstructed the original image using the correct keys and the parameters of the optical system. The computer simulation results are shown in figure 11.14. One of the three interferograms of figure 11.14(a) is shown in figure 11.14(c). The image of recovered object from 256 × 253 × 54.2% measurements using our proposed method is shown in figure 11.14(d). The computer simulation results show that this compressive holographic imaging system with security features is implemented by using a completely optical scheme in the Fresnel domain.

11.6 Prospects in compressive image security

Optical image security is an emerging technology developed by combining modern optics and image security. This technology can take advantage of features such as multi-dimensional processing for handling a large amount of data, more degrees of freedom, high speed, parallel processing, and large storage capacity to combine optical information processing technology with image security applications. However, one of the challenging problems in this field is due to the large volume of data available in optical image security. This problem has become an important factor in restricting its development. The emerging CS theory provides a new direction for solving this problem. Combining the optical image encryption technique or optical image hiding technique with the CS technique, the method introduces an all-optical solution to sensing the original object, encrypting or hiding the object and compressing the object in the analog domain. Compared with current optical image security systems, the presented method can overcome limitations due to the large data volume of holograms. The utilization of CS in optical image security processing will improve the overall security of the system due to an additional encryption layer. The method can be applied to gray-scale images and binary images to perform image encryption or hiding, and compression in an all-optical system.

In addition, the proposed method can perform high-speed measurements because of the use of DMD in the system. It can safely obtain a 3D scene during the imaging process. Hence, it can achieve highly sensitive imaging and overcome the limitation of array imaging under weak light conditions. In the near future, this method is expected to be used widely for applications in other technical fields such as biomedical imaging, aviation imaging, real-time video secure transmission, the weak signal detection, cloud computing, future all-optical networks and the IoT or Internet of Everything (IoE).

Acknowledgments

This work was supported by the Natural Science Foundation of Guangdong Province, China (2015A030313384); the Science and Technology Program of Guangzhou, China (201607010275) and the National Science Foundation for Young Scientists of China (No. 11404115).

Author contributions

LJ designed the system, analysed the data and wrote the manuscript. LM analysed the data and wrote the manuscript. DXF, LJS, ZT, LHB, JBW, YCX and WSY helped with the manuscript preparation and revision. All of the authors discussed the results and commented on the manuscript.

References

[1] Donoho D L 2006 Compressed sensing *IEEE Trans. Inf. Theory* **52** 1289–306

[2] Bioucas-Dias J M and Figueiredo M A T 2008 A new TwIST: two-step iterative shrinkage/thresholding algorithms for image restoration, image processing *IEEE Trans.* **16** 2992–3004

[3] Refregier P and Javidi B 1995 Optical image encryption based on input plane and Fourier plane random encoding *Opt. Lett.* **20** 767

[4] Peng X, Cui Z and Tan T 2002 Information encryption with virtual-optics imaging system *Opt. Commun.* **212** 235–45

[5] Cai L Z, He M Z, Liu Q and Yang X L 2004 Digital image encryption and watermarking by phase-shifting interferometry **43** 3078–84

[6] Chen L and Zhao D 2007 Color information processing (coding and synthesis) with fractional Fourier transforms and digital holography *Opt. Express* **15** 16080–9

[7] Situ G and Zhang J 2004 Double random-phase encoding in the Fresnel domain *Opt. Lett.* **29** 1584

[8] Li J, Li J, Shen L, Pan Y and Li R 2014 Optical image encryption and hiding based on a modified Mach–Zehnder interferometer *Opt. Express* **22** 4849–60

[9] Chen W, Javidi B and Chen X 2014 Advances in optical security systems *Adv. Opt. Photon.* **6** 120–55

[10] Patten R E, Hennelly B M, Kelly D P, O'Neill F T, Liu Y and Sheridan J T 2006 Speckle photography: mixed domain fractional Fourier motion detection *Opt. Lett.* **31** 32–4

[11] Frauel Y, Naughton T J, Matoba O and Tajahuerce E 2006 Three-dimensional imaging and processing using computational holographic imaging *Proc. IEEE* **94** 636–53

[12] Alfalou A, Brosseau C, Abdallah N and Jridi M 2013 Assessing the performance of a method of simultaneous compression and encryption of multiple images and its resistance against various attacks *Opt. Express* **21** 8025–43

[13] Alfalou A, Mansour A, Marwa A and Christian B 2011 Optical compression scheme to simultaneously multiplex and encode images *Optical and Digital Image Processing: Fundamentals and Applications* ed G Cristobal, P Schelkens and H Thienpont (Wiley) 463–83

[14] Javidi B *et al* 2016 Roadmap on optical security *J. Opt.* **18** 083001

[15] Di H, Zheng K, Zhang X, Lam E Y, Kim T, Kim Y S, Poon T C and Zhou C 2012 Multiple-image encryption by compressive holography *Appl. Opt.* **51** 1000–9

[16] Deepan B, Quan C, Wang Y and Tay C J 2014 Multiple-image encryption by space multiplexing based on compressive sensing and the double-random phase-encoding technique *Appl. Opt.* **53** 4539

[17] Huang R, Rhee K H and Uchida S 2014 A parallel image encryption method based on compressive sensing *Multimedia Tools Appl.* **72** 71–93

[18] Lang J and Zhang J 2015 Optical image cryptosystem using chaotic phase-amplitude masks encoding and least-data-driven decryption by compressive sensing *Opt. Commun.* **338** 45–53

[19] Zhang A, Zhou N and Gong L 2013 Color image encryption algorithm combining compressive sensing with Arnold transform *J. Comput.* **8**

[20] Veena V K, Lal G J, Prabhu S V and Kumar S S 2012 A robust watermarking method based on compressed sensing and Arnold scrambling *Int, Conf. Machine Vision and Image Processing* pp 105–8

[21] Candès J E 2008 The restricted isometry property and its implications for compressed sensing *C. R. Math.* **346** 589–92
[22] Guo P and Devaney A J 2004 Digital microscopy using phase-shifting digital holography with two reference waves *Opt. Lett.* **29** 857–9
[23] Meng X F, Cai L Z, Xu X F, Yang X L, Shen X X, Dong G Y and Wang Y R 2006 Two-step phase-shifting interferometry and its application in image encryption *Opt. Lett.* **31** 1414
[24] Li J, Pan Y Y, Li J S, Li R and Zheng T 2014 Experimental study of two-step phase-shifting digital holography based on the calculated intensity of a reference wave *J. Opt. Soc. Korea* **18** 230–5
[25] Li J, Li H, Li J, Pan Y and Li R 2015 Compressive optical image encryption with two-step-only quadrature phase-shifting digital holography *Opt. Commun.* **344** 166–71
[26] Nomura T and Javidi B 2000 Optical encryption using a joint transform correlator architecture *Opt. Eng.* **39** 2031–5
[27] Barrera J F, Vargas C, Tebaldi M, Torroba R and Bolognini N 2010 Known-plaintext attack on a joint transform correlator encrypting system *Opt. Lett.* **35** 3553–5
[28] Li J, Zheng T, Liu Q Z and Li R 2012 Double-image encryption on joint transform correlator using two-step-only quadrature phase-shifting digital holography *Opt. Commun.* **285** 1704–9

Chapter 12

Simultaneous encryption and arithmetic coding for performing image compression

Nihel Neji, Maher Jridi, Ayman Alfalou and Nouri Masmoudi

12.1 Introduction

Cryptography is a set of techniques for protecting a message by means of a secret code. It makes the information unintelligible and ensures the authentication of the partners. It restricts access to data, which guarantees confidentiality and access control. The techniques of encryption and decryption are based on mathematical notions related to the security of information [1]. With the scalable power of computers and the great advances in mathematics, cryptography has made enormous progress, which broadens its field of use. In the literature, we distinguish several classes and techniques of encryption. Cryptography is essentially based on the arithmetic used to transform a visible message into a coded message. The transformations used are mathematical functions called cryptographic algorithms to encrypt messages with a parameter called a key. In order to transmit a message in a confidential way, we apply a transformation that makes it incomprehensible; this is called encryption. This change results in unreadable text from plain text. Each encryption system is composed of an encoding algorithm and one or more security keys based on mathematical theories to render the communication inviolable.

Different approaches exist that we can differentiate according to where we perform the encryption during video compression: before or after compression, before or after transformation and quantification, during entropy coding or after. If encryption is performed before compression, we see a large negative influence on the video compression performance.

Thus, if the totality of the visual information must be hidden, this approach is not the right choice since we can easily guess the outlines of objects. Another solution consists of encrypting the data of the image and encoding modified information, for example, a permutation of the positions of the pixels is proposed in [2]. For lossy

doi:10.1088/978-0-7503-1457-2ch12

video compression standards like H.264 we consider if decryption after lossy compression is possible and how we keep compression performance that is easier in the compression chain [3–5].

12.2 Prior work of video encryption methods

Researchers have presented schemes to perform encryption in different domains. In this section, a review of the literature is presented separately for each embedding domain.

12.2.1 Embedding in the video codec structure

Embedding can be performed in the video compression chain. A modification of the Intra prediction mode has been proposed in [6–12]. In [13], the author proposed to perform permutations of the information sent on inter-predicted macroblocks having the same number of motion vectors. Some proposed encryption of the syntax elements signs [10, 14–16], and others spoke about the exponential Golomb suffixes codes encryption [6]. Embedding can also be performed in some transform domains before quantization e.g. DCT, DFT, DWT. In [17] the authors proposed a robust watermarking algorithm for H.264/AVC that consists of choosing the coefficients to be watermarked based on frequency sensitivity, luminance masking and contrast masking. Some researchers have proposed algorithms to embed hidden messages in QTCs of H.264/AVC. In [18], Gong and Lu embedded watermarks in a H.264/AVC video by modifying the quantized DC coefficients in luma residual blocks. Noorkami and Merserau have presented a technique to embed messages in both intra and inter frames in all non-zero QTCs [19]. In [8], they proposed to encrypt the DC coefficients for intra-macroblocks. Thus, the authors of [20] encode non-zero coefficients that undergo Exp-Golomb encoding before binary arithmetic coding. In [21, 22], the authors speak of the encryption of the least significant bits. But all these techniques of selective encryption do not make it possible to provide maximum security. Many unsafe algorithms are employed by adding constants with the coefficients [23], which reduces the compression efficiency.

12.2.2 Embedding in the entropy coder

Before the entropy coding of each 4×4 block, the non-zero coefficients are mapped to a zigzag scan. A modification in the mapping order has been proposed in [10, 24]. Many works also talk about joint compression and encryption for the two types of entropy encoders CAVLC and CABAC. With CAVLC, a large number of data is encoded with Exp-Golomb codes, which can be frequently encrypted [9, 15, 25]. The code words of the syntax elements can be used as 'run before' in [26]. If CABAC is used for data encoding, only bits coded in bypass mode, such as the MVD, must be encrypted in order to retain the encoder performance [27, 28]. Security is based on the uncertainty of the arithmetic decoding state, which means that the syntax element to decode and contextual models state [29] must be guessed, as well as various decoding variables (CodIRange and codIOffset represented with 16 bits) which make decryption very complex for an attacker.

12.2.3 Embedding in the bitstream

Some methods have suggested embedding messages in the compressed bitstream [30–34]. In [35], Li *et al* propose to perform robust watermarking for H.264/AVC by embedding a hidden message in a new syntax element named the reference index. Partial H.264 stream encryption can reduce the amount of data to be encrypted, leading to improvements in the broadcast system performance [36]. We can limit to the IDR encryption or the heading encryption of a NALU [37], which requires an unconventional decoder but may be insufficient to ensure security [36]. An approach to reduce the computational complexity of selective encryption of the bitstream with secure encryption can be found in [38]. In [39], it is proposed that one could use an AES for the least important parts of the bitstream. Since the bitstream consists of NALUs, one can apply a simple encryption on the integer format like packet encryption in JPEG2000 [40]. There are also encryption systems that eliminate tagging sequences [36]. Encryption of the NALUs while retaining the headers of the units has been proposed in [36, 41, 42]. Several propositions [43–45] encrypt fractions of the H.264/AVC binary stream with classical algorithms such as AES. In addition, there is a standard for secure broadcasting of RTP data, including H.264: AVC/SVC [46].

12.3 Simultaneous encryption and video compression

12.3.1 Principle of compression

The compression scheme chosen in this part is shown in figure 12.1. This is a simplified compression scheme that can be used for image compression or part of the video compression chain. After the block decomposition of the image, we carry out the transformation and the quantification of the coefficients of the image. Then we perform a zig-zag mapping to arrange the coefficients in a specific way and end with an entropy encoding of the quantified symbols. We use a BAC entropy encoder for binarized video data.

However, we can apply one or more encryption techniques in the compression string that apply in various levels in the chain. As illustrated in figure 12.1, the video encoder can be interrupted at any level to perform encryption. There are 3 different floors:

- Encryption of raw data prior to compression;
- Encryption of the transformed and quantified coefficients realized after the quantification;

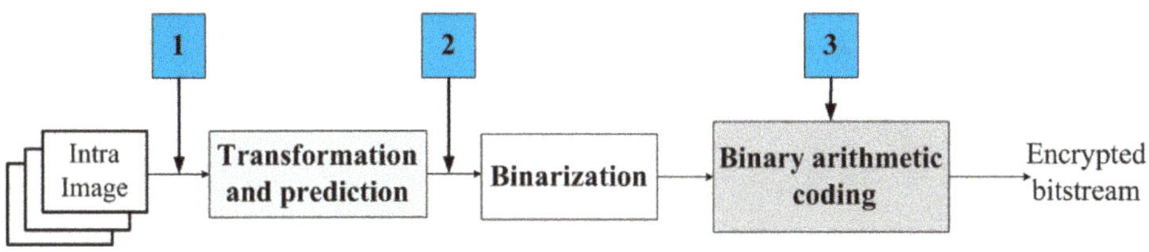

Figure 12.1. Adopted video compression chain model with data encryption in different levels: (1) encryption of raw data (2) encryption of transformed and quantified coefficients (3) simultaneous encryption and entropy coding.

- Simultaneous encryption of the entropy coding applied to the binary information of the image.

12.3.2 Raw data compression and encryption (RD C&E)

As a first level of security, we have integrated synchronous streaming encryption. This encryption consists in combining the message clearly with a secret binary sequence. The secret binary sequence comes from a pseudo-random number generator or a biometric key. In our compression chain we used a biometric key for this first level of security. This key is generated from a biometric sample. At the beginning, we extracted the minimum value and the average value of the pixels contained in the biometric key to make a pseudo-random image. The size of the random key is the same as the Intra image to be encrypted. Next, we apply the encryption as shown in figure 12.2.

Indeed, this technique is very simple. It is based on an 'exclusive OR' between the image to be encrypted and the key that must be in binary. Each bit of raw data is combined with a bit of the secret encryption key according to the truth table of the Boolean XOR function. Thus, if the information bit and the key bit are identical, the bit of the cryptogram emitted by the algorithm will be 0, otherwise it will be 1. For decryption, the bits are again combined one by one with a bit of the same biometric key used in the encryption. Our key is absolutely private in this approach. The XOR encryption algorithm is presented in table 12.1.

The key used for this algorithmic description is an image of the same size as the original that contains randomly distributed 0s and 1s. At the development level of

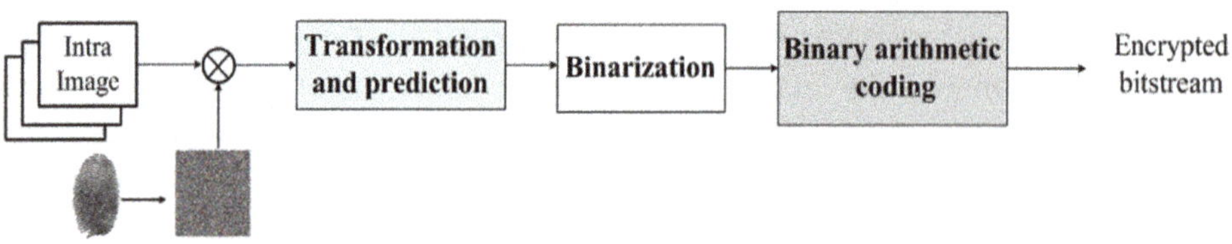

Figure 12.2. Encryption based on the biometric key applied to raw data before video compression.

Table 12.1. Synchronous flow encryption algorithm with a biometric encryption key.

```
 i ← 1
While Image is different from image end DO
c ← next image bit //Reading binary image pixels
write (c XOR Key [i]) //Application of XOR between image
and key bits
i ← i + 1
End While
 End
```

our method we used a fixed length key and we reused the key as many times as necessary to cover the total length of the information to be encrypted. This is necessarily related to the video resolution used. Figure 12.3(a) shows the reconstructed image after synchronous stream encryption and compression. The correlation between figure 12.3(a) and the original image is illustrated in figure 12.3(b). We find this time a PSNR = 11.828 dB, a peak to correlation energy (PCE) = 0.0071 and a compression ratio (CR) = 23.152%.

The value of the PCE is very low compared to the PCE found in the case of compression without encryption (PCE = 0.0845). This is due to the fact that the correlation method is sensitive to contours that have remained visible after decoding. Moreover, the compression ratio is low compared to the value of 83.278% obtained without encryption, for the same quantization step. This is because the random key was applied to the raw data. Spatial redundancies between neighboring pixels and blocks of neighboring pixels have been eliminated. Consequently, the number of coefficients quantified at zero has been reduced and the arithmetic coding is no longer effective.

We thus see all the disadvantages of making an encryption followed by a compression.

12.3.3 Transformed and quantified coefficients compression and encryption (TQC C&E)

In this section we have performed biometric key synchronous stream encryption after the transformation and quantification of the video data, as shown in figure 12.4. The encryption algorithm used is equivalent to that of the algorithm I except that the or-exclusive applies to the transformed and quantified data.

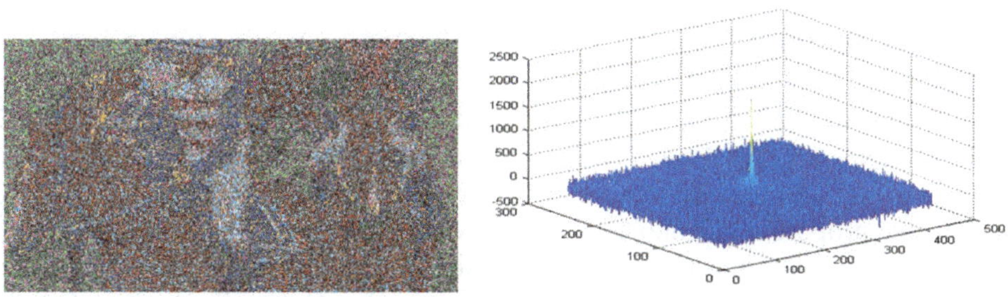

(a) (b)

Figure 12.3. Decoded picture after compression and encryption of raw data.

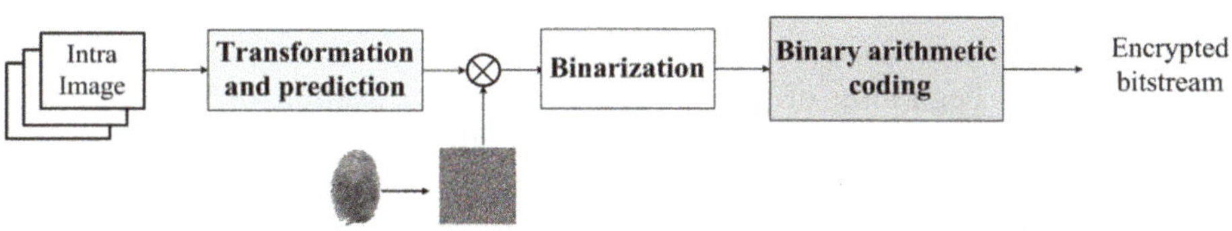

Figure 12.4. Encryption based on the biometric key applied to transformed and quantified coefficients.

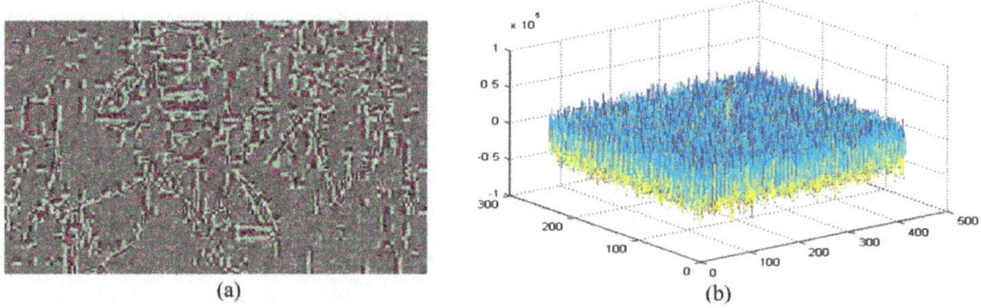

Figure 12.5. Decoded picture after compression and encryption of transformed and quantified coefficients.

When we adopt this optimization, we use the same encryption key as before and we get a PSNR = 11.764 dB, a PCE = 0.0073 and a CR = 1.381%. The results are shown in figure 12.5. On the level of the quality of the reconstructed image (figure 12.5(a)), the PSNR is close to what we found with (RD C&E) (11.61≈11.58), whereas the reconstructed image has a level of visual encryption more interesting than that of (RD C&E).

It is for this reason that it is interesting to use another objective measure of the quality of the images. The compression ratio deteriorated completely compared to the configuration without encryption (83.278% 1.50). This is because the use of the random key has reduced the number of consecutive zeros in the high frequencies of the spectrum and therefore the BAC is no longer effective. To remedy this problem, we will present a simultaneous compression and encryption approach to improve the compression ratio.

12.4 Joint compression and encryption (JCE)

12.4.1 RAC algorithm

The arithmetic coding used in our chain is a lossless compression technique which represents the data to be coded as a fractional value in the scalar segment [0, 1] [47, 48]. Many approaches dedicated to encoding and reconstructing encrypted data with encoders of still and moving images are described in the literature, ([49–51] etc). In this part we propose a technique of simultaneous compression encryption with the aim of changing the generated bitstream without modifying the performance of the encoder, especially in terms of the compression ratio. Our method is based on the randomized arithmetic coding (RAC) algorithm and is described in figure 12.6. The operating principle of this RAC encryption relies to a large extent on that of the binary arithmetic coder (BAC). The block diagram of figure 12.7 shows the different blocks which constitute the proposed method. The random number is generated by the rand function of Matlab.

Indeed, at the beginning of the process there is acquisition of the data to be compressed and to be encrypted simultaneously with the encryption key denoted R in figure 12.7. The statistics on the binary inputs are carried out in order to determine the size of the data to be compressed as well as the number of 1 and 0 in the string to be compressed. With this, an initialization of the bands lower 'low' and

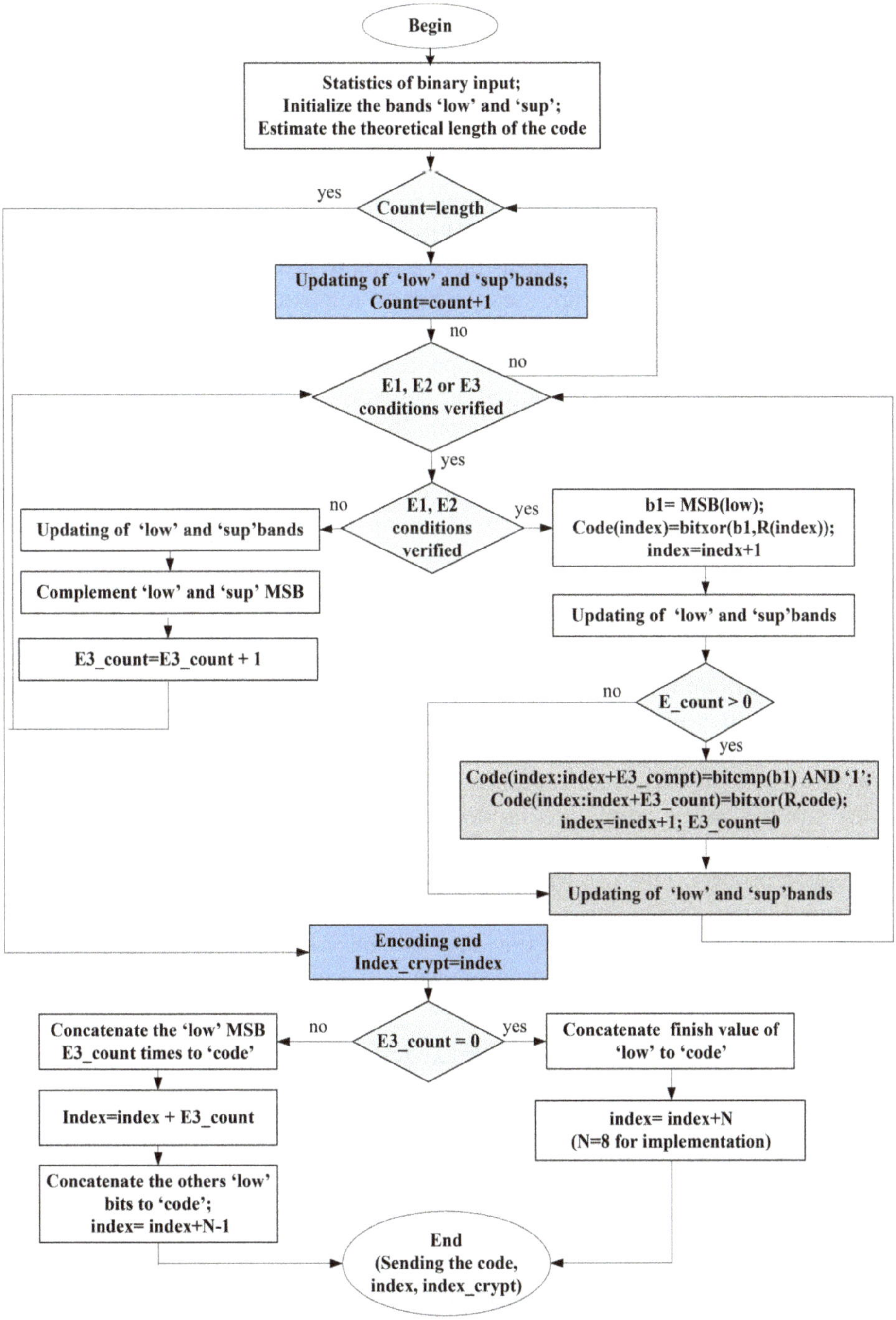

Figure 12.6. Principle of the proposed simultaneous encryption and compression.

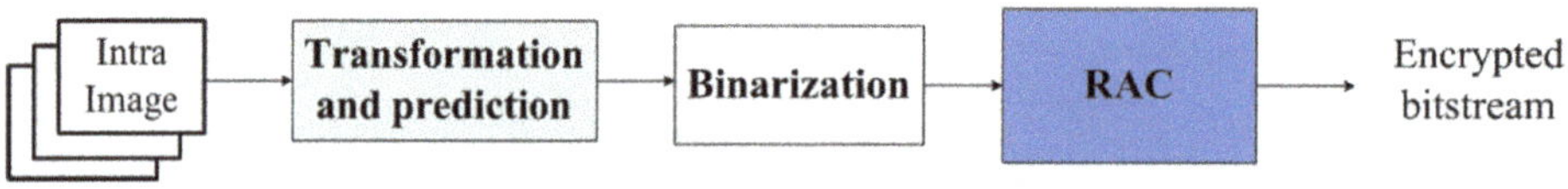

Figure 12.7. Simultaneous encryption and entropy coding.

higher 'sup' is performed in addition to the estimation of the theoretical size of the compressed data. Then, for any element of the binary string to be compressed, there is an update of these bands. As long as there is data to be compressed (the end of the file is not reached), the algorithm uses the bands 'low' and 'sup' to be in one of the three conditions E1, E2 or E3.

If the condition E3 is satisfied, there is update of the bands, complementation of the MSB of the bands and incrementation of the counter 'E3_compt'. Otherwise condition E1 or E2 is satisfied. It is precisely at this point that the encryption part can begin. The MSB (bit b1) of the low band is recovered, which makes it possible to find the code for this index, which is the combination with an or-exclusive between b1 and the value of the random vector R taken at this index. We made the choice to do the encryption at this level to increase the robustness of the encryption. It is important to note that the number of times the condition E3 is satisfied is given by the counter E3_compt.

As long as the value of this counter is non-zero, we realize an exclusive value between the value of the complement of b1 taken E3_compt times (according to the BAC algorithm) and the corresponding bits in the random vector R. These operations continue until the end of the data to be compressed. That is when we go to the end-of-encoding process. Here, we note the index of the compressed data at this precise moment, which corresponds to the intermediate size of the code.

The variable 'index_crypt' is a useful datum of the encryption process since it indicates the size of the random vector R that was used during the crypto-compression process. This size will be very useful for the decryption of information. In the end, the variable 'index' will indicate the size of the compressed data and the variable 'index_crypt' indicates the size of the random vector used for the encryption. At the end of the process, we can find the following relationship between these two counters:

$$index = \begin{cases} index_{encrypt} + N \rightarrow si - E3_{count} = 0 \\ index_{encrypt} + E3_{count} + N - 1 \rightarrow else \end{cases}. \tag{12.1}$$

Recall that $N = 8$ for a numerical implementation. Once the encoding is finished, the algorithm returns the compressed and encrypted code and the 'index' and 'index_crypt' variables that represent the size of the code and key used for encryption, respectively.

12.4.2 LFSR structures for pseudo-random numbers generation

Concerning the generation of pseudo-random numbers, we have opted for structures called LFSR (linear feedback shift register). This choice is justified by the very low complexity of hardware implementation of the LFSR and by the excellent statistical quality of the product numbers. More precisely, a binary LFSR of length L is

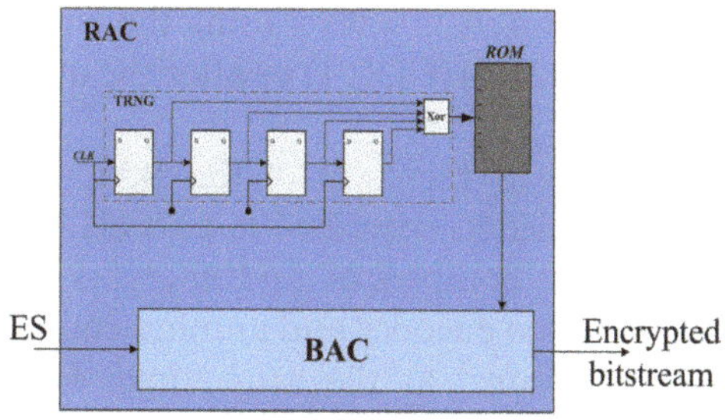

Figure 12.8. RAC block diagram.

composed of a register with L flip-flops. This register is controlled by an external clock which defines the operating periods of the LFSR. During each period of time, each flip-flop shifts its contents to the right. The delay caused by the flip-flop propagation times is used to construct a random number. Figure 12.8 shows the block diagram of the entropy coder integrating the encryption.

The initialization values of the flip-flops play an important role for the statistical property of the generated numbers. Indeed, these initialization values define the feedback polynomial of an LFSR and must remain secret during a stream encryption, even if the Berlekamp–Massy algorithm allows one to reconstruct it from $2L$ bits of the encrypted sequence [52, 53]. It is for this reason that they are often used as a basic module in dedicated pseudo-random generators, but within a more complex device. For example, to increase the nonlinearity of the code, the LFSRs can be used in parallel. This is the case of the Geffe algorithm where the cascading of several LFSRs of different lengths followed by a logical combination of each LFSR makes it possible to obtain more robust generated codes [54].

12.4.3 Decryption process

Not only does the proposed method compress data with an identical performance to the case without encryption, it also ensures that the information is perfectly decrypted. When the encrypted and compressed data are sent, the value of the variable index_crypt is also sent. Since the users authorized to see the contents of the image share a key of sufficient length, they have only to adapt the length of the key with respect to the value of index_crypt by taking the first elements of the key.

For these first elements which correspond to index_crypt, one applies an exclusive or bit-by-bit between the code and the key R, and for the other elements no action is envisaged. Finally, we use the arithmetic decoding algorithm BAD. Figure 12.9 gives a synthetic idea of this decryption algorithm.

Once the decryption is complete, the reverse zigzag order, the inverse quantization and the IDCT are applied to find the decrypted image.

12.5 Compression and double encryption (CDE)

Encryption can act on one level, but can also act on several levels simultaneously. In fact, we want to see if the concatenation of two encryption methods in various blocks of the compression chain supersedes the previously cited deficiencies. Here we introduce two levels of encryption. A first level using the biometric key in one of the stages (before the DCT (figure 12.10) (CDE 1) or after the quantization phase (figure 12.11) (CDE 2)).

The second security level is at the level of the BAC entropy encoder whose compression algorithm is modified as described in part (JCE). The crypto-compression chain then simultaneously compresses and encrypts the data in two levels. Indeed, an encryption in the video encoder and another encryption in the module of the entropy encoder are performed at the same time. Figures 12.12 and 12.13 represent the images decoded after compression and double encryption with a

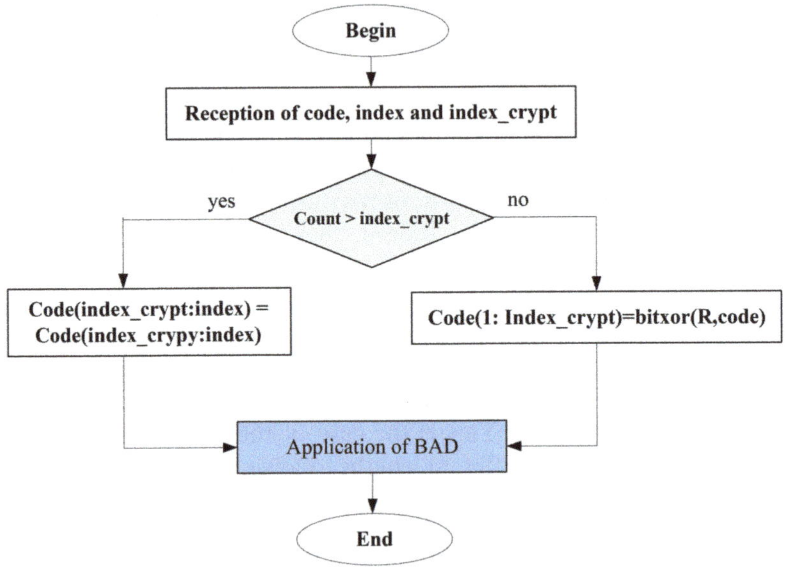

Figure 12.9. The proposed decryption algorithm.

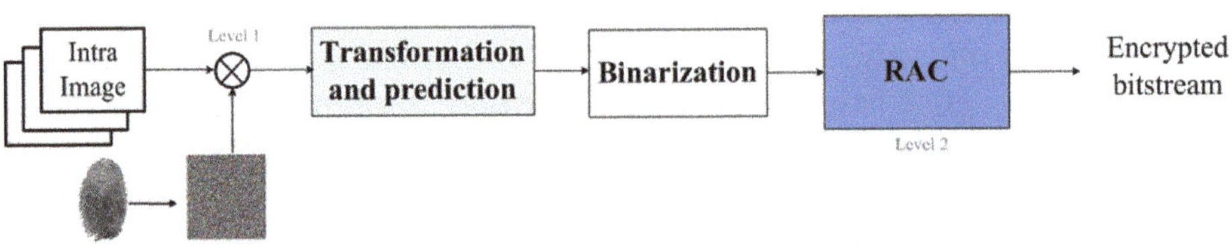

Figure 12.10. Double encryption with a biometric key applied to the raw data.

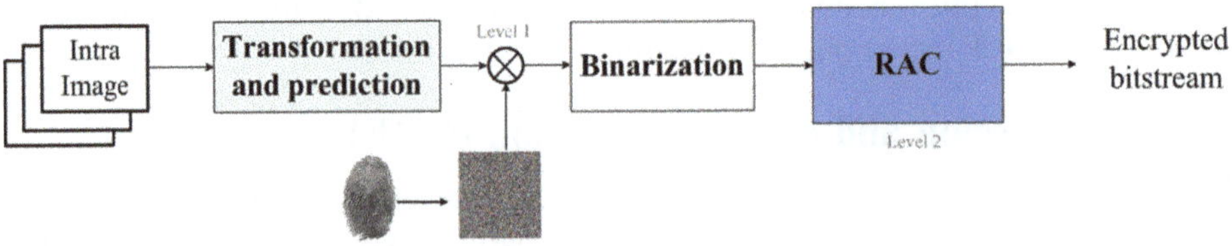

Figure 12.11. Double encryption with a biometric key applied to the transformed and quantified coefficients.

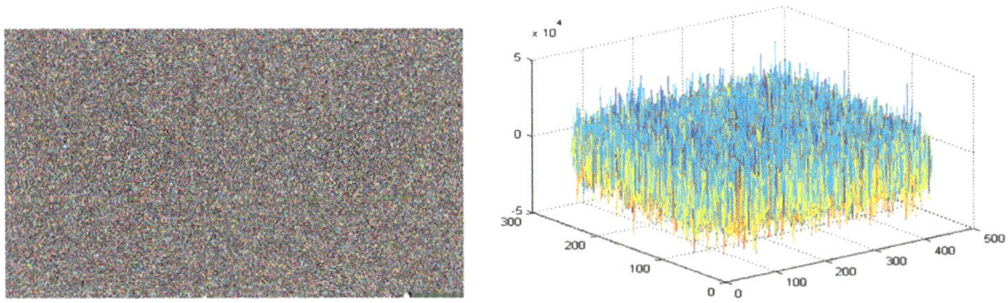

(a) (b)

Figure 12.12. Image decoded after compression and double encryption with a biometric key applied to the raw data.

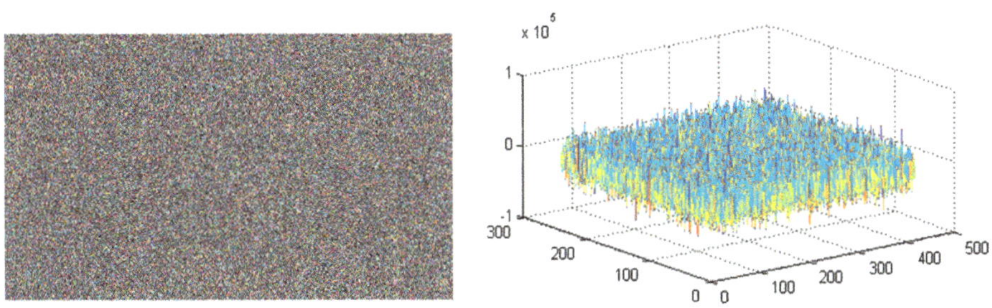

(a) (b)

Figure 12.13. Image decoded after compression and double encryption with a biometric key applied to the transformed and quantified coefficients.

biometric key applied to the raw data and to the transformed and quantified coefficients, respectively.

When encryption with a biometric key is applied to the raw data, we obtain PSNR = 11.385 dB, PCE = 0.0073 and RC = 23.152% (figure 12.12). And if encryption with a biometric key is applied to the transformed and quantified coefficients, we obtain PSNR = 10.358 dB, PCE = 0.0071 and RC = 1.381% (figure 12.13). These two techniques give us two slightly different PCEs, a difference of 1 dB at the PSNR level and a compression ratio less than that found with the RAC.

12.6 Simulation results

We performed simulations with the Matlab tool of the crypto-compression method used. The random number is generated by the rand function of Matlab. The encrypted image of this proposed method is presented in figure 12.14. We obtained a PSNR = 12.081 dB, a PCE = 0.0858 and a CR = 83.278%.

The PSNRs and PCEs are very close to those obtained in (TQC C & E) with a much greater compression ratio (+88%). This compression ratio is identical to the ratio obtained in the case without encryption. This is the maximum value we can achieve. It is the purpose of the proposed RAC algorithm to ensure the encryption

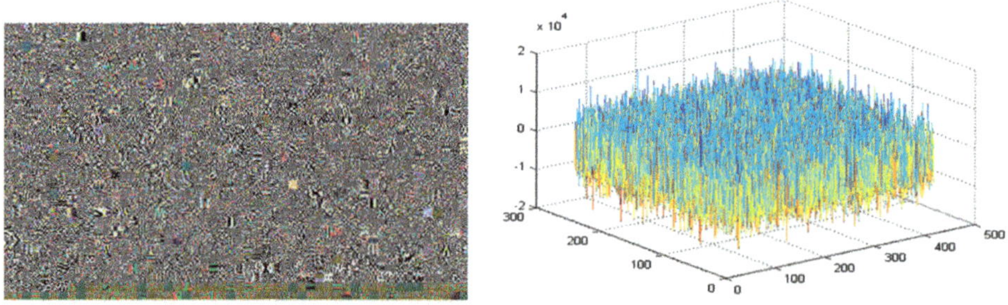

(a) (b)

Figure 12.14. Decoded image after compression and RAC encryption.

without changing the statistics of the sequence to be compressed. Several simulations were made with various and varied test sequences. The aim of changing the test sequences is to see the behavior of the proposed method in relation to the quantification or the very nature of the sequence to be tested. To evaluate and compare the different compression and encryption techniques, the QCIF and SD formats were tested in a 4:4:4 to 10-bit YUV version. We have varied the quantization steps QP between low, medium and large values. The test sequences are the QCIF sequences: (Forman, Carphone, Suzie and Mobile) and the SD sequences: (BasketballPass, BlowingBubbles, BQSquare and RaceHorses). The average change in PSNR as a function of QP is presented in the figures 12.15 and 12.16.

Joint compression and encryption (JCE) gives us the best compression ratio. Measurements of PSNRs and PCEs are very similar with several methods. RAC encryption gives the best results. JCE gives us the highest PSNR, then we put the other methods (RD C&E) (TQC C&E) (CDE 1) (CDE 2) in descending order. The compression ratios for the same sequences cited before are shown in figures 12.17 and 12.18.

The joint compression and encryption (JCE) assigns us the best CR. We obtain CRs, respectively, decreasing by (RD C&E) and (TQC C&E). We note that the variation of QP does not significantly affect the values of this metric. We obtain very close PCEs with (RD C&E), (JCE) and (TQC C&E) and much lower than the PCEs of compression without encryption (CWE), as shown in tables 12.2 and 12.3.

The execution time of the joint compression and encryption is shown in table 12.4. The timing ratio (%) represents the ration of compression and encryption time and compression time.

It is much lower compared to the double compression. The second level of encryption is not efficient and the performance in terms of execution time is very low.

12.7 Security analysis

The security of each crypto-compression system is linked to the ability to guess the values of the encrypted data. The security analysis of the proposed RAC encryption

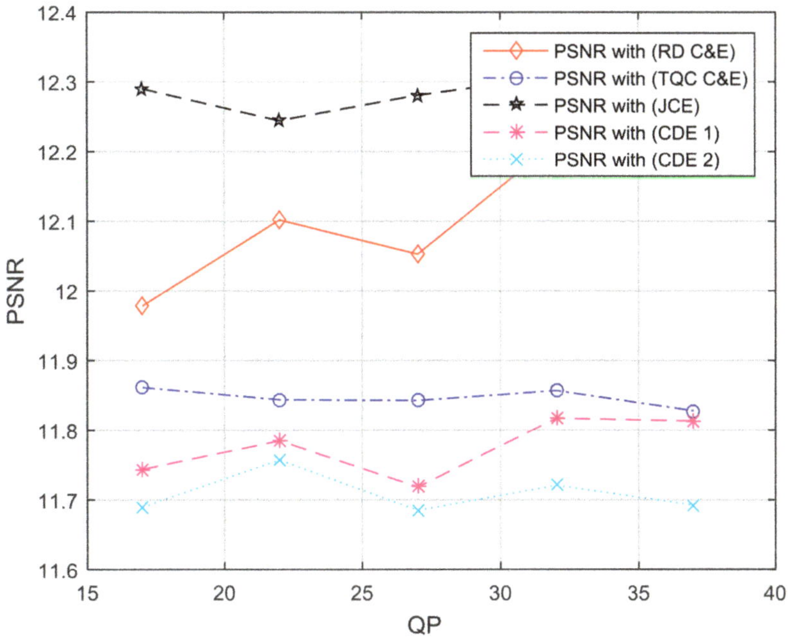

Figure 12.15. PSNR variation of QCIF sequences for different QP values.

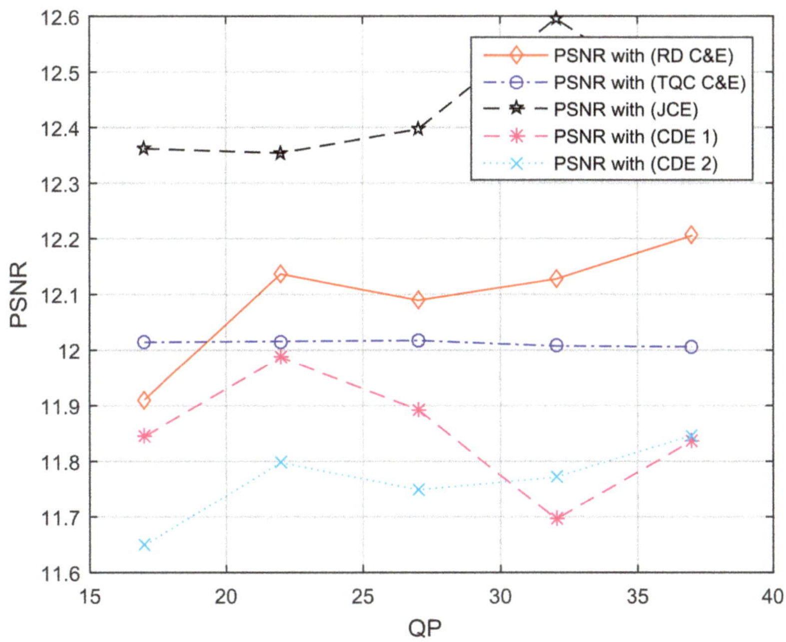

Figure 12.16. PSNR variation of QD sequences for different QP values.

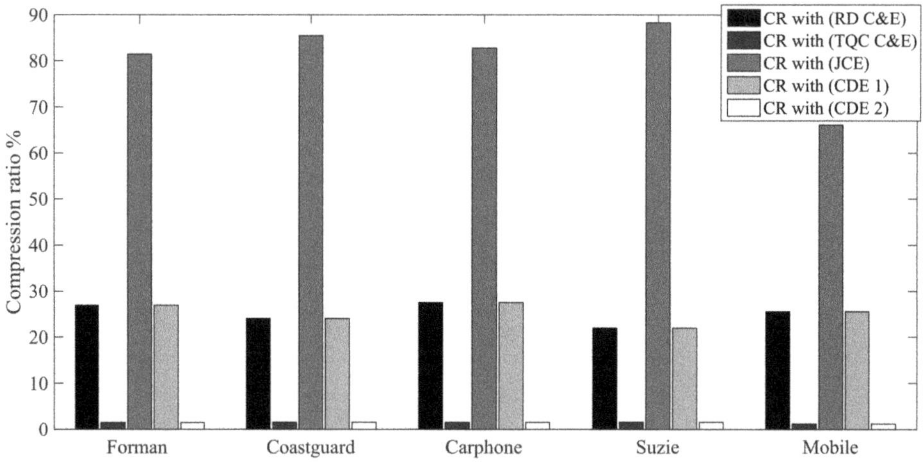

Figure 12.17. Compression ratio of QCIF sequences for QP = 32.

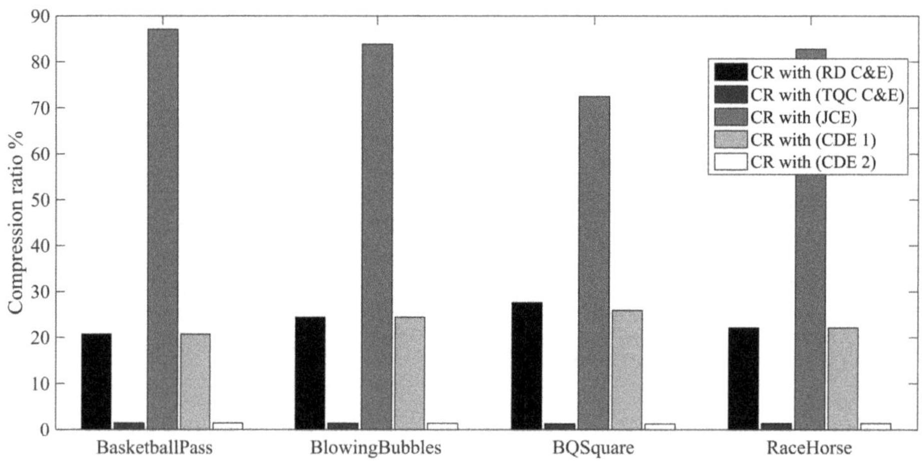

Figure 12.18. Compression ratio of SD sequences for QP = 32.

scheme includes the histogram analysis, the correlation analysis, the sensitivity of the original image, the key to attacks and the addition of noise.

12.7.1 The histogram analysis

We present the result of the RAC encryption image. The histogram of a plain image is presented in figure 12.19(b). The ciphered image is presented in figure 12.20(a), and its histogram resembles evenly a normal distribution as shown in figure 12.20(b). In statistics, a normal law is one of the laws of probability most adapted to model natural phenomena coming from several random events. Thus, in the context of cryptanalysis, it becomes very difficult to decrypt the image on the basis of statistical analyzes of the image to be decrypted.

Table 12.2. PCE for QCIF sequences for different QP values.

		37	32	27	22
Forman	(CWE)	0.6626	0.8479	0.6626	0.6626
	(RD C&E)	0.5065	0.7189	0.5065	0.5065
	(TQC C&E)	0.5739	0.5807	0.5739	0.5739
	(JCE)	0.5739	0.5807	0.5739	0.5065
	(CDE 1)	0.4390	0.7189	0.5065	0.5739
	(CDE 2)	0.5739	0.7189	0.5065	0.6414
Coastguard	(CWE)	0.3325	0.3370	0.3370	0.3325
	(RD C&E)	0.1174	0.1099	0.1097	0.1174
	(TQC C&E)	0.1242	0.1099	0.1097	0.1242
	(JCE)	0.1174	0.1099	0.1097	0.1242
	(CDE 1)	0.1310	0.1099	0.1136	0.1310
	(CDE 2)	0.1242	0.1099	0.1136	0.1174
Carphone	(CWE)	0.1917	0.1294	0.1294	0.1917
	(RD C&E)	0.0367	0.0167	0.0167	0.0367
	(TQC C&E)	0.0423	0.0235	0.0235	0.0423
	(JCE)	0.0395	0.0167	0.0235	0.0451
	(CDE 1)	0.0423	0.0235	0.0167	0.0395
	(CDE 2)	0.0451	0.0167	0.0167	0.0423
Suzie	(CWE)	1.0321	0.9207	1.0321	1.0321
	(RD C&E)	1.0052	0.8199	1.0052	0.9104
	(TQC C&E)	0.7557	0.7087	0.7557	0.7557
	(JCE)	1.0052	0.6809	1.0052	0.9104
	(CDE 1)	0.9104	0.7087	0.6009	0.7557
	(CDE 2)	0.4461	0.6253	0.7557	1.0652
Mobile	(CWE)	0.0404	0.0400	0.0404	0.0404
	(RD C&E)	0.0017	0.0016	0.0017	0.0016
	(TQC C&E)	0.0018	0.0013	0.0018	0.0018
	(JCE)	0.0017	0.0010	0.0017	0.0017
	(CDE 1)	0.0019	0.0016	0.0019	0.0018
	(CDE 2)	0.0020	0.0016	0.0018	0.0017

So, the histogram of the shuffled image is different from that of the plain image due to the significant substitution effect introduced in the scrambler process and shows no statistical resemblance to the plain image.

12.7.2 Correlation analysis

In an ordinary image, each pixel is highly correlated with its adjacent pixels. Unlike the cipher image, the correlation of the adjacent pixels is sufficiently low. However, the test of the correlations of the adjacent pixels in the encrypted image determines the superior confusion and diffusion properties. The correlation of the adjacent pixels is

Table 12.3. PCE for SD sequences for different QP values.

		37	32	27	22
BasketballPass	(CWE)	0.4340	0.4340	0.4340	0.4340
	(RD C&E)	0.4340	0.4340	0.4340	0.4340
	(TQC C&E)	0.3966	0.3966	0.3966	0.3966
	(JCE)	0.3966	0.3966	0.3966	0.4340
	(CDE 1)	0.3966	0.3966	0.3966	0.4340
	(CDE 2)	0.4340	0.3966	0.3592	0.4340
BlowingBubbles	(CWE)	0.0675	0.0675	0.0675	0.0675
	(RD C&E)	0.0675	0.0675	0.0675	0.0675
	(TQC C&E)	0.0660	0.0660	0.0660	0.0660
	(JCE)	0.0660	0.0675	0.0646	0.0675
	(CDE 1)	0.0675	0.0660	0.0675	0.0660
	(CDE 2)	0.0646	0.0646	0.0646	0.0632
BQSquare	(CWE)	0.0186	0.0186	0.0186	0.0186
	(RD C&E)	0.0186	0.0186	0.0186	0.0186
	(TQC C&E)	0.0337	0.0337	0.0337	0.0337
	(JCE)	0.0337	0.0337	0.0337	0.0487
	(CDE 1)	0.0186	0.0487	0.0337	0.0186
	(CDE 2)	0.0487	0.0487	0.0487	0.0337
RaceHorses	(CWE)	0.0845	0.0845	0.0845	0.0845
	(RD C&E)	0.0845	0.0845	0.0845	0.0845
	(TQC C&E)	0.0858	0.0858	0.0858	0.0858
	(JCE)	0.0871	0.0858	0.0871	0.0871
	(CDE 1)	0.0858	0.0858	0.0871	0.0845
	(CDE 2)	0.0845	0.0845	0.0858	0.0871

Table 12.4. Timing ratio for QCIF sequences.

	Timing ratio (%)		
Sequences	(JCE)	(CDE 1)	(CDE 2)
Foreman	0.1	54.59	6.84
Coastguard	0.803	51.052	3.791
Carphone	3.39	52.209	9.013
Suzie	0.351	58.175	7.827
Mobile	0.898	48.447	6.867

calculated with the use of equations (12.2) and (12.3) to calculate the correlation coefficients in the horizontal, vertical and diagonal directions:

$$\rho_{x,y} = \frac{\mathrm{cov}(x, y)}{\sqrt{D(x) \times D(y)}} \tag{12.2}$$

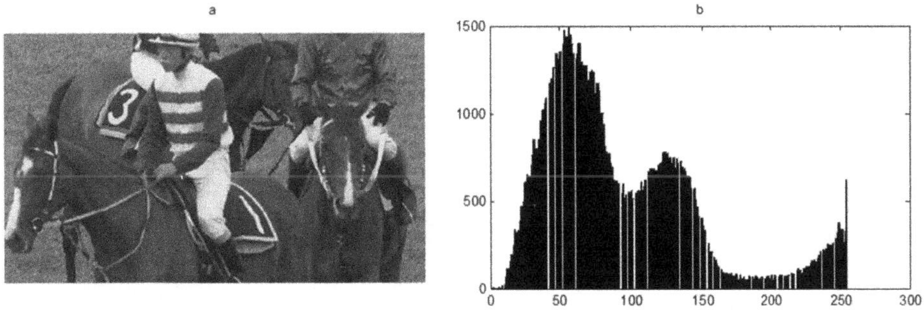

Figure 12.19. (a) Original image, (b) histogram of the original image.

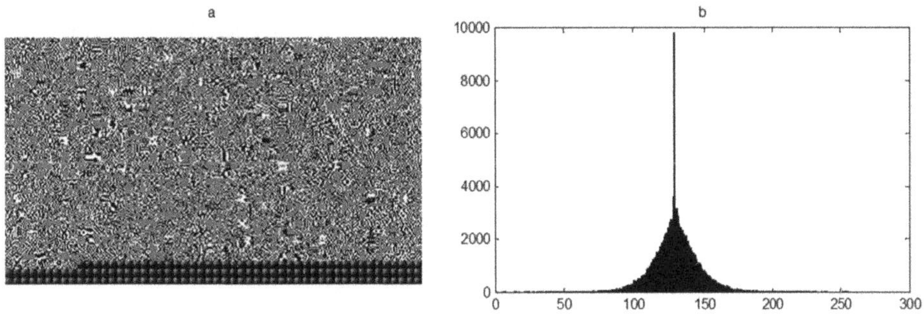

Figure 12.20. (a) Crypto-compressed image, (b) histogram of (a).

$$\text{with} \begin{cases} \operatorname{cov}(x, y) = \frac{1}{N}\sum_{i=1}^{N}(x_i - E(x)) \times (y_i - E(y)) \\ D(x) = \frac{1}{N}\sum_{i=1}^{N}(x_i - E(x))^2 \\ E(x) = \frac{1}{N}\sum_{i=1}^{N}x_i \end{cases} \tag{12.3}$$

where x and y are grayscale values of two adjacent pixels in the image, $E(x)$ is the expectation of x, $D(x)$ is the estimation of the variance of x and $\operatorname{cov}(x, y)$ is the estimation of the covariance between x and y.

Figure 12.20 shows the correlation of neighboring pixels in the horizontal direction of the images taken in figures 12.19(a) and 12.20(a).

By observing the point cloud in the plaintext image of figure 12.21(a), we can see that there is a linearity relation between adjacent pixels. This neighborhood property disappears with the encrypted image. The zoom on this distribution illustrated in figure 12.21(c) shows that any pixel whose intensity is between 110 and 150 can take any adjacent pixel value. This makes the decryption of images difficult. To give a quantitative analysis of what we advance, we can rely on the equations of the correlation coefficient explained in equation (12.2) (table 12.5).

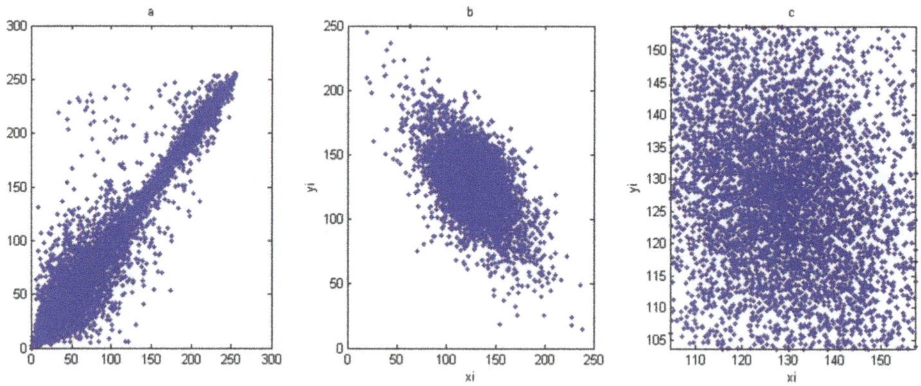

Figure 12.21. Correlations of adjacent pixels horizontally: (a) in the original image, (b) in the encrypted image obtained by RAC, (c) zooming on the encrypted image.

Table 12.5. Correlation of the original images and the corresponding encrypted ones.

		Correlation coefficients	
Sequences	Direction	Original images	Encrypted images
Race Horses	Horizontal	0.942	−0.496
	Vertical	0.933	−0.524
	Diagonal	0.895	0.303
Tulipe	Horizontal	0.960	−0.531
	Vertical	0.945	−0.541
	Diagonal	0.922	0.381

As illustrated in table 12.5, we can see that for natural images, the correlation coefficient is very close to 1, which is the maximum value. The closer the correlation coefficient is to 1, the more closely the neighboring pixels are correlated. In the case of the encrypted image, the correlation coefficient has values close to 0.5, which shows the non-correlation of the neighboring pixels. This value could have been lower, but the pixel distribution shown in figure 12.20(b) shows a peak in the histogram around the value 130. The standard deviation around these points is not very high and makes the correlation coefficient higher (but still less important than in the original image). Note that the negative sign emphasizes the downward relationship between the neighboring pixels. This means that a correlation coefficient which equals −1 also shows a strong correlation between neighboring pixels.

12.7.3 Sensitivity of the original image

A test of the sensitivity of the original image to attacks makes it possible to conclude on the robustness of the proposed algorithm. The idea of this test is to understand what happens when a single bit of the original image is changed. Let P1 be the

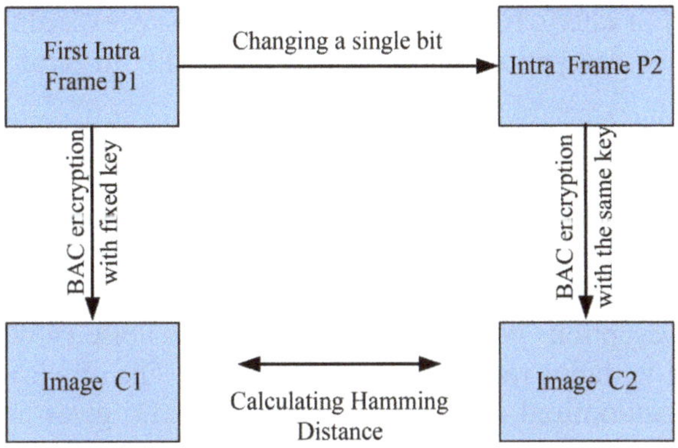

Figure 12.22. The adopted sensitivity test scheme.

original image. If a single bit of image P1 changes, image P2 is obtained. Next, the encryption of P1 and P2 by the same method and the same key gives the encrypted images C1 and C2. Figure 12.22 illustrates the adopted sensitivity test scheme.

The sensitivity of the original image to attacks amounts to calculating the number of different bits between C1 and C2 divided by the total number of bits. This is called the Hamming distance and is expressed mathematically by this equation:

$$d_{\text{Hamming}}(C1, C2) = \frac{\sum_{k=1}^{L\times H\times 8} C1(k) \oplus C2(k)}{L \times H \times 8}. \tag{12.4}$$

L and H denote the size of the input image encoded on 8 bits and designates the operator or exclusive. The various tests carried out show that on average the change of a bit in the original image makes it possible to have a Hamming distance of 0.24, which is equivalent to the change of 24% of the output image.

12.7.4 Sensitivity of the key

The test of the sensitivity of the encryption key to attacks much resembles that of the previous section. We use the images P1 and P2 obtained in the previous section by a change of a single bit. Then we apply the RAC encryption, but this time with a single bit change to the key level. Thus, image P1 is encrypted with a key and image P2 is encrypted with a second key that differs from the first one of a single bit. The block diagram in figure 12.23 illustrates the sensitivity measurement scenario of the encryption key to attacks.

In the same way, we calculate the Hamming distance between C1 and C2. We find that the average distance is 0.25% which means that the rate of change of the pixels is 25%. This demonstrates the robustness of the proposed encryption method, not to mention that it is a method that simultaneously performs the compression. However, we note that a decryption with a bad key always gives a noisy image. Also, if the correct encryption key is obtained, a decryption with an incorrect index_crypt does not allow the picture to be displayed in clear text. This constitutes an additional level of safety. If the value of the index_crypt exceeds the size of the file to be decrypted, the decryption algorithm responds with an error.

12.7.5 Noise addition

Generally, addition of noise is responsible for the degradation and distortion of the image. The cipher-image is also degraded by noise addition, resulting in difficulties in image decryption. We thus obtain a reconstituted image C1 after compression and decrypted after encryption. We add a Gaussian bench noise (AWGN) with a zero mean and a noise variance ranging from 0.02 to 0.4. The application of the RAC followed by the randomized arithmetic decoding (RAD) gives us a reconstructed and decrypted image C2 after applying the noise. The synoptic of this attack is detailed in figure 12.24. To test the robustness of the noise encryption scheme, we use

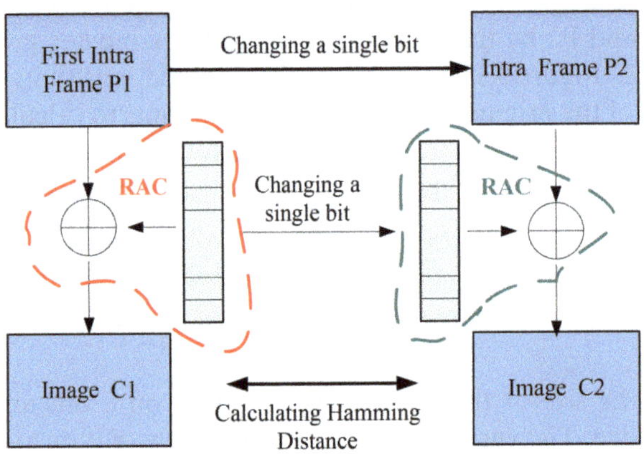

Figure 12.23. Scenario for measuring key sensitivity to attacks.

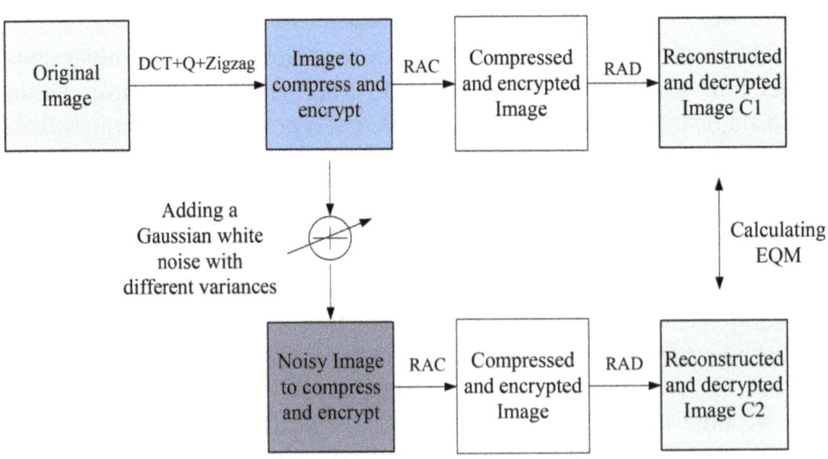

Figure 12.24. Sensitivity of the encryption scheme to noise.

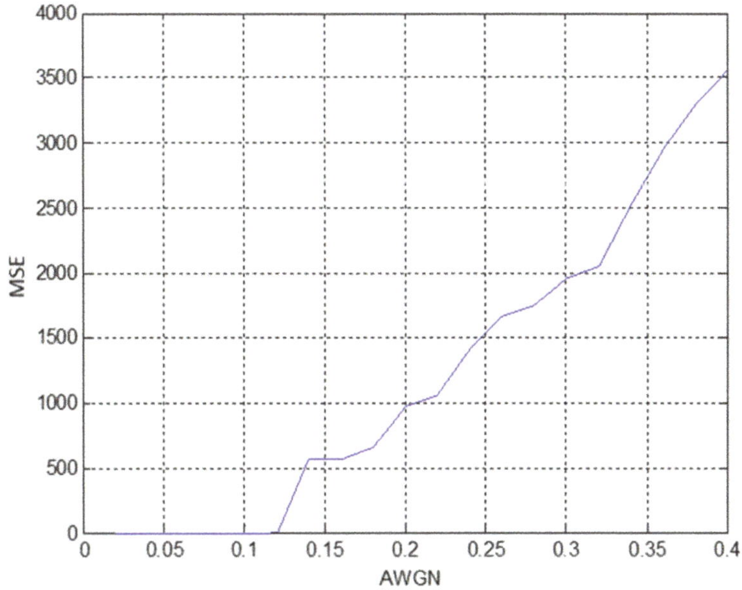

Figure 12.25. Evaluation of the MSE for different values of Gaussian noise.

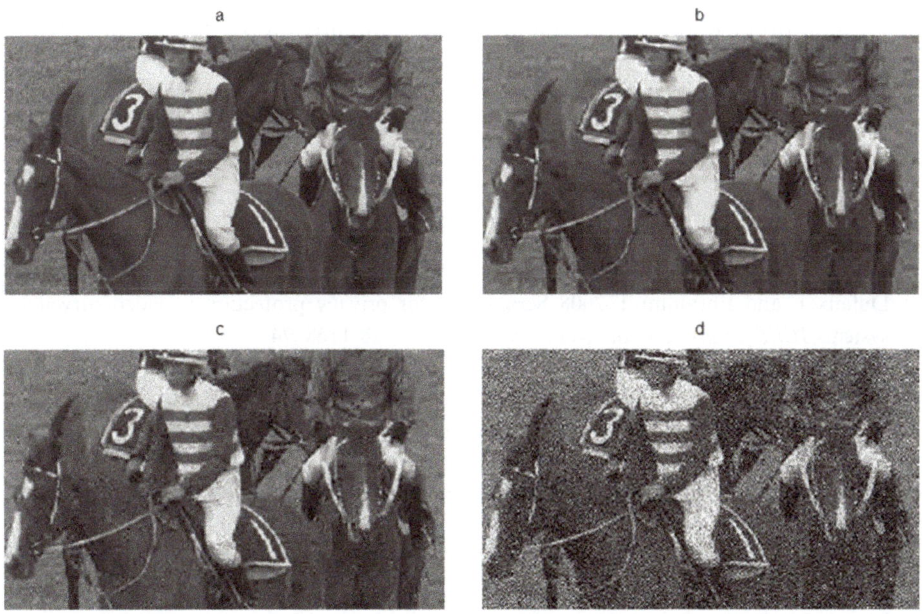

Figure 12.26. Results of noise resistance: (a) Original image; (b) encrypted and decrypted image with the secret key without the presence of noise; (c) image encrypted and decrypted with the secret key with the presence of variance noise equal to 0.3; (d) image encrypted and decrypted with the secret key with the presence of variance noise equal to 0.4.

a distance measurement between the two images C1 and C2. This measure can be achieved by evaluating the mean square error (MSE) between C1 and C2.

We have plotted the different values of the MSE with respect to the variance of the noise as shown in figure 12.25. The step of the noise power is set to 0.02. It is evident that the higher the variance of the noise, the greater the difference between C1 and C2.

These tests show that we can decrypt the images and we can recognize them despite the interferences, and that the crypto-compression system can tolerate a certain level of noise. To prove what we are advancing, figures 12.26(a) and (b) show the original image and the image C1, respectively. For values of noise variance powers equal to 0.3 and 0.4, figures 12.26(c) and (d), respectively, show the reconstructed and decrypted images.

12.8 Conclusion

In this chapter, we presented a simultaneous compression and encryption procedure implemented in the standard compression of still and animated images. We tried to show the performance and limitations of the modified video encoder generating a stream of protected data. If we compare the extracts of the sequences with the original in the different methods proposed, we do not recognize the details.

Our combination of encryption and compression allows us to save computing time by keeping the initial compression rate. The significant difference in the compression ratios makes it possible to see a certain advantage in joint compression encryption (JCE). It presents the best compromise between CR, PCE, PSNR and computing time since it is a single level of encryption that requires less computation.

References

[1] Buchmann J 2004 *Introduction to Cryptography* (Springer) p 335

[2] Carrillo P, Kalva H and Magliveras S 2008 Compression independent object encryption for ensuring privacy in video surveillance *Proc. ICME* pp 273–6

[3] Dufaux F and Ebrahimi T 2008 Scrambling for privacy protection in video surveillance systems *IEEE Trans. Circuits Syst. Video Technol.* **18** 1168–74

[4] Dufaux F and Ebrahimi T 2008 H.264/AVC video scrambling for privacy protection *Proc. IEEE ICIP* pp 1688–91

[5] Lian S, Liu Z, Ren Z and Wang H 2007 Commutative encryption and watermarking in video compression *IEEE Trans. Circuits Syst. Video Technol.* **17** 774–8

[6] Noorkami M and Mersereau R M 2007 Digital video watermarking in P-frames *Proc. SPIE Conf. Security Steganography Watermarking Multimedia Contents IX*

[7] Ahn J, Shim H J, Jeon B and Choi I 2004 Digital video scrambling method using intra prediction mode *Proc. Adv. Multimedia Inf. Process* pp 386–93

[8] Lian S, Liu Z, Ren Z and Wang H 2006 Secure advanced video coding based on selective encryption algorithms *IEEE Trans. Consum. Electron.* **52** 621–9

[9] Lian S, Sun J, Liu G and Wang Z 2008 Efficient video encryption scheme based on advanced video coding *Multimedia Tools Appl.* **38** 75–89

[10] Su P C, Hsu C W and Wu C Y 2010 A practical design of content protection for H.264/AVC compressed videos by selective encryption and fingerprinting *Multimedia Tools Appl.* **52** 529–49

[11] Dai Y, Zhang L and Yang Y 2003 A new method of MPEG video watermarking technology *Proc. Int. Conf. on Communication Technology* **2** 1845–7
[12] Qiu G, Marziliano P, Ho A T S, He D and Sun Q 2004 A hybrid watermarking scheme for H.264/AVC video *Pattern Recognition, Int. Conf.* **4** 865–9
[13] Li Y, Liang L, Su Z and Jiang J 2005 A new video encryption algorithm for H.264 *Proc. 5th ICICS* pp 1121–4
[14] Thomas N, Lefol D, Bull D and Redmil D 2007 A novel secure H.264 transcoder using selective encryption *Proc. IEEE ICIP* pp 85–8
[15] Lian S, Liu Z, Ren Z and Wang Z 2005 Selective video encryption based on advanced video coding *Proc. PCM Multimedia, Adv. Multimedia Inform. Process.* pp 281–90
[16] Won Y G, Bae T M and Ro Y M 2006 Scalable protection and access control in full scalable video coding *Proc. 5th IWDW* pp 407–21
[17] Golikeri A, Nasiopoulos P and Wang Z 2007 Robust digital video watermarking scheme for H.264 advanced video coding standard *J. Electron. Imaging* **16** 0430008
[18] Gong X and Lu H 2008 Towards fast and robust watermarking scheme for H.264 video *Proc. IEEE Int. Symp. on Multimedia* pp 649–53
[19] Noorkami M and Mersereau R 2008 Digital video watermarking in P-frames with controlled video bit-rate increase *IEEE Trans. Inf. Forensics Secur.* **3** 441–55
[20] Shahid Z, Chaumont M and Puech W 2009 Fast protection of H.264/AVC by selective encryption of CABAC *Proc. IEEE ICME* pp 1038–41
[21] Magli E, Grangetto M and Olmo G 2006 Conditional access to H.264/AVC video with drift control *Proc. IEEE ICME* pp 1353–6
[22] Grangetto M, Magli E and Olmo G 2007 Conditional access to H.264/AVC video by means of redundant slices *Proc. IEEE ICIP* **6** pp 485–8
[23] Lee H J and Nam J 2006 Low complexity controllable scrambler/descrambler for H.264/AVC in compressed domain *Proc.ACM Multimedia* pp 93–6
[24] Dufaux F and Ebrahimi T 2008 H.264/AVC video scrambling for privacy protection *Proc. IEEE ICIP* pp 1688–91
[25] Bergeron C and Lamy-Bergor C 2005 Compliant selective encryption for H.264/AVC video streams *Proc. IEEE Workshop MMSP* pp 1–4
[26] Mian C, Jia J and Lei Y 2007 An H.264 video encryption algorithm based on entropy coding *Proc. 3rd Conf. IIH-MSP* pp 41–4
[27] Zou D and Bloom J 2010 H.264 stream replacement watermarking with CABAC encoding *Proc. IEEE ICME* pp 117–21
[28] Stutz T and Uhl A 2011 A survey of H.264 AVC/SVC encryption *IEEE Trans. Circuits Syst. Video Technol.* **22** 325–39
[29] Richardson I E G 2003 *H.264, MPEG-4 Video Compression: Video Coding for Next Generation Multimedia* (New York: Wiley)
[30] Noorkami M and Mersereau R 2005 Compressed-domain video watermarking for H.264 *Proc. IEEE Int. Conf. on Image processing* pp 890–3
[31] Lu C, Chen J and Fan K 2005 Real-time frame-dependent video watermarking in VLC domain *Signal Process., Image Commun.* **20** 624–42
[32] Kim S M, Kim S B, Hong Y and Won C 2007 Data hiding on H.264/AVC compressed video *Image Analysis and Recognition. ICIAR 2007* ed M Kamel and A Campilho (Lecture Notes in Computer Science vol 4633) (Berlin: Springer)

[33] Mobasseri B and Raikar Y 2007 Authentication of H.264 streams by direct watermarking of CAVLC blocks *Proc. Security, Steganography and Watermarking of Multimedia Contents IX SPIE* **6505** 65051W
[34] Zou D and Bloom J 2009 H.264/AVC substitution watermarking: a CAVLC example *Proc. SPIE Media Forensics and Security XI* **7254** 9
[35] Li J, Liu H, Huang J and Zhang Y 2009 A robust watermarking scheme for H.264 *Digital Watermarking. IWDW 2008* ed H J Kim, S Katzenbeisser and A T S Ho (Lecture Notes in Computer Science vol 5450) (Berlin: Springer)
[36] Hellwagner H, Kuschnig R, Stutz T and Uhl A 2009 Efficient in-network adaptation of encrypted H.264/SVC content *Signal Process., Image Commun.* **24** 740–58
[37] Almasalha F, Agarwal N and Khokhar A 2008 Secure multimedia transmission over RTP *Proc. 8th IEEE ISM* pp 404–11
[38] Massoudi A, Lefèbvre F, De Vleeschouwer C, Macq B and Quisquater J J 2008 Overview on selective encryption of image and video, challenges and perspectives *EURASIP J. Inform. Secur.* **2008** 18
[39] Fan Y, Wang J, Ikenaga T, Tsunoo Y and Goto S 2007 A new video encryption scheme for H.264/AVC *Proc. Adv. Multimedia Inf. Process.* pp 246–55
[40] Engel D, Stutz T and Uhl A 2009 A survey on JPEG2000 encryption *Multimedia Syst.* **15** 243–70
[41] Arachchi H, Perramon X, Dogan S and Kondoz A M 2009 Adaptation-aware encryption of scalable H.264/AVC video for content security *Signal Process., Image Commun.* **24** 468–83
[42] Zou Y, Huang T, Gao W and Huo L 2006 H.264 video encryption scheme adaptive to DRM *IEEE Trans. Consumer Electron.* **5** 1289–97
[43] Iqbal R, Shirmohammadi S and El-Saddik A 2006 Secured MPEG-21 digital item adaptation for H.264 video *Proc. ICME* pp 2181–4
[44] Iqbal R, Shirmohammadi S and El-Saddik A 2007 A framework for MPEG-21 DIA based adaptation and perceptual encryption of H.264 video *Proc. SPIE Multimedia Comput. Netw.* **6504** 1–12
[45] Kuschnig R, Kofler I, M.Ransburg M and Hellwagner H 2008 Design options and comparison of in-network H.264/SVC adaptation *J. Vis. Commun. Image Represent.* **19** 529–42
[46] ISMA 2007 *Encryption and Authentication Specification 2.0* Internet Streaming Media Alliance
[47] Langdon G G 1984 An introduction to arithmetic coding *IBM J. Res. Dev.* **28** 135–49
[48] Sayood K 2005 *Introduction to Data Compression* (San Mateo, CA: Morgan Kaufmann)
[49] Pande A 2011 Algorithms for secure multimedia delivery over mobile devices and mobile agents *Ubiquitous Multimedia and Mobile Agents: Models and Implementations* (IGI Global)
[50] Grangetto M, Magli E and Olmo G 2006 Multimedia selective encryption by means of randomized arithmetic coding *IEEE Trans. Multimedia* **8** 905–17
[51] Varalakshmi L M and Florence G 2010 Performance enhancement of binary randomized arithmetic coding for multimedia encryption *Int. J. Recent Trends Eng. Technol.* **3**
[52] Berlekamp E R 1968 *Algebraic Coding Theory* (New York: McGraw-Hill)
[53] Massey J L 1969 Shift-register synthesis and BCH decoding *IEEE Trans. Inf. Theory* **15** 122–7
[54] Geffe P R 1973 How to protect data with ciphers that are really hard to break *Electron.* 99–101

IOP Publishing

Advanced Secure Optical Image Processing for Communications

Ayman Al Falou

Chapter 13

A comparative study of CFs, LBP, HOG, SIFT, SURF, and BRIEF for security and face recognition

Yassin Kortli, Maher Jridi, Ayman Al Falou and Mohamed Atri

13.1 Introduction

In recent years, developing surveillance systems (SS) for security has been one of the most active research fields in most applications. These systems are used to adjust, enhance, and improve security. The face recognition system, plays an efficient and very important role for several applications including access in military, civilian, control, and system security, and can be employed for optical surveillance systems [1, 2]. In addition, face recognition systems are usually used to adjust, enhance, and automate the identification process for security and safety. This system can be used to assist and help police to understand suspect behavior and launch adequate measures to tackle it.

Face recognition based on vision systems is usually performed by two processes: the detection of the face and then the recognition. The face detection method uses many algorithms to detect and locate a face in an image obtained by the system. The face recognition process includes a feature extracted from an unknown face in order to compare it with all the template faces in the database to decide the identity of this unknown face ('acceptance or rejection'). Moreover, a face recognition system presents three major processing modules [3]: preprocessing and detection, feature extraction, and recognition. The recognition (or decision) is done by matching the features stored in the system datasets with the features extracted from the unknown face.

Face recognition is a task that humans perform naturally and effortlessly in their daily lives. The high availability of powerful and inexpensive computers and embedded computing systems has generated an enormous interest in the automatic

doi:10.1088/978-0-7503-1457-2ch13

processing of digital images and videos in many applications, including biometric identification, monitoring, human–machine interaction and multimedia data management. The recognition of the face can be used without direct contact. It can be used anywhere and in various applications. However, it suffers from various changes in the same face and large image size. In the literature, it can be found that some techniques are used exclusively for the process of detection, and others are dedicated to the recognition process with high robustness and discrimination [4]. For the past two decades, various face recognition methods have been proposed to reduce the number of calculations needed and improve the recognition rate. These proposed methods can be categorized into three significant categories:

- Local feature approaches
- Subspace learning approaches
- Correlation filters approaches

Local feature methods are mainly proposed for face recognition and all show the best performance in this task. Many methods are implemented for representing the face such as Gabor filters [5], local ternary patterns (LTP) [5], local binary patterns (LBP) [6], histogram of oriented gradient (HOG) [7], scale invariant feature transform (SIFT) [8], binary robust independent elementary features (BRIEF) [9], and speeded-up robust features (SURF) [10], etc.

Unlike the local approaches, subspace learning approaches or global approaches are another approach used for face recognition. These approaches include Fisher linear discriminant analysis (LDA) [11], principal components analysis (PCA) [12], and locality preserving projections (LPP) [13]. Subspace learning approaches are methods to obtain a discriminative subspace. However, these approaches generally ignore the local details. This characteristic makes the subspace learning approaches ineffective for face recognition.

Correlation approaches like the Vander Lugt correlator (VLC), and joint transform correlator (JTC) [2, 14–15] are used for face recognition applications. These methods are mainly used due to their ability to instantly detect, discriminate and estimate the target objects.

However, these techniques have many problems to surmount. For example, the same person's facial images can look profoundly different due to changes in orientation of the head, illumination conditions, and expression [2]. It is difficult for any recognition system to obtain a correct decision because of lighting conditions and unexpected noise interference [15]. To build a reliable system that can deal with all of these problems, we test and evaluate some face recognition techniques in order to choose the best technique regarding computation time and algorithmic complicity.

The rest of the chapter is organized as follows: in section 13.2, we provide an overview of face recognition techniques in order to select some of them. Section 13.3 is dedicated to the description and implementation of the many face recognition systems. Section 13.4 presents the experimental setup and discusses the results obtained under various conditions. The last section concludes the work and discusses the direction of future research.

13.2 Related work

In recent years, face recognition systems in real-time using monocular vision have become a reputed area of study in security surveillance systems. In the literature related to computer vision, feature extraction allows the detection and classification of faces. This is possible by the use of symmetry and edges on the features to detect faces. Two main classes are applied to identify faces: detection (extraction) and recognition (identification) modules [15]. The choice made during one step can influence the other step. The detection module after a preprocessing step consists in obtaining the face region for each image. The recognition module comprises a description and extraction of the feature vectors of the face region obtained in the first step, and then a comparison with all the template faces in the database. In this section, we provide an overview of the many methods related to face detection and recognition.

However, there are many methods for face detection, and recognition tasks rely on describing the local features of an image, the location of the main points or points of interest in each image, and specifying the region pixels featured around the point of interest.

In addition, local feature methods and correlation methods can be divided into two categories: local appearance-based methods and interest point based methods. The first methods include feature analysis (e.g. local binary patterns (LBP), histogram of oriented gradient (HOG) and direct correlation methods). These methods divide the face into blocks (or regions) for extraction of the local features, or they directly extract the local features using correlation. The second methods are composed of points of interest methods, contours or edges methods, and corners methods. These methods detect and extract the point of interest of an image and extract the local features localized on these points (e.g. scale invariant feature transform (SIFT), binary robust independent elementary features (BRIEF), and speeded-up robust features (SURF)).

13.2.1 Local feature-based methods

There are geometrical methods also called methods to feature, to local characteristic, or analytic. In these methods, the face is represented by a set of characteristic vectors of low dimensions, rather than by a single large dimension vector. Local methods are concerned with the critical points of the face such as nose, mouth, eyes; and will generate more details. The advantage of these local methods is that they take into account the particularity of the face as a natural form to be recognized and a reduced number of parameters. Also, these methods describe local features by local orientation information, histograms, geometric properties, and correlation. The most well-known methods in these approaches are the HOG [7], LBP [16], and correlation filters [2].

Local feature descriptors show a good performance in some applications in pattern recognition. Other advantages of these descriptors are that they are robust to scaling, rotation, and translations.

- ***Histogram of oriented gradients (HOG):*** the HOG descriptor is one of the best accepted processes to describe image local texture and it has one of the best features of shape and edge information. The HOG descriptor can describe the shape of the face by distribution of the edge direction or light intensity gradient. The process of this technique done by dividing the whole face image into cells (small region or area), a histogram of pixel edge direction or direction gradients is generated from each cell, and finally, the histograms of the whole cells are combined to extract the feature of the face image. It displays great success in face recognition [7].
- ***Local binary patterns (LBP):*** LBP is a great general texture descriptor used for feature extraction from an image, and has been widely used in a lot of applications, such as face recognition, facial expression recognition, texture segmentation, and texture classification. The most important advantages of this descriptor are its low computational complexity for describing local scene regions (good texture structure description), its invariance to monotonic grayscale value changes and convenient multi-scale extension. At each pixel, the LBP descriptor presents binary comparisons of pixel intensities between its eight surrounding pixels and the center pixel. Also, this descriptor has been successfully used in pattern recognition [16]. In this work, for HOG and LBP descriptors, the matching task can be done by distance measures such as chi-squared distance and histogram intersection. These distances are used to obtain the correspondence between two images [17]. The HOG and LBP descriptor inherits the advantages of chi-squared distance, which effectively handles the unbalanced nature of HOG and LBP histograms [18].
- ***Correlation filters (CFs):*** correlation filters are used for pattern recognition applications due to their ability to instantly detect, discriminate and estimate target objects. The correlation filters are based on an optical configuration called the 4f-setup. This configuration is done by three planes: the input plane, the Fourier plane and the correlation plane. A convergent lens separates each plane. The first lens performs the Fourier transform of the target image. The second lens, after the multiplication of the correlation filter (obtained from the reference image) by the spectrum of the target image (correlation filter * spectrum of the target image), performs the inverse Fourier transform to get the correlation result, characterized by the central correlation peak. The matching process is used to classify the face image when the features are selected and extracted. A wide variety of matching algorithms are used in pattern recognition. However, for this matching process to be reliable and fast, it is necessary to establish an efficient technique. To evaluate the correspondence of the correlation, the peak-to-correlation energy (PCE) is used to decide the degree of similarity between the reference and target images [1].

13.2.2 Interest-point based methods

Many interest-point based methods rely on describing the features of an image, the location of the main points or points of interest and describe the region pixels featured around the point of interest. SIFT [8], BRIEF [9], and SURF [10] are methods based on the detection of interest points to describe the feature of an image.

- ***Scale invariant feature transform (SIFT):*** SIFT is one of the most popular descriptors based on point of interest that uses the intensity without color information. For this purpose, the difference of Gaussian (DoG) is used to identify the interest points in the image region which are invariant to orientation, scale, illumination, and zoom. This descriptor has been widely used in pattern recognition and classification systems such as face recognition, visual mapping, and image stitching. However, when the scale or the size of the image of the dataset increases significantly, the disadvantage of the SIFT descriptor is its high computational cost [8].
- ***Speeded-up robust features (SURF):*** the SURF descriptor deduced from the SIFT descriptor, even outperforms or approximates the SIFT descriptor at robustness, distinctiveness, and repeatability, and also can be compared and calculated more quickly [19]. In this descriptor, the scale changes by altering the size of the box filter instead of changing the image size. Therefore, it significantly minimizes the number of operations for the simple box convolutions. It is used mainly for classification, 3D reconstruction, image registration, face recognition, and extraction points of interest. Also, to describe the local feature, the SURF descriptor uses a 64-dimensional feature vector, while the SIFT uses 128 that increases the processing speed. Besides, the SIFT descriptor is more adapted to describe faces affected by illumination deformations, scaling, translation, rotation [3]. In a matching case, a Euclidean-distance based nearest-neighbor matching can be used for matching vector features. In addition, the best nearest-neighbor technique depends on the descriptor used for the extraction of the feature vector. Also, the distance ratio between the two neighborhoods must be required to be less than a predefined threshold, in order to eliminate the candidates which are considered as noise. However, a Euclidean-distance based nearest-neighbor matching can be used for matching vector features. In the literature, in the high-dimensional features, the fast library for approximate nearest neighbors (FLANN) and the randomized k-d forest are the most efficient techniques that have been used for matching. In this work, for SIFT and SURF descriptors, the FLANN is used to obtain the correspondence between two images [20].
- ***Binary robust independent elementary features (BRIEF):*** BRIEF is a binary descriptor that is simple and fast to compute. This descriptor is based on the differences between the pixel intensity that is similar to the family of binary descriptors such as BRISK and FREAK in terms of evaluation. To reduce noise, the BRIEF descriptor smooths the image patches. After that, the differences between the pixel intensity is used to represent the descriptor. This descriptor has achieved the best performance and accuracy in pattern recognition. Moreover, for binary features, for example, BRIEF or FAST, the FLANN, and randomized k-d forest techniques are not adequate. Then, the Hamming distance is used to compare the binary features in a matching case. The Hamming distance, instead of the Euclidean distance, is computed

by performing the sum of XOR between two binary descriptors followed by a bit count on the result [20].

Generally, for the descriptors based on interest points, the choice of associated image descriptors and the properties of the underlying interest points affect the performance of the matching tasks. In this study, six of the most frequently-used features, HOG, LBP, VLC, SIFT, SURF, and BRIEF are implemented and compared.

13.3 Methods implementation

In this section, we describe our implemented methods for face recognition. We apply six algorithms of feature extraction for the three approaches: the VLC for the correlation approaches and the LBP, HOG, SIFT, SURF, and BRIEF for the local approaches. For each image obtained from the PHPID datasets, a preprocessing step is applied to enhance the feature extraction in the image. After the extraction phase, we obtain feature vectors for each target and reference image. Next, we use a metric parameter to compare these images (target and reference images). Finally, our system determines the decision, depending on the comparison scores. The flowchart of our proposed system is presented in figure 13.1.

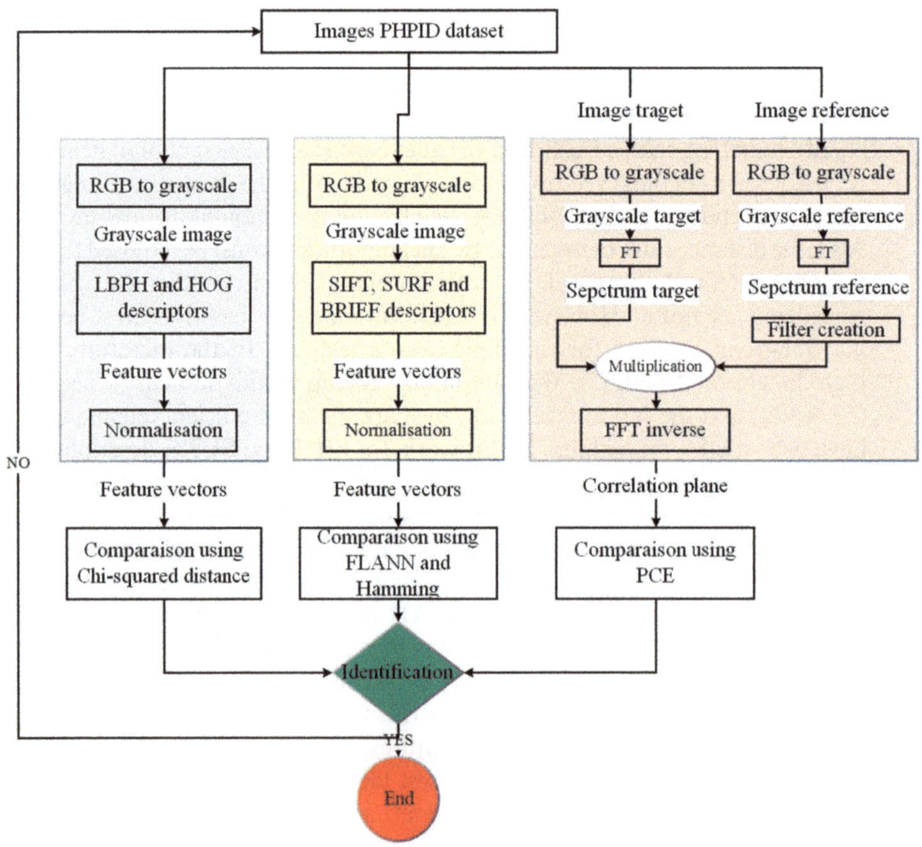

Figure 13.1. Flow diagram of the implemented face recognition algorithms.

For local face deformations in terms of insensitive and discriminative matching, the face structure or content needs to be described within a suitable descriptor. This descriptor must be indicated by the main orientations and their red squares of a set of interest points extracted from a face. There are many descriptors of facial features in the literature. In the following parts, we discuss the most common descriptors. The experimental results for each descriptor, for face descriptions under different lighting conditions, are presented below.

13.3.1 Histogram of oriented gradients (HOG)

The HOG descriptor is one of the most popular descriptors used for feature extraction in the local region by taking orientation histograms of the edge intensity. The feature vector computation by the HOG descriptor proceeds as follows:

- *Calculation of the gradient:*

  At first, to obtain the HOG descriptor, the gradient image is calculated by an appropriate filter mask (e.g. Laplacian, Sobel, and Prewitt filters) in order to extract the edge gradients and orientations. Next, depending on their gradients and orientations, a grid of histograms is created, where every histogram votes for the respective gradients using its length into a small spatial area called a 'cell'. Then, we divide the local image into regions called cells, and then calculate the amplitude of the first order gradients of each cell in both the horizontal and vertical direction (e.g. equations (13.1) and (13.2)). The most common method is to apply a 1D mask [−1 0 1].

$$G_x(x, y) = I(x + 1, y) - I(x - 1, y) \tag{13.1}$$

$$G_y(x, y) = I(x, y + 1) - I(x, y - 1) \tag{13.2}$$

  where, $I(x, y)$ is the pixel value of the point (x, y), $G_x(x, y)$ and $G_y(x, y)$ denote the horizontal gradient amplitude and the vertical gradient amplitude, respectively.

  The amplitude and gradient direction of the pixel (x, y):

$$G(x, y) = \sqrt{G_x(x, y)^2 + G_y(x, y)^2} \tag{13.3}$$

$$\theta(x, y) = \tan^{-1}\left(\frac{G_y(x, y)}{G_x(x, y)}\right) \tag{13.4}$$

- *Each block 16 * 16 is divided into 2 * 2 cells:*

  16 * 16 Gaussian weight blocks adjust the amplitude of the gradient, each range of bins is $[\frac{k\pi}{9}, \frac{(k+1)\pi}{9}, k = 0, 1, \ldots, 8]$. The amplitude of the gradient and the orientation of each pixel in the cell are voted in 9 bins with tri-linear interpolation. The tri-linear interpolation is important, as it makes the image block of the edge zone better allocated to improve the stability of the features.

- *Concatenate the histograms of all blocks:*

The histograms of the four cells of each block concatenate on a vector of 36D characteristics which are normalized by L2-Hys to reduce the influence of the local variation of the illumination and contrast of the foreground.

The HOG descriptor has been one of the best accepted processes to describe image local texture, and it has one of the best features of shape and edge information. The HOG descriptor can describe the shape of the face by the distribution of edge direction or light intensity gradient. It displays great success in face recognition.

13.3.2 Local binary patterns

LBP are one of the high-texture descriptor approaches, and have been widely used in different applications. The LBP descriptor was initially revealed for face recognition and texture classification, due to its robustness to binary coding of threshold intensity values. Generally, the LBP descriptor works in a 3 × 3 pixel matrix ($p1...p8$) of an image. The pixels of this matrix are thresholded with the value of the center pixel p_0 (i.e., use the intensity value of the center pixel $i(p_0)$ as a reference for thresholding) to produce the binary code. If a neighbor pixel's value is lower than the center pixel value, it is given a zero, if not, one. Finally, this code binary is multiplied by powers of two and then summed to obtain a local binary pattern descriptor value for the center pixel. The LBP is defined in a matrix of size 3 × 3 and is given by equation (13.5).

$$LBP = \sum_{p=1}^{8} 2^p s(i_0 - i_p), \quad \text{with} \quad s(x) = \begin{cases} 1 & x \geq 0 \\ 0 & x \leq 0 \end{cases} \tag{13.5}$$

where i_c and i_p are the intensity value of the center pixel and neighborhood pixels, respectively. Also, an illustration of the LBP descriptor can be seen in figure 13.2.

The most influential benefits of LBP are its low computational complexity, convenient multi-scale extension, and invariance to monotonic gray-scale changes. LBP descriptors have three patterns: rotation invariant uniform pattern, rotation invariant pattern, and uniform pattern. LBP is defined as a set of binary commands comparing pixel intensities between the central pixel and its eight surrounding pixels in the image. In fact, the LBP descriptor was used in this report for face authentication or verification. In this paper, for HOG and LBP descriptors, the

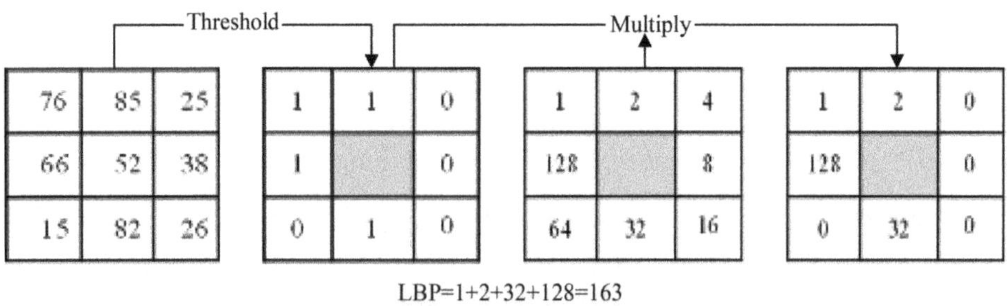

Figure 13.2. The LBP descriptor.

matching task is performed with the chi-squared distance. This distance is used to find a similarity between two histograms [17]. The HOG and LBP descriptors inherit the advantages of chi-squared distance, which effectively handles the unbalanced nature of HOG and LBP histograms [18]. The chi-squared ($\chi 2$) was successfully used for local descriptors matching, shape classification, object categories and texture classification [21], and boundary detection.

Generally, in the natural histograms the difference between the small bins is more significant than the difference between the large bins. The chi-squared ($\chi 2$) is a histogram distance that takes this into account. To compare two histograms $S_1 = (u_1, \ldots\ldots u_m)$ and $S_2 = (w_2, \ldots\ldots w_m)$ the chi-squared ($\chi 2$) distance can be defined as:

$$\chi 2 = D(S_1, S_2) = \frac{1}{2}\sum_{i=1}^{m}\frac{(u_i - w_i)^2}{u_i + w_i} \tag{13.6}$$

where m is the dimensionality of the spatially enhanced histograms.

The $\chi 2$ histogram distance is performed to test the difference between the observed frequencies and distribution.

13.3.3 Correlation filters

The correlation filters are widely used in face recognition tasks. There are two main approaches for correlation filters: the joint transform correlator (JTC) [22, 23] and the Vander Lugt Correlator (VLC). In this chapter, we will use the second approach, the VLC, for face recognition. The VLC has been mainly used for face recognition applications due to its ability to instantly detect, discriminate capability and estimate target objects [15]. In addition, the similarity matching based on the PCE criterion is used to obtain the recognition results.

The VLC technique, as shown in figure 13.3, has three planes: the input plane P1, the Fourier plane P2, and the correlation plane P3. Two converging lenses (L1 and L2) are used to pass from one plane to the other. The first lens realizes the FT of the target image. The second realizes the FT^{-1} of the multiplication result (spectrum of target image * the correlation filter).

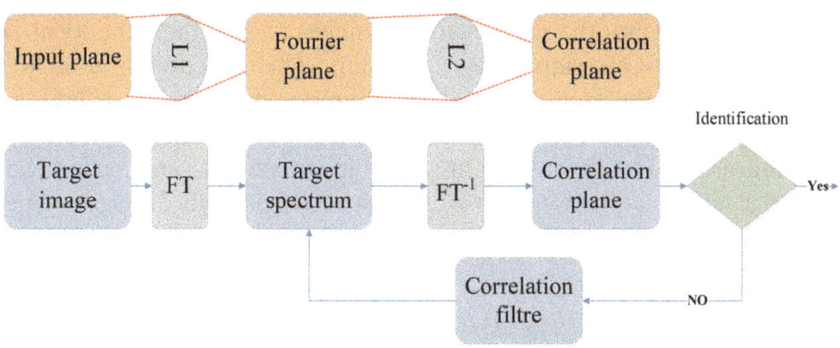

Figure 13.3. Vander Lugt correlator.

The flowchart of the VLC technique is presented as follows: firstly, a Fourier transform is applied to the target image to get a target spectrum. After that, a multiplication between the target image spectrum and a given POF filter is performed. Then by an inverse Fourier transfom, we obtained in the output the correlation plane. Next, it provides the correlation result recorded on the correlation plane; this multiplication is effected by inverse Fourier transform.

The correlation result, described by the peak intensity, is used to determine the similarity degree between the target and reference images. Figure 13.3 illustrates the VLC technique. To evaluate the correlation, the PCE is defined as the energy in the correlation peaks intensity normalized to the overall energy of the correlation plane.

$$PCE = \frac{\sum_{i,j}^{N} E_{\text{peak}}(i, j)}{\sum_{i,j}^{M} E_{\text{correlation}_{\text{plane}}}(i, j)} \tag{13.7}$$

where M and N are the size of the correlation plane and the size of the peak correlation spot, respectively.

To enhance the matching process, Horner and Gianino [24] proposed a phase-only filter (POF). The POF can produce correlation peaks marked with an enhanced discrimination capability. The POF is an optimized filter defined as:

$$H_{\text{POF}}(u, v) = \frac{S^*(u, v)}{|S(u, v)|} \tag{13.8}$$

where $S^*(u, v)$ is the complex conjugate of the FT of the reference image.

The correlation peak indicates clearly when the face image was in its corresponding face image (in figure 13.4(a)). Besides, the correlation peak is less prominent when the face image was correlated with the other face image, and the peaks were less prominent in figure 13.4(b).

Higher correlation peaks were associated with lower noise levels and smaller ones with higher noise levels.

13.3.4 Scale invariant feature transform (SIFT)

The scale-invariant feature transform (SIFT) descriptor is presented by [25]. This descriptor establishes the scale space of images and extracts the local extrema with the difference-of-Gaussian (DoG) function. This function is used to detect the interest points in the face image. The difference of Gaussian (DoG) in scale space is applied in order to detect the extrema and obtain the feature points (local extrema of the DoG) that are invariant to orientation and scale. The DOG function is the difference between two Gaussian kernels with different scales. It is defined as:

$$\begin{aligned} D(x\ y) &= (G(x, y, k\sigma) - (G(x\ y, \sigma))*I(x\ y) \\ &= L(x, y, k\sigma) - L(x, y, \sigma) \end{aligned} \tag{13.9}$$

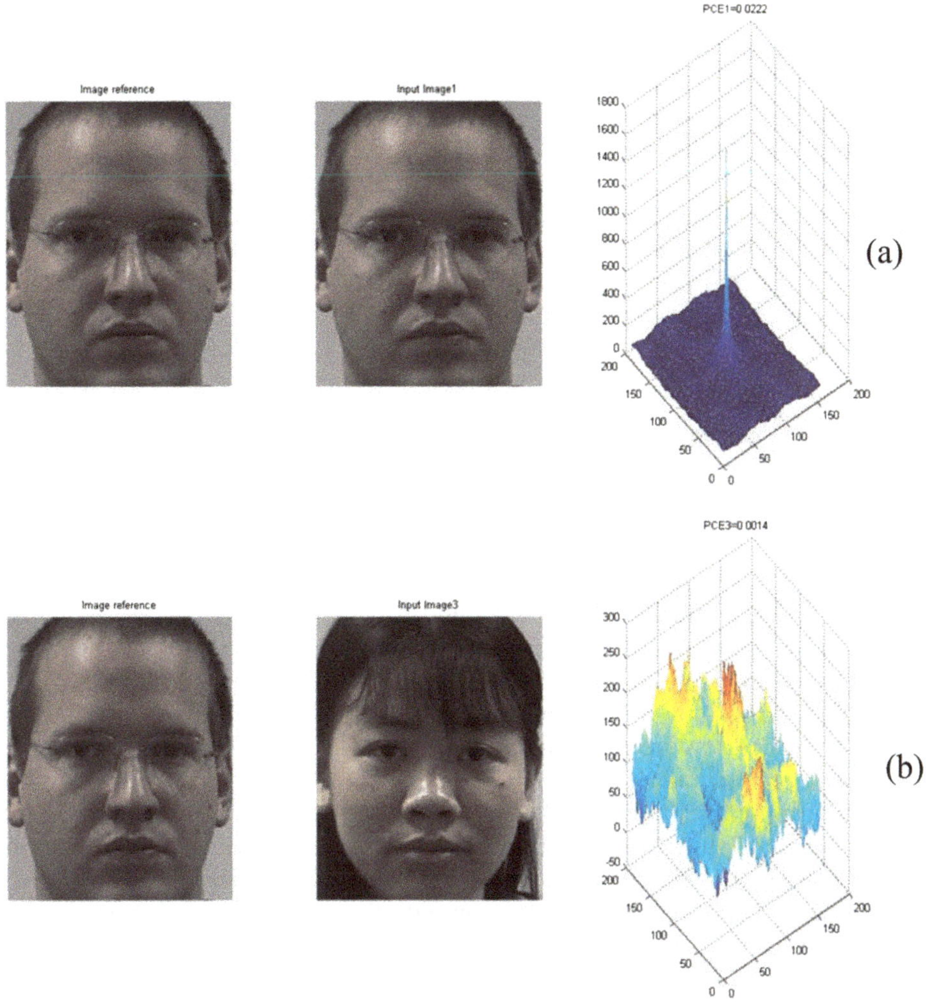

Figure 13.4. Results of the correlation technique.

with

$$G(x, y, k\sigma) = \frac{1}{2\pi\sigma^2} e^{-\frac{x^2+y^2}{2\sigma^2}}$$

where $I(x, y)$ is the gray level of pixel (x, y), σ is the scale space factor, k is the ratio factor between the two neighboring scales, $G(x, y, \sigma)$ is the Gaussian kernel function, and $L(x, y, \sigma)$ is the image scale space.

Then, a feature vector is extracted at each interest point. Then, to provide invariance against rotation, the local orientation of the image is estimated using the local image properties over a neighborhood and, over a number of scales around the point of interest. Next, based on local image information at the characteristic scale, a descriptor is computed for each detected point. In addition, this method performs a histogram of gradient orientations of sample points in a region around the interesting point, determines the interest points that have the highest orientation value to construct a feature vector, and uses these directions as the dominant direction of the interesting point.

Typically, by putting the key point in the center, an adjacent 16 × 16 region is performed. SIFT divides this region into 4 × 4 sub-regions with 8 orientation bins in each. The vector obtained by the SIFT descriptor has 128 elements since there are 4 × 4 = 16 histograms each with 8 bins. Thus, meaningful descriptors are extracted from the image that are robust, highly distinctive, and resistant to change in a camera's viewpoint and illumination. Figure 13.5 shows the representation of the SIFT descriptor for a 16 × 16 pixel patch and a 4 × 4 descriptor array.

13.3.5 Speeded up robust features (SURF)

SURF is a technique used generically for the detection of features, but also for description. This descriptor has been constructed on the basis of a SIFT descriptor. Also, it is very robust, and several times faster at obtaining results than SIFT. The SURF descriptor has been used mainly to extract points of interest, to track objects, to locate and recognize faces or people, and to reconstruct 3D scenes. Also, this descriptor is invariant to scale and rotation. Its algorithm outputs a descriptor vector and also finds key point locations.

The SURF descriptor has three main steps including interest point detection, local neighborhood description, and matching. Initially, the SURF descriptor used an intermediate representation of the face image, to improve the speed of the SURF algorithm. This representation is called the 'integral image', as shown in figure 13.6. It contains the sum values of the grayscale pixel of the face image. Its representation reduces a computational time advantage as it only depends on three integer operations.

When the integral image is used to compute the surface integral from the face image, it is necessary to read only four-pixel values so as [26]. Next, to detect the interesting point, SURF employs the determinant of the Hessian blob detector. In

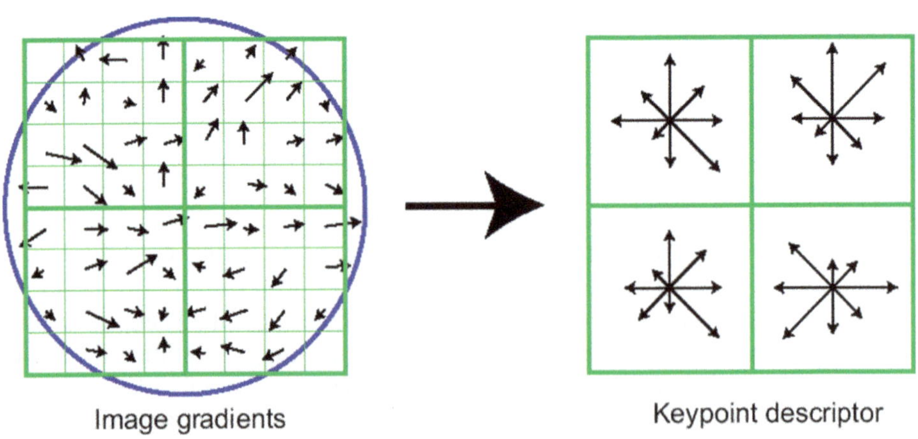

Figure 13.5. SIFT descriptor for a 16 × 16 pixel patch and a 4 × 4 descriptor array.

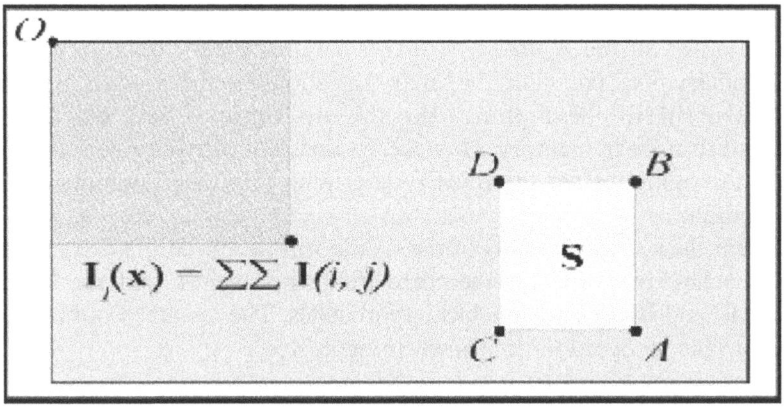

Figure 13.6. Evaluation of the integral image.

the image I, $x = (x, y)$ is the given point, the Hessian matrix $H(x, \sigma)$ in x at scale σ, can be defined as (e.g. equation (13.10)):

$$H(x, y) = \det\begin{pmatrix} \frac{\partial^2 f}{\partial x^2} & \frac{\partial^2 f}{\partial x \partial y} \\ \frac{\partial^2 f}{\partial x \partial y} & \frac{\partial^2 f}{\partial y^2} \end{pmatrix}. \tag{13.10}$$

The approximated determinant of the Hessian describes the blob response in the image. To approximate the image partial derivatives, SURF utilizes a second order Gaussian filter, since this filter allows both space and scale analysis. The responses to the Haar wavelets are used for orientation assignment, before the interest point descriptor is formed from the wavelet responses in a given surrounding interest point neighborhood. Finally, similarity matching based on the FLANN distance (Euclidean distance) is used to obtain the recognition results for the SIFT and the SURF techniques.

13.3.6 Binary robust independent elementary features (BRIEF)

BRIEF belongs to the family of binary descriptors that include BRISK and LBP, and only performs simple binary comparison tests and uses Mahalanobis or Hamming distance instead of Euclidean distance. Thus, it is only necessary to compare the intensity between two-pixel positions located around the detected interest points for building a binary descriptor. This allows one to obtain a representative description at very low computational cost. Also, to represent an image patch as a binary string, this descriptor relies on a relatively small number of intensity difference tests. Generally, a binary descriptor for a patch of pixels of size $S \times S$ is performed by concatenating the results of the following test:

$$\tau = \begin{cases} 1 & \text{if } I(P_j) > I(P_i) \\ 0 & \text{otherwise} \end{cases} \tag{13.11}$$

where $I(P_i)$I(pi) denotes the (smoothed) pixel intensity value at P_i, and the selection of the location of all the P_i uniquely defines a set of binary tests. In addition, the number of binary pixel pairs and the binary threshold are the two setting parameters used with the BRIEF descriptor. Also, this descriptor is very efficient both to compute and to store in memory. However, it does not provide rotation invariance. Finally, in this study we use the BRIEF descriptor with the Hamming distance for face recognition.

We present the software results of the six algorithms of feature extraction for the three approaches: the VLC for the correlation approaches, and the LBP, HOG, SIFT, SURF, and BRIEF for the local approaches. The results of our implemented methods for face recognition are shown in figure 13.7.

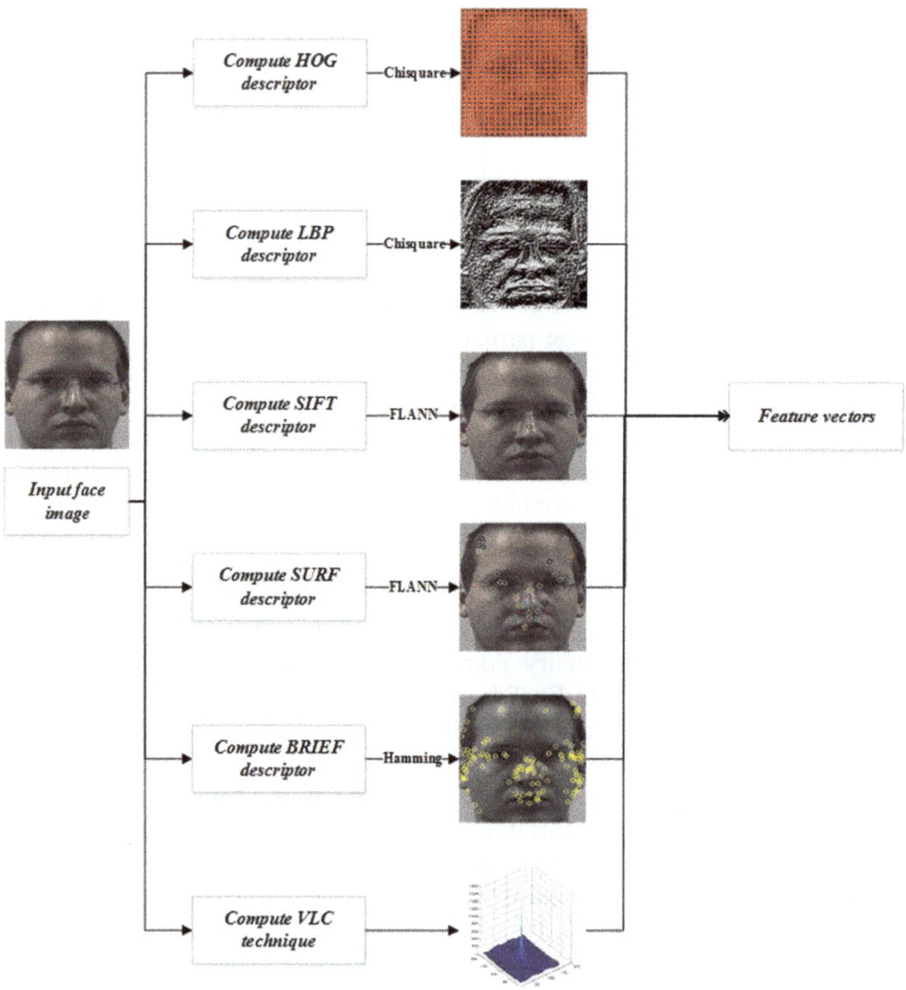

Figure 13.7. The software results of our implemented methods for face recognition.

13.4 Results analysis

We implemented the proposed methods in C++ and OpenCV for testing a wide variety of images of faces. These experiments have been successfully evaluated on Intel PC platform equipped with a CPU Core i3 3.2 GHz.

13.4.1 Pointing head pose image database (PHPID)

In order to evaluate the performance of the six algorithms, we use the Pointing Head Pose Image Database (PHPID) [27]. In addition, in the face recognition research community, the PHPID is one of the most widely used databases. It contains 2790 monocular face images of 15 sets of people with tilt angles from −90 to +90 degrees and variations of pan. Every set contains two series of 93 different poses (93 images) of the same person. In this study, we have two series per person in order to be able to train and test our algorithms on unknown and known faces. The face images in the PHPID database have various skin colors and some wear glasses. In addition, this database differs in the degree of illumination, expression, and variation of pose present in the images. Some of the images used for evaluation are shown in figure 13.8.

13.4.2 Experimental settings

Before the evaluation of the different methods, a preprocessing step is performed. In order to extract the region of interest of the position of the face, a geometrical normalization of the faces is used to resize the image to 154 × 188. The preprocessing step is performed in order to reduce local shadowing, highlights, and the effect of the variation in illumination.

Figure 13.8. Examples of the PHPID face images used in our tests.

For the HOG and LBP methods, we used the same size of the region of interest located by applying the grayscale conversion and normalization steps. The chi-squared distance is applied for recognition. Concerning the VLC method, we used the same sized window of the region of interest to extract the correlation result and the peak-to-correlation energy (PCE) for recognition.

For the SIFT, SURF, and BRIEF methods, we used the same descriptors (SIFT, SURF, and BRIEF) to detect and describe the feature vectors and we performed the FLANN distance (SIFT and SURF) and Hamming distance (BRIEF) for recognition.

13.4.3 Evaluation techniques

In order to evaluate the performance of these six techniques for face recognition, two major evaluation plots are used, the PR (Precision (P) and Recall) curve and the ROC (receiver operating characteristics) curve. The PR curve determines the relation between the detection precision and the detection rate (recall), while the ROC curve determines the relation between the true positive rate (TPR) and the false positive rate (FPR). In our study, we need a two class recognition (i.e. true or false face). The true positive means the true positive to be accepted by the technique, while the false positive means the false face to be accepted as the true face. The detection rate and recall have the same meaning as the term true positive. In addition, we summarize a detailed description for evaluation in table 13.1. However, we use a specific threshold for a given case to decide the positive or negative result for recognition. We could achieve high TPR and FPR, and vice versa with low threshold. Then, in this study, we use specific thresholds to represent each point on the ROC curve.

The definition of the TPR, the FPR and precision are:

$$TPR = Recall = TP/(TP + FN) \tag{13.12}$$

$$FPR = FP/(FP + TN) \tag{13.13}$$

$$P = TP/(TP + FP). \tag{13.14}$$

13.4.4 Results and discussion

In this study, three steps are presented for each method: preprocessing, feature extraction using the LBP, HOG, and VLC, SIFT, SURF, and BRIEF, classifying

Table 13.1. The definition of parameters used in this study.

Recognition	Accepted (positive)	Rejected (negative)
Desired face	True positive (TP)	False negative (FN)
Undesired face	False positive (FP)	True negative (TN)

with chi-squared, FLANN, and Hamming distances or with PCE. Thus, in order to demonstrate the effectiveness and feasibility of the best method for face recognition, we discuss and compare the six face recognition algorithms in terms of recognition, implementation, and application under various conditions and environments.

The PHPID dataset provides characteristic points such as mouth positions, nose, and eyes. We used eye labels for the cropped face images and resized them to 154 × 188 pixels. In addition, we applied a grayscale conversion as a preprocessing step in order to alleviate the large illumination variation contained in the PHPID dataset. We computed the TPR and FPR by changing the threshold values to draw the ROC curves for 30 faces without rotation. We used two subsets containing 30 faces of 15 people with zero-degree rotation under various illumination conditions. The mean values of FPR and TPR of the PHPID dataset with fixed thresholds over all the two subsets for performance evaluation are shown in table 13.2.

We used the PHPID dataset with rotation (vertical and horizontal) 45° to 45° in this case study, where the dataset is divided into three subsets for validation, with two subsets containing 42 pairs of genuine matches and 42 pairs of impostor matches of the same person and the other subset containing 42 pairs of another person for verification. The mean values of FPR and TPR of the PHPID dataset with rotation and fixed thresholds over all the three subsets are shown in tables 13.3–13.5.

To evaluate our techniques implemented for face recognition, we plot the ROC curve using TPR and FPR for performance evaluation in figure 13.9.

When reading an ROC curve, the points farther away from the diagonal line ($y = x$) and closer to the point of coordinates (0, 1) are better, as that indicates lower FPR (fewer incorrect classifications) and higher TPR (more correct classifications).

We calculated the TPR and FPR by varying the threshold values to plot the ROC curves. The ROC curve for the six techniques shows that LBP and VLC outperform HOG, SIFT, SURF, and BRIEF in face recognition tasks.

More specifically, VLC achieves 61% TPR at 46% FPR, LBP achieves 59% TPR at 41% FPR, whereas HOG, SIFT, SURF and BRIEF obtain 55%, 40%, 48% and 58% TPR at 9%, 52%, 44% and 95% FPR, respectively (table 13.6). Figure 13.9 shows ROC curves for the six methods, VLC and LBP achieve a better performance than HOG, SIFT, SURF, and BRIEF.

Table 13.2. The results under different thresholds for 30 faces (2 desired faces and 28 undesired faces) for various techniques.

Methods/parameters	TP	FN	FP	TN	TPR	FPR	P
VLC	25	5	34	386	0.83	0.91	0.42
LBP	23	7	31	389	0.76	0.92	0.42
HOG	20	10	130	290	0.66	0.69	0.13
SIFT	28	2	9	411	0.93	0.97	0.75
SURF	23	7	10	410	0.76	0.97	0.69
BRIEF	26	4	21	399	0.86	0.95	0.55

Table 13.3. The results under different thresholds for 126 faces (84 desired faces and 42 undesired faces) for VLC and BRIEF techniques.

VLC (*T*=PCE)								BRIEF (*T*= Hamming distance)							
T	TP	FN	FP	TN	TPR	FPR	P	*T*	TP	FN	FP	TN	TPR	FPR	P
***T* = 0.0010**	84	0	42	0	1	1	0.65	***T* = 160**	79	5	40	2	0.94	0.95	0.66
***T* = 0.0012**	83	1	42	0	0.98	1	0.65	***T* = 150**	66	18	37	5	0.78	0.88	0.64
***T* = 0.0014**	75	9	39	3	0.89	0.92	0.65	***T* = 140**	51	33	31	11	0.65	0.73	0.62
***T* = 0.0016**	68	16	25	17	0.80	0.59	0.73	***T* = 130**	45	39	28	14	0.53	0.66	0.61
***T* = 0.0018**	56	28	13	29	0.66	0.30	0.81	***T* = 120**	42	42	27	15	0.5	0.64	0.60
***T* = 0.0020**	53	31	4	38	0.63	0.09	0.92	***T* = 110**	40	44	27	15	0.47	0.64	0.59
***T* = 0.0022**	31	53	3	39	0.36	0.07	0.91	***T* = 100**	36	48	27	15	0.42	0.64	0.57
***T* = 0.0024**	23	61	3	39	0.27	0.07	0.88	***T* = 90**	36	48	27	15	0.42	0.64	0.57
***T* = 0.0026**	18	66	3	39	0.21	0.07	0.85	***T* = 80**	36	48	26	16	0.42	0.61	0.58
***T* = 0.0028**	13	71	2	40	0.15	0.04	0.86	***T* = 70**	34	50	26	16	0.40	0.61	0.56

Table 13.4. The results under different thresholds for 126 faces (84 desired faces and 42 undesired faces) for SIFT and SURF techniques.

SIFT (*T*= FLANN distance)								SURF (*T*= FLANN distance)							
T	TP	FN	FP	TN	TPR	FPR	P	*T*	TP	FN	FP	TN	TPR	FPR	P
***T* = 450**	74	10	41	1	0.88	0.97	0.64	***T* = 0.4**	78	6	38	4	0.92	0.90	0.67
***T* = 440**	64	20	38	4	0.76	0.90	0.62	***T* = 0.3**	66	18	31	11	0.78	0.73	0.68
***T* = 430**	45	39	33	9	0.53	0.72	0.57	***T* = 0.2**	53	31	23	19	0.63	0.52	0.69
***T* = 420**	32	52	25	17	0.38	0.59	0.56	***T* = 0.1**	39	45	18	24	0.46	0.42	0.68
***T* = 410**	21	63	19	23	0.25	0.45	0.52	***T* = 0.09**	38	46	15	27	0.45	0.35	0.71
***T* = 400**	18	66	14	28	0.21	0.33	0.56	***T* = 0.08**	36	48	13	29	0.42	0.30	0.73
***T* = 390**	14	70	10	32	0.16	0.23	0.58	***T* = 0.07**	35	49	13	29	0.41	0.30	0.72
***T* = 380**	11	73	7	35	0.13	0.16	0.61	***T* = 0.05**	27	57	12	30	0.32	0.28	0.69
***T* = 370**	10	74	3	39	0.11	0.07	0.76	***T* = 0.02**	13	71	5	37	0.15	0.11	0.72
83/***T* = 360**	9	75	2	40	0.10	0.04	0.81	***T* = 0.005**	2	82	1	41	0.023	0.02	0.66

13.4.5 Computational time

In this section, we compare the complexity of the most widely used methods for face recognition: HOG, LBP, VLC, SIFT, SURF and BRIEF descriptors. The face recognition times for different methods are shown in table 13.7.

Regarding the computational complexity of the six methods implemented in this study inside the identification (or recognition) phase, the VLC, which is based on the fast Fourier transform (FFT), and HOG, which calculates histograms by using the angle orientation of the edges, are the less complex methods, as compared to the other methods including LBP, which has simple thresholding and encoding, and SIFT, SURF, and BRIEF which consist of two steps such as key points detection description.

Table 13.5. The results under different thresholds for 126 faces (84 desired faces and 42 undesired faces) for LPB and HOG techniques.

LBP *(T* = Chi-square distance)								HOG *(T* = Chi-Square distance)							
T	TP	FN	FP	TN	TPR	FPR	P	*T*	TP	FN	FP	TN	TPR	FPR	P
***T* = 1.3**	83	1	41	1	0.98	0.97	0.66	***T* = 2500**	61	23	4	38	0.72	0.09	0.93
***T* = 1.2**	83	1	39	3	0.98	0.92	0.68	***T* = 2400**	61	23	4	38	0.72	0.09	0.93
***T* = 1.1**	80	4	30	12	0.95	0.71	0.72	***T* = 2300**	57	27	3	39	0.67	0.07	0.95
***T* = 1**	70	14	21	21	0.83	0.5	0.76	***T* = 2200**	48	36	2	40	0.57	0.04	0.96
***T* = 0.9**	59	25	14	28	0.70	0.33	0.80	***T* = 2100**	43	41	1	41	0.51	0.02	0.97
***T* = 0.8**	47	37	5	37	0.55	0.11	0.90	***T* = 2000**	39	45	0	42	0.46	0	1
***T* = 0.7**	32	52	3	39	0.38	0.07	0.91	***T* = 1900**	34	50	0	42	0.40	0	1
***T* = 0.6**	24	60	0	42	0.28	0	1	***T* = 1800**	30	54	0	42	0.35	0	1
***T* = 0.5**	8	76	0	42	0.09	0	1	***T* = 1700**	25	59	0	42	0.29	0	1
***T* = 0.4**	3	81	0	42	0.03	0	1	***T* = 1600**	23	61	0	42	0.27	0	1

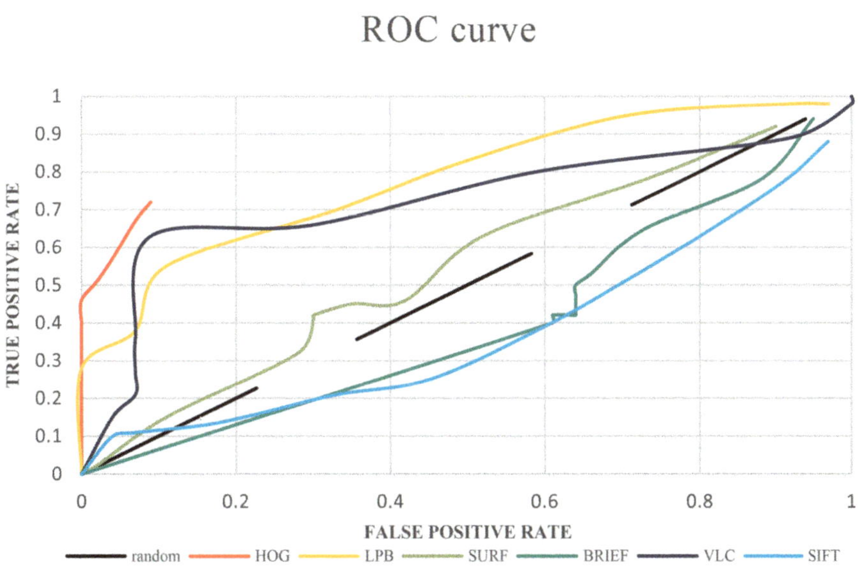

Figure 13.9. The ROC curves using TPR and FPR, the vertical axis shows the true-positive rate TPR and the horizontal axis shows the false-positive rate FPR.

Considering the VLC method, LBP requires a computational complexity which is 2 times greater (the calculation of VLC features and the construction of the correlation plane is very fast when compared to the computation of a simple thresholding and the binary code to get the feature vectors), SIFT, SURF, BRIEF require a computational complexity 4 times greater, 2 times greater, and 3 times greater, respectively (the detection and the description of the interest points are the most complex processes due to the techniques used in these descriptors (e.g. DOG,

Table 13.6. The average values of TPR, FPR, and P for different techniques.

Parameters Methods	TPR	FPR	P
VLC	0.61	0.46	0.75
LBP	0.59	0.41	0.80
HOG	0.51	0.09	0.89
SIFT	0.40	0.52	0.63
SURF	0.48	0.44	0.69
BRIEF	0.58	0.95	0.72

Table 13.7. Face recognition times for different methods.

Methods	HOG	LBP	VLC	SIFT	SURF	BRIEF
Time (seconds)	0.039	0.099	0.046	0.41	0.116	0.304

wavelet transform, Harr filter, histogram of gradients). Also, the HOG and the VLC methods probably have the same time, but the VLC method outperforms in recognition rates on the PHPID database.

13.5 Conclusion and future work

This chapter establishes a comparative study of face recognition of various vision algorithms used with feature mining. The changes in angle and lighting makes the recognition of faces very complex, and they often differ in subtle ways. Therefore, a face recognition system should include several elements such as change of illumination, change in pose, change in facial expression, a moustache, beard, mask in front, occlusion due to the scarf and scaling factor. Experiments on the PHPID dataset, a challenging database of real-world human faces, demonstrated that the implemented methods yield the best results in various conditions. VLC and LPB convey more efficiency and discriminative power than other descriptors such as HOG, SURF, SIFT and BRIEF, and for feature extraction that does not involve any quantization in the computation of descriptors. The results clearly show that VLC and LBP outperformed the other methods studied on all the subsets of the PHPID dataset. Finally, we conclude that VLC and LBP are the best methods in terms of implementation and application in real time and for future work.

References

[1] Alfalou A, Brosseau C, Katz P and Alam M S 2012 Decision optimization for face recognition based on an alternate correlation plane quantification metric *Opt. Lett.* **37** 1562–4

[2] Elbouz M, Bouzidi F, Alfalou A, Brosseau C, Leonard I and Benkelfat B E 2013 Adapted all-numerical correlator for face recognition applications *Proc. SPIE* **8748** 874807-1

[3] Lu X, Duan X, Mao X, Li Y and Zhang X 2017 Feature extraction and fusion using deep convolutional neural networks for face detection *Math. Probl. Eng.* **2017** 1376726

[4] Solanki D V and Kothari A M 2014 A survey on face recognition techniques and its applications *J. Image Process. Pattern Recog. Progr.* **1** 16–21

[5] Abhishree T M, Latha J, Manikantan K and Ramachandran S 2015 Face recognition using Gabor filter based feature extraction with anisotropic dffusion as a pre-processing technique *Proc. Comput. Sci.* **45** 312–21

[6] Rahim M A, Azam M S, Hossain N and Islam M R 2013 Face recognition using local binary patterns (LBP) *Glob. J. Comput. Sci. Technol.*

[7] Huang C and Huang J 2017 A fast HOG descriptor using lookup table and integral image *arXiv:*1703.06256

[8] Feng Y, An X and Liu X 2016 The application of scale invariant feature transform fused with shape model in the human face recognition *Advanced Information Management, Communicates, Electronic and Automation Control Conf. (IMCEC)* (IEEE) pp 1716–20

[9] Calonder M, Lepetit V, Strecha C and Fua P 2010 Brief: Binary robust independent elementary features *Computer Vision–ECCV* **2010** 778–92

[10] Verma R and Kaur R 2014 Enhanced character recognition using surf feature and neural network technique *Int. J. Comput. Sci. Info. Technol.* **5** 5565–70

[11] Singh N A, Kumar M B and Bala M C 2016 Face recognition system based on SURF and LDA technique *Int. J. Intell. Syst. Appl.* **8** 13

[12] Vishwakarma N 2016 Face recognition system using principal component analysis (PCA) **5** 1759–66

[13] Liu Y, Pados D A and Yeh C H 2016 Two-stage tensor locality-preserving projection face recognition *2016 IEEE Second Int. Conf. on Multimedia Big Data (BigMM)* (IEEE) pp 214–8

[14] Alfalou A and Brosseau C 2010 Understanding correlation techniques for face recognition: from basics to applications *Face Recognition* (InTech)

[15] Wang Q, Alfalou A and Brosseau C 2017 New perspectives in face correlation research: a tutorial *Adv. Opt. Photon* **9** 1–78

[16] Král P and Vrba A 2017 Enhanced local binary patterns for automatic face recognition *Preprint* arXiv: 1702.03349

[17] Ren J, Jiang X and Yuan J 2015 A chi-squared-transformed subspace of LBP histogram for visual recognition *IEEE Trans. Image Process* **24** 1893–904

[18] Rami H, Hamri M and Masmoudi L 2014 Tracking object using exponential forgetting factor in the LBP code *J. Theor. Appl. Inf. Technol.* **63** 3

[19] Du G, Su F and Cai A 2009 Face recognition using SURF features *Proc. SPIE* **7496** 749628-1

[20] Işık Ş 2014 A comparative evaluation of well-known feature detectors and descriptors *Int. J. Appl. Math. Electron. Comput* **3** 1–6

[21] Zhang J, Marszalek M, Lazebnik S and Schmid C 2005 Local features and kernels for classifcation of texture and object categories: an indepth study *Technical Report* RR-5737 INRIA Rhône-Alpes.

[22] Butt J A and Wilkinson T D 2006 Binary phase only filters for rotation and scale invariant pattern recognition with the joint transform correlator *Opt. Commun.* **262** 17–26

[23] Leonard I, Alfalou A and Brosseau C 2012 Spectral optimized asymmetric segmented phase-only correlation filter *Appl. Opt.* **51** 2638–50

[24] Horner J L and Gianino P D 1984 Phase-only matched filtering *Appl. Opt.* **23** 812–6
[25] Lowe D G 2004 Distinctive image features from scale-invariant keypoints *Int. J. Comput. Vis* **60** 91–110
[26] Svab J, Krajnik T, Faigl J and Preucil L 2009 FPGA based speeded up robust features *IEEE Int. Conf. on Technologies for Practical Robot Applications (TePRA 2009)* (IEEE) pp 35–41
[27] Gourier N, Hall D and Crowley J L 2004 Estimating face orientation from robust detection of salient facial structures *FG Net Workshop on Visual Observation of Deictic Gestures* vol 6

Lightning Source UK Ltd.
Milton Keynes UK
UKHW05n1723040618
323584UK00004B/75/P

9 780750 314558